ENVIRONMENTAL STUDIES AND ETHICS

ENVIRONMENTAL STUDIES AND ETHICS

For B.A., B.Sc. and B.Com. Courses

(A compulsory course for undergraduate students of all streams – Arts, Science and Commerce – as per the recommendation of University Grants Commission, with additional information for the Universities of Jharkhand)

GOURI SURESH
Head of the Department of Botany & Former Co-ordinator,
Environment and Water Management
Jamshedpur Women's College, Jamshedpur, Jharkhand

TECHSAR™
NEW DELHI

Published and Printed by

Techsar Pvt. Ltd.
4435-36/7, Ansari Road, Daryaganj
New Delhi 110 002 (India)
E-mail: tech_info@techsarworld.com
Website: www.techsarworld.com

ISBN 978-81-89866-26-6

Reprint 2025

Foreword

The environment includes both natural and social processes and the relation between people and the physical biological systems of which we are a part. Sustainability is the maintenance of natural and human systems with reference to social, economic and environmental priorities. In view of this, the environment becomes an integral part of human civilization.

Over the past few centuries man has been exploiting natural resources without any consideration of the dangerous consequences of such a wanton act. Environmental degradation has reached an all time high. Uncontrolled exploitation of natural resources along with environmental pollution of various types is leading to a grave global catastrophe. If something tangible is not done now, life on earth may be put to great hazard.

In such a situation, it is the duty of every right thinking citizen to conserve nature and hand it over to future generations safe and sound. In this context, making environmental studies a compulsory subject in undergraduate courses by the UGC is a welcome step in the right direction.

Guided by this scenario, Ranchi University has introduced a paper on 'Environmental Studies and Ethics' in undergraduate curriculum to drive home the message that understanding environmental issues not only enhances better living but is also a good social responsibility. The textbook on 'Environmental Studies and Ethics' written by Dr. Gouri Suresh covers the contents of the Ranchi University syllabus. The book is well written and is presented in simple language for the benefit of students. I congratulate Dr. Suresh for this commendable work and wish her well.

A.A. Khan
Vice-Chancellor
Ranchi University, Ranchi

PREFACE

Life exists in all kinds of situations – on top of mountains, in valleys; in water bodies as well as deserts; on the hard surfaces of rocks and stones as well as on the soft surfaces of plants themselves; outside and inside living organisms; on the soil or under the soil….. The list can be made longer and longer…..All these living organisms are influenced by the environment in which they live; the environment is in turn influenced by the organisms too. The two entities are thus interdependent.

The study of the relationships between living organisms and the environment is called 'Ecology'. The original term was 'Oikologie' and it was introduced by Reiter in 1868. 'Oikos' means home and 'logos' is study. Thus Ecology describes the home life of organisms, i.e. it is the study of organisms at home, as they exist in their natural surroundings. 'Oikologie' was later changed to Ecology.

Ecology was studied in the early days separately as Plant Ecology and Animal Ecology. However, as the artificial nature of this classification became evident, this branch of natural science was renamed as Environmental Science or Environmental Studies.

Man is part of nature. So the consequences of the activities of human beings on nature – both positive and negative – are of great significance in maintaining the harmonious equilibrium of the earth. Growth of civilization and social structures must go hand in hand with sustainable development.

Today, the need for creating awareness in society regarding our environment has become crucial for all round development. In such a situation, making Environmental Studies a compulsory paper for undergraduate students of all streams is a highly laudable step. Recommendation to this effect by the University Grants Commission is a most welcome measure. As a lover of nature, I hereby express my heartfelt thanks to the UGC.

This book is an attempt at explaining the topics suggested by UGC in simple language for the benefit of the students. The style of presentation is such that students can remember the points easily. Another objective of the book is to create environmental awareness among the students and through them, in the society as a whole.

We are facing many environmental problems in our day to day lives. Some of these are not under our control. But we can do quite a lot to solve some other problems. Water scarcity and accumulation of solid wastes in our surroundings are two examples. Therefore, these topics have been dealt with extensively in this book. Methods of rain water harvesting and solid waste management have been described in detail with relevant pictures. Formulae to calculate Annual Rainwater Harvesting Potential and sizes of settling tanks have been given in Appendix 1. This is meant to be a textbook, no doubt, yet the author hopes that it will also motivate readers to start rainwater harvesting projects and solid waste management programmes in their respective localities.

The contents have been arranged as per the core module syllabus recommended by the University Grants Commission, with modifications to suit the requirements of the Universities of Jharkhand.

At the end of each unit, a large number of questions have been given. These are of various types, ranging from multiple choice questions to long essay type questions. Answers to the multiple choice questions have been given in Appendix 2. A brief idea of websites and search engines with examples can be seen in Appendix 3. This is to give students a brief concept of how to go about surfing the internet to get information on any topic. Tips to students on improving memory, concentration power, etc. and suggestions regarding how to score higher marks in examinations are the contents of Appendix 4.

I have great pleasure in thanking the publisher for their offer and prompt publication.

Constructive comments and suggestions for improving any aspect of this book in the next edition are most welcome from students as well as fellow teachers and other readers.

Jamshedpur
December 2006

THE AUTHOR

Acknowledgements

Love for nature was inclucated in me as a child by my father, late Dr. Raja Varma. He was compassionate not only to his patients, but also to all around him, indeed all parts of nature, to be precise. I learnt to love nature and nature's creations from my father. My mother, late Mrs. Muktha Varma, was an epitome of love and kindness. Thus, first and foremost, I thank my beloved parents for making me competent enough to write this book.

Dr. A.A. Khan, Honourable Vice Chancellor, Ranchi University is well known as a keen educationist and a great reformer. In spite of his busy schedule, he has spent valuable time to write the foreword of my book. I am indebted to our respected Vice Chencellor for this kind gesture.

I wholeheartedly thank to the Principal of Jamshedpur Women's College, Dr. Shukla Mahanty for the support and cooperation she has provided to me. She is always an inspiration to all of us, staff and students alike.

An environmentalist at heart, I have always been propagating the principle of sustainable development among my colleagues, friends and students. Dr. B.K. Mehta, Lecturer in the Faculty of Commerce of our college, suggested that I write a textbook on 'Environmental Studies and Ethics' for the final year undergraduate students of Ranchi University, based on the slightly modified UGC syllabus. Now when I have finished writing the book, I'd like to take the opportunity to thank Dr. Mehta for his invaluable support in initiating the process.

Colleagues of mine in Jamshedpur Women's College in all the three Streams—Arts, Science and Commerce—have been most supportive of me that I don't have words to express my gratitude to them.

I am sure the blessings of my seniors and good wishes of my peers and juniors in the University and different colleges will always be with me, and I thank them all for their goodwill and cooperation.

I also wish to thank all those who have directly or indirectly encouraged and helped me by any means.

It is pleasure to extend my sincere thanks to I.K. International Publishing House Pvt. Ltd. for their offer and prompt publication. I hope this book will be widely beneficial to the students and teachers of Ranchi University as well as other Universities, especially those of Jharkhand.

Gouri Suresh

CONTENTS

Unit 1

Definition, Scope and Importance of Environmental Studies

The origin of life on the earth was decided and controlled by a set of environmental conditions. Through these past millions of years, environment has played a major role in the sustenance of life, and it will continue to be so for all times. Let us examine how it all began. About two thousand million years ago, the earth was a ball of hot molten mass, revolving in the space around the sun. This fiery matter cooled through millions of years, forming a shell of rocks on the surface. The core remained hot and turbulent and forced molten rock out through the crust. Some lighter rocks remained on the earth's surface and became land masses. In some other parts, the crust was pulled inwards. These became the beds of the seas. The inside of the earth kept on cooling and shrinking. This gave rise to huge ridges and furrows on the earth. These are the mountains and valleys that we see today.

During the cooling of the earth, water vapours and other gases were emitted into the sky continuously. The water vapours condensed into thick clouds in the sky. They were so thick that sun's rays could not reach the earth. Rain from the clouds fell downwards endlessly, but before touching the hot earth, was converted into steam again. This was repeated many times. Finally, when the earth's crust cooled down sufficiently, rain fell down touching the earth, forming the first seas and rivers and other water bodies.

The thickness of the clouds reduced and for the first time, sun's rays fell on the earth. Thus came up the possibility of life on the earth, because life, as we know today, cannot exist without the sun.

Land was dry and was made of rocks in the beginning. There was no soil. Volcanic gases were still present in the air. Therefore, life must have originated in water. Most scientists believe that the first 'living' molecules to be formed in water, which could divide on their own, must have been nucleic acids. Proteins and other complex molecules must have originated later.

The exact origin of life is a mystery even today. Yet we can reason that the first formed tiny droplets of protoplasm, the physical basis of life, were neither plant nor animal. It is assumed that this happened more than a thousand million years ago. The first living things were simpler than cells. Gradually, these structures evolved into cells. These simple cells branched off along two courses of evolution. One division became animals, the other became plants. A third group of cells remained simple in structure and function. These are the microbes or the micro-organisms of the present times.

Both plants and animals developed from a common stock. They show many similarities with each other. Yet they are so very different from each other. Evolution did not stop there. Diversification proceeded in many directions. This resulted in the formation of innumerable varieties of living things. The biodiversity that we see today is thus the result of millions of years of development and diversification.

Plants, animals and microorganisms are thus the living beings in nature. In other words, they make the biotic part of nature. The non-living or the abiotic part includes the physical and chemical components of nature. The biotic and the abiotic components together make our environment. They are in perfect harmony with each other and form a stable system. Therefore it is our duty to protect the environment, of which we are a part.

DEFINITION OF ENVIRONMENTAL STUDIES

Environment can be defined as the external conditions influencing the growth and development of living organisms in any way. Study of any one or more aspects of the environment in relation to their effect on living organisms and also the effects of living organisms on the environment is Environmental Studies. Unlike other streams of studies, Environmental Studies pertains to our everyday lives. It has direct effect on our health and overall well-being. It is basically an applied field of study.

SCOPE OF ENVIRONMENTAL STUDIES

The living organisms of the earth interacting with the non-living environment form a vast, gigantic system. This is the largest, most nearly self-sufficient biological system we know of. *This is the biosphere or the eco-*

sphere. In the biosphere, biotic and abiotic components are constantly acting, reacting and interacting with one another. This results in structural and functional changes in the biosphere.

It is difficult to study the biosphere as a whole. Therefore, we study nature by dividing it in smaller parts forests, deserts, grasslands, crop-lands, ponds, lakes, rivers, oceans, etc. These systems — the ecosystems — are separated from one another only by space and time. In other words, the plant, animal and microbial communities in different areas and at different time periods (seasons or years or centuries) are different, yet they all are linked with each other through air, soil and water. The molecules of carbon dioxide being absorbed by a plant or the molecules of oxygen inhaled by us may have come from a far-off area. They also have been used and reused number of times in the past. Thus plants, animals and the non-living components of nature are linked to each other functionally.

There are no practical demarcations between the ecosystems. All of them are interdependent in various ways. For example, land plants absorb water from the soil, but this water ultimately comes from rain; and rain is the result of condensation of clouds, which are formed by evaporation of water from the water-bodies. Thus land and water are closely linked not only physically, but also functionally.

The plants, animals and microorganisms of any region influence and interact with one another, and together constitute a biome or biotic community. In a biome, the living beings act on the environment and are acted upon by it. There are innumerable examples of such interdependencies. Some of them are cited here. Trees provide shelter to birds and birds help in pollination and dispersal of fruits and seeds of many plants. Animals provide carbon dioxide to the plants, which need it for photosynthesis. Plants give oxygen to the animals which require it for respiration. Microorganisms act on the dead bodies of plants and animals and convert them into carbon dioxide, water and minerals which are thus returned to the air, water bodies and soil in the environment so that they can be used again. The minerals from the soil enter the plants through their roots. Animals get minerals mainly from plants. In this way, the living and the non-living are closely connected to each other in nature. The energy needed to support life on the earth comes from the sun. This radiant energy is absorbed by the non-living as well as the living components of the biosphere. Heat energy is released into the environment by both the components.

Man is an animal too. He depends on other animals, plants and microorganisms and all the non-living components of the environment in many and varied ways. Therefore if the equilibrium of the natural systems is disturbed by any means, the resulting harmful effects will have to be endured by all components, including man.

To study various aspects of the biosphere, contributions from workers in many different fields are essential. All that we know today in this field is the result of studies in various branches of biology, chemistry, physics, geology, engineering, resource management, anthropology, sociology, disaster management, medicine, economics, philosophy, statistics, computer science, meteorology etc. Thus the scope of Environmental Studies is widespread.

IMPORTANCE OF ENVIRONMENTAL STUDIES

We are surrounded by the environment. Whatever happens in our environment has a bearing on us. Therefore, studying all the issues concerning environment, understanding the problems related to environment, find-

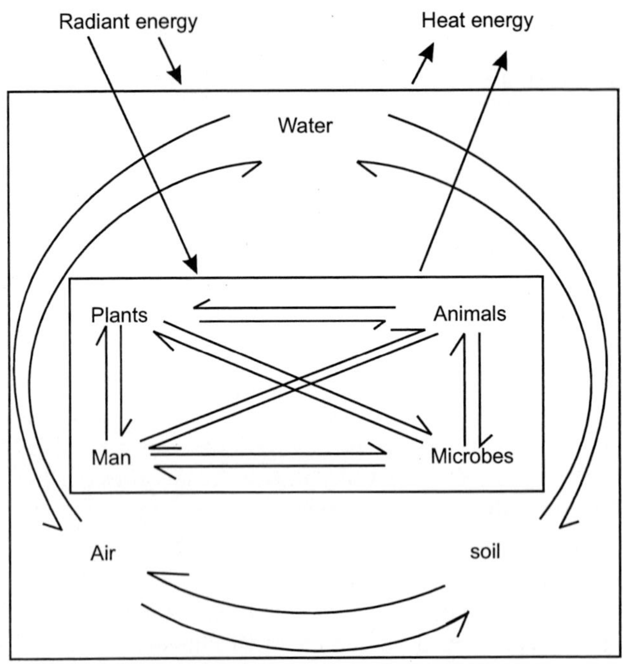

Interrelationships in the Biosphere.

ing solutions to these problems, and implementing the solutions are of utmost importance to our well-being and also to that of the living and non-living world around us.

The primary natural resources available to us are air, water and soil, forests, minerals, food and energy resources etc. Some of these are non-renewable, like soil, minerals etc. while some others are renewable, like forests, water etc. However, even the renewable sources need time to be regenerated. However, if trees in a forest are cut down faster than saplings can grow into trees, the forest will be destroyed completely in no time. So we have to be very careful while taking out resources from nature.

In India, our ancestors had fully realized the importance of conserving natural resources. Indian sages of the Vedic age lived in hermitages (*ashrams*) in forests. The best form of spiritual, scientific and cultural researches was done in these hermitages. The learned saints dwelling in the forests provided guidance to the people as to how to live a righteous life and how to treat Mother Nature. This is why ancient Indian culture is referred to as 'Aranya Sanskriti-(*aranya* means forest).

The masses were advised not to harm any natural resource; rather they were taught to respect nature. People looked to nature, the provider of air, water, light, food and so many other resources, with a sense of gratitude. Prayers to this effect can be seen in different Vedas. Some examples:

1. "O Earth! Let thy hills, snow-clad mountains and forests be pleasant."
2. "Minerals and other elements dwell in your naval part (central core). Please offer these to us."
3. "Your land and the clouds which pour rainwater on it nurture us like our parents. May the shelter of these purify and strengthen us."

The Vedas also guide us as to what we must do to protect our environment. The meanings of some verses go thus:

- "We should establish harmony with water and medicinal plants."
- "It is in the interest of mankind to plant more and more trees as these safeguard the water resources."

Indian culture was thus based on harmonious co-existence with the environment (See Unit 2 for more references).The same culture was followed later during the times of Budha and Mahavir also.

It was in the modern times, from approximately two centuries ago, that man's greed overtook his needs. He became so self-centred that he stopped thinking about the environment. Man cleared forests on a large scale for various purposes. The immediate effects of this blatant deforestation were soil erosion, floods and droughts.

One of the prime concerns of today is this continuing *deforestation.* Planting trees in as many places as possible will help to reduce the severity of this problem to some extent. Plantation forests cannot match natural forests in species variety of trees or other plants or wildlife. Yet, they help to reduce air pollution, to purify the atmosphere and to conserve precious soil. Therefore tree plantation programmes must be organized regularly. This can be done by students in groups all over the country. While planting trees, care must be taken in the selection of species. For example, arid (dry) zones should be planted with draught resistant trees like Shisham (*Dalbergia sissoo*), Ber (*Zizyphus jujuba*), *Cassia* etc. In this way, the suitability and adaptability of the species and the needs of the different zones should be looked into during tree plantation programmes.

As *population is increasing in geometrical progression*, there is a resultant shortage of land for cultivation. In some parts of the country, forest land is cleared for cultivation even now. Researches in genetics, plant breeding, biotechnology, tissue culture etc. are being carried out on a large scale. As a result of this, many new varieties of high yielding crop plants have been found out. But most of these need more fertilizers. Application of chemical fertilizers in huge quantities is degrading the land. Pesticides are polluting air, water and soil. Hybrid seeds have become popular, traditional seeds are being neglected, leading to gradual loss of many wild type genes and gene combinations. It has thus become necessary to convince the farmers of the importance of using traditional seeds to maintain the fertility of the soil. Otherwise, more and more areas in the country will have to face the problems as faced by the farmers of Punjab and Haryana today. In these two states large tracts of land have been degraded due to unsustainable farming using hybrid seeds, large quantities of fertilizers and pesticides. Mono culture, i.e. cultivation of one particular crop year after year has been another reason for the loss of fertility of the soil. Intense irrigation with the help of dams is bad for soil too, as it makes soil saline and uncultivable. *Green Revolution thus has its limitations.* It must therefore be practised carefully and only in areas suitable for such agricultural practices.

Dams of many types and capacities have been built on different rivers in India. They are irrigating thousands of hectares of cropland, and are providing us with electricity. But major dams have always been a heatedly debated topic. This is mainly because the tribal and other people who are evacuated from the dam site are never given proper rehabilitation packages. Villages are wiped out in floods when vast areas in the dam site are inundated due to blocking the path of rivers. Moreover, major dams are very expensivc to be built. They cause flash floods and change breeding patterns of fish. In fact, all plants and animals in the river are disturbed when dams are built. The developed countries have stopped making major dams.

It is not that dams are not required. They are reservoirs of water, and help in irrigation and power generation and flood control. The point is that how worthy it is to construct major dams, while many minor dams can serve the purpose. Let us examine *points for and against major dams*. The advantages projected are that they help in

1. Storing water for irrigation
2. Flood control
3. Power generation

Now, these purposes can be served by other means. For example, rainwater conservation will be of great help in this regard. Except some states like Rajasthan, Delhi, Gujarat etc., most parts of India get adequate rain.

1. If the rainwater is conserved properly by constructing ponds and lakes in low lying areas, this stored water can be used for many purposes. The huge amounts needed for constructing one major dam is more than sufficient for constructing a few minor dams or thousands of ponds, reaching a far more number of villages that need water.
2. Rainwater harvesting by this method will also reduce the amount of water that flows to rivers. The occurrence of floods is thus automatically reduced.
3. Hydroelectric power (hydel power) can be generated in minor dams also. The quantity of power generated will of course be less, but then there will be many minor dams. Electricity from many power stations can be more easily and cheaply transmitted to surrounding areas; losses in transmission will be less too.

Besides all these, one major disadvantage of constructing major dams is the human factor. Thousands of villagers are always displaced, and rehabilitation is never satisfactory. Moreover, vast forest areas have to be cleared of trees for making a major dam. This loss is irreparable. This is why the World Commission on Dams (WCD) of 2000 suggested that *projects on dams should be undertaken only after all other alternatives have been assessed and the consent of communities to be affected is obtained.*

Rainwater harvesting can do wonders for India. What is needed is awareness. Most of the people do not know the potential of rainwater harvesting. It can be done with ease in rural as well as urban areas (See Water Harvesting in Unit 6 for details). In this era of rapid urbanization, rising population and growing industrialization, availability of water is becoming more and more scarce all over the world, especially in the developing countries. The cities have become concrete jungles. As the green areas decrease giving way to cement, mosaic and marble, surface runoff increases greatly. Much less water is percolating underground. Whatever water is available under ground is overexploited too.

In such a situation, we must guide the runoff from rooftops and footpaths and other hard surfaces into storage reservoirs or underground water tanks. *Rain is the primary source of water.* Forgetting this truth, we depend on secondary sources like rivers, lakes and wells for our daily needs. This trend must be reversed. Water harvesting was an Indian way of life in the distant past. We must revive our traditional methods of water harvesting and also implement new methods.

At the Centre for Science and Environment (CSE), New Delhi, most of the rainwater is captured and recharged into underground aquifers or stored. It is not allowed to drain away into streets. The Centre uses various methods for water harvesting. Annual Rainwater Harvesting Potential (ARHP) at CSE is 366,600 litres. The Rashtrapathi Bhavan has taken many measures for harvesting rainwater.

Many children in India are harvesting rainwater successfully in their respective schools. Some of them are Bishop Cotton School, Simla; St. John's School, Jalpaiguri; Mussorie Public School, Mussorie; The School (KFI), Chennai; Shri Ram School, Gurgaon; Pallikoodam, Kottayam, Kerala etc. The same can very well be done by college students too. Colleges in general have large flat concrete roofs; all the rain falling on these roofs can be harvested through suitable methods.

Keeping the environment clean is the duty of every citizen. We cannot and should not depend on the government for everything. In the previous paragraph, the wasteful style of living of people in the rich countries was discussed. But we have to learn many things from them too. For example, nobody throws paper or plastic bags or toffee wrappers on the roadsides there. In our country, this is a common occurrence. People keep their homes clean, but in general, are seldom bothered about their surroundings. This attitude has to be changed. Bits of paper, toffee wrappers, etc. can easily be put in pockets or hand bags, taken home and disposed of properly. For managing household wastes, composting and vermicomposting can be done. (See Unit 5 for details)

Using plastic carry bags must be discouraged too. Cloth or jute bags can be used while shopping. These are biodegradable, while plastic is non degradable. Plastic will not decay for years and will pollute the soil. People throwing used plastic bags here and there are leading to another lesser known problem. Cows are eating them and are developing tumors. There have been cases of removing kilograms of plastic bags from the stomachs of cows. It is not practically possible to stop using plastic; it has its own advantages. In the case of the small size plastic bags, it must be realized that those black in colour are the thinnest, being the most recycled and the most dangerous. Food put directly into these bags becomes toxic; may even be fatal. Therefore, these must not be used at all. White ones are the thickest, those of other colours slightly thinner. These differences in thickness shows which ones have been recycled more number of times.

Segregation of wastes at source is a very good practice. Degradable garbage like kitchen wastes, paper, dry leaves etc. must be put in one bin while non-degradable wastes like plastics, metal, broken glass, crockery etc. in another bin. This way, the contents of the first bin can be used at home in the making of organic manure while the contents of the second bin can be deposited in the municipality waste collectors, or in their absence, in a common dumping area from where rag pickers can take them, sort them and send them to recycling plants. The wastes that cannot be recycled usually find a place in landfills.

Water pollution is a major concern in all countries, but more so in the developing countries. Many industries in India discharge their waste materials into rivers and other water bodies. People also use water bodies for dumping garbage. This should never be done. Drinking polluted water causes many diseases. In India and other less developed countries, waterborne diseases like diarrhea, dysentery, jaundice, cholera etc. and diseases spread by mosquitoes like malaria, filarial, encephalitis (brain fever) etc. are taking a heavy toll year after year only due to the lack of proper waste management. In 2006, many cases of dengue fever and the uncommon chikungunia were reported from many parts of the country. In some localities, they even reached epidemic proportions. These diseases are also related to sanitary conditions in the environment.

The value of natural resources must be realized fully by everyone. Just like water, other resources like food, minerals, wood, petroleum products etc. should not be wasted. Rich and developed nations like the US follow a wasteful style of living. Large quantities of food, paper, plastics etc. go waste in the US. They do have facilities for waste management; still wasting resources must be considered a crime, environmentally and also philanthropically. It is an irony that tonnes of food is wasted in some parts of the world while millions are starving in other parts. Wasting food is a practice with some people in India too. While having food, one must think of how many peoples' efforts have gone into the making of it, starting from the croplands to the dining table.

Biodiversity is precious. The possible gene combinations are so many in nature; the number of chemicals produced by living organisms based on them is numerous. There are thousands of plants, animals and microbes whose uses we do not know yet. Therefore, any species becoming extinct is a loss for us. As such, some species become extinct as time passes, due to various natural factors. We should not add to their numbers by our activities. A species lost is lost for ever.

Wild species are most essential for plant and animal improvement. They contain the wild or the dominant genes which are used in breeding experiments. New diseases appear now and then in plants and animals. Resistance to these diseases is seen in the wild species. Such resistant genes have been used to create disease-resistant plants and animals all over the world. Wild stocks cannot be maintained in parks and zoos, as the attention they receive there results in rapid loss of variability. Therefore, *entire habitats must be protected.* To prevent deterioration, they must be given the status of national parks, sanctuaries etc. At the same time, zoos, aquaria, botanical and zoological parks etc. are also essential, for they help urban people to know about the wild plants and animals and the variety in nature, thereby making them more environmentally oriented. Both these types of protection, known as *in situ* and *ex situ* conservation respectively, also help to maintain the biodiversity of nature.

Conserving nature for future generations is the duty of the present generation. We are duty bound to prevent extinction of species. If we do not take care, our future generations will not get a chance to even see many of the creations of nature. This will be a great injustice to them. We inherited a beautiful nature with a great variety of plants and animals from our ancestors. We are reaping all benefits from it. Now we cannot destroy the same and deny our descendents what we inherited.

Ecotourism combines conservation with development. The revenue generated by ecotourism can be reinvested in conserving the nature. It also gives employment opportunities to traditionally remote and marginalized communities. It increases environmental awareness and creates a desire in the people for developing more and more tourist spots. It builds public support for conservation and encourages private sector participation in conservation efforts. However, care must be taken that areas opened for ecotourism do not get degraded by the activities of the tourists themselves.

Nature and natural scenes give us great inner peace. Nature has a great aesthetic value. We forget all our worries while enjoying the beauty of nature. An unpolluted river, lush green forests, a clean blue sea, a majestic mountain, muticloured flowers, birds and butterflies—all give us tranquillity. Deformation and pollution of the beautiful nature is a crime in itself.

Pollution, overexploitation of natural resources and lack of water harvesting and proper waste management practices are the chief environmental problems facing mankind today. All these problems can be rectified to a great extent by creating environmental awareness among the general public. Students, as an enlightened group of energetic youngsters, can play a major role in bringing about this awareness in society.

NEED FOR PUBLIC AWARENESS OF ENVIRONMENTAL STUDIES

Pollution of air, water and soil, and mindless utilization of natural resources have raised environmental degradation to dangerous level across the world. One of the methods to reverse this trend, undoubtedly, is creating environmental awareness and educating the people. What is needed is a holistic approach, involving all strata of the society.

The need for creating public awareness of environmental issues has been discussed at national and international levels. The goal is to make people aware of and concerned about environmental problems, and work individually and collectively towards solving current problems and preventing the occurrence of new ones.

People's attitudes towards their environment determine to a large extent how their lifestyle affects the environment. *If people are aware of the environment and its significance, they will act in an environmentally responsible manner and support the environmental programmes.* Education, training and supplying information to the people are the tools to boost this up. Information must be presented to different target groups so as to serve the local need for the immediate solution of the problem. Activities must be organized in every school, college, panchayat, village, town and city. It is in this context that NGOs assume a important role as they are in constant touch with the local communities.

Mass media like radio, television, newspapers, documentary films, hoardings, etc. have a great influence on the people. Media can spread awareness with ease. More visible changes are given sufficient coverage by the media. For example, we regularly get information on population growth and disappearing forests. The media also help in giving the public information on how to construct rainwater harvesting systems, where to get eco-friendly products, etc.

To have a more positive impact, the media will need to understand the issues and develop a long-range outlook. Mass extinction is a very important issue, but as there is not much change in a day, it is not considered newsworthy. Such long-term changes should also receive the attention due to them.

Population explosion is perhaps the biggest problem faced by our country today. This is leading to rapid environmental degradation. According to the World Bank Report 1984, in North India, it is the population growth that has led to severe deforestation, resulting in devastating floods. Desertification, pollution and all other environment related problems increase greatly by unchecked population growth. So, spreading awareness about the need for family planning is of utmost importance.

The need to increase literacy can never be overemphasized. Non-formal methods of education like Adult Education, Total Literacy Mission etc. have been taken up by both central and state governments and also various NGOs in India. These have been partially successful.

Environmental awareness among the common people is seen more in places where literacy standards are high. This is but natural. Kerala can be cited as a typical example. People protest vehemently if any industry is polluting air, water or soil. Similarly, they are conscious of problems of deforestation, rehabilitation etc. on construction of major dams. The Silent Valley project had to be dropped by the government as it would have led to the deforestation of vast tracts of forest land with great biodiversity. In November 2006, there was a statewide protest when just ten families left the premises of Mullaperiyar Dam fearing evacuation due to rising water level and a possible breach in the dam. This is undoubtedly the result of education and awareness. Therefore, literacy campaigns must be conducted on a war footing in India. It will help in creating awareness about not only environmental issues, but also the importance of sanitation, hygiene and health, family planning and many other issues, all related to environment and overall welfare of the general public.

Students can take part in various activities to create environmental awareness among the general public. Some of these are:

- Organize activities like seminars, workshops, competitions in essay writing, elocution, slogans, posters and quiz etc. at college and inter-college levels
- Conduct public meetings, festivals, rallies, exhibitions, demonstrations, folk songs and dances, street theatres, science fairs etc. on various themes
- Distribute pamphlets/handbills on environmental issues
- Celebrate Van Mahotsav
- Participate in local literacy campaigns
- Write relevant articles and publish them in newspapers, magazines etc.
- Help neighbours and other members of the society in setting up community waste management and rainwater harvesting projects
- Discuss with friends and relatives topics like resource management, pollution control, saving water, paper, energy etc.
- Set practical examples by using cloth and jute bags instead of plastic, walking short distances instead of using automobiles etc.
- Encourage civic sense and hygiene by not spitting on walls and roadsides, not urinating in public places, not throwing garbage on roads and inside trains and buses, not smoking in public etc.
- Join local NGO's who popularize environmental issues
- Read newspapers and magazines regularly and be up to date with global environmental issues
- Visit local zoos, parks, sanctuaries etc.
- Celebrate important days on the Environmental Calendar, like World Environment Day, Wildlife Week, World Food Day, World Popualtion Day, etc.

There are many institutions working for the betterment of environment all over the world. In India, the ministry of Environment and Forests (MoEF) has many divisions and departments, with various centres and institutes working under it. Some of them are listed below:

1. Centre for Environment and Education, Ahmedabad

2. CPR Ayar Environment Education Centre, Chennai
3. BB Pant Himalayan Paryavaran Evam Vikas Sansthan, Almora
4. Tropical Botanic Garden and Research Institute, Thiruananthapuram
5. Centre for Ecology, Research and Training, Bangalore
6. National Museum of Natural History, New Delhi
7. Salim Ali Centre for Ornithology and Natural History, Coimbatore
8. Botanical Survey of India, Kolkata
9. Zoological Survey of India, Kolkata
10. Centre for Environmental Management of Degraded Ecosystems, New Delhi
11. Centre for Mining Environment, Dhanbad
12. Central Pollution Control Board
13. Wildlife Institute of India, Dehradun
14. Indian Institute of Forest Management, Bhopal etc.

These institutes and centres do valuable work in:

- Spreading environmental awareness and promoting environmental education
- Publish magazines and journals
- Train environmental educators
- Conduct programmes on conservation and biodiversity
- Perform Environment Impact Assessment work
- Produce research scientists
- Train personnel in eco-development, wildlife biology, habitat management, disaster management etc.
- Procure information on India's wildlife wealth
- Prepare extensive lists of the wetlands of the country
- Prepare resource material on Environmental Education
- Prepare books, posters, video films etc. and conduct workshops, seminars, conferences etc.
- Establish Nature Clubs and Eco clubs etc.

Many awards are also given annually to citizens/institutions by MoEF, like:

- Indira Gandhi Paryavaran Puraskar
- Pitambar Pant National Environment Fellowship Award for Biodiversity
- Rajiv Gandhi Wildlife Conservation Award
- Indira Priyadarshini Vriksha Mitra Award

- Dr. Salim Ali and Dr. Sankala Fellowship etc.

Many projects have been set up in India to protect wildlife. Some of these are the Project Tiger, Project Elephant etc. Crocodile breeding is done in the Madras Crocodile Bank Trust. Survey of plant resources is done by BSI while corresponding surveys of animals is done by ZSI, both located at Kolkata. We also have the Ganga Action Plan, the Yamuna Action Plan etc.

Many other national organisations are also doing valuable work in the feld of environment. Some examples are:

- State Pollution Control Boards (SPCBs)
- Industrial Toxicology Research Centre (ITRC)
- National Environmental Engineering Research Institute (NEERI) etc.

In addition to this, hundreds of non-governmental organisations are also involved in various projects. Some well-known NGOs in India are:

- World Wide Fund for Nature-India (WWF-I)
- Kalpavriksh, Pune
- Kerala Sastra Sahitya Parishad (KSSP)
- Bombay Natural History Society (BNHS)

The names of many individuals who have spread the importance of conservation of environment stand out clearly in our country. The contributions of Medha Patkar, famous for her Narmada Bachao Andolan, Sunderlal Bahuguna of Chipko Movement, Salim Ali, the great ornithologist (expert on birds), M.S. Swaminathan, the agricultural scientist, Rajendra Singh, the Magsaysay award winner for water harvesting, M.C. Mehta, the famous environmental lawyer etc. are unforgettable.

The good work done by them and so many others has to be carried forward with a renewed zeal. There are thousands involved in projects creating awareness and propagating environmental education in our country. Yet the task cut out for us Indians is an immense one as we have to educate one hundred crore people. There is need for as many more environmentalists as possible. And it is here that young, creative, motivated students can render invaluable service to the nation.

QUESTIONS

A] Very short answer type questions:

a) Multiple choice type: Tick the right answer:

1. The largest ecosystem on the earth is:

   (*a*) The oceans (*b*) Rainforests

(*c*) Mountains (*d*) The earth itself

2. Salim Ali was an expert in:
   (*a*) Agricultural Sciences (*b*) Rainwater harvesting
   (*c*) Ornithology (*d*) Environmental law
3. NEERI is:
   (*a*) an Indian Environmental Research Institute
   (*b*) an NGO
   (*c*) an international organization
   (*d*) a Pollution Control Board
4. Which of the following waste material is degradable?
   (*a*) Plastic (*b*) Metal
   (*c*) Glass (*d*) Paper
5. Which of the following diseases is not spread by water?
   (*a*) Diarrhoea (*b*) Malaria
   (*c*) Cholera (*d*) Dysentery
6. Botanical Survey of India is located in
   (*a*) New Delhi (*b*) Mumbai
   (*c*) Kolkata (d) Chennai

b) Define the following:
   1. Environment
   2. Environmental Studies
   3. Biosphere
   4. Biome

c) Name the following:
   1. Five disciplines related to Environmental Studies
   2. Five primary natural resources
   3. Five centres/institutions under MoEF, Government of India
   4. Three Environmental awards given in India
   5. Three diseases caused by drinking contaminated water

B] Short Answer type questions:

a) Differentiate between:
   1. Renewable and non-renewable natural resources
   2. *In situ* and *ex situ* conservation

b) Write short notes on:
   1. Deforestation
   2. Effects of population explosion on environment
   3. Ecotourism
   4. Role of media in creating environmental awareness
   5. Functions of MoEF and the centres and institutions working under it

6. Environmental awareness and literacy
7. Segregation of waste at source
8. Use of wild species in crop improvement

C] Long answer type questions:

a) Write long notes on any two:
   1. Water pollution
   2. Biodiversity
   3. The advantages and disadvantages of major dams
   4. Rainwater harvesting
   5. The advantages of minor dams

b) Discuss in detail the scope of Environmental Studies.

c) What can students do to promote environmental awareness in society?

d) Give an account of the importance of Environmental Studies.

e) Public awareness of Environmental Studies is the need of the hour. Discuss.

Unit 2

Man, Environment and Society

There has been life on earth for over one thousand million years. Mammals, the most evolved of all animals, which carry and nourish their young inside their bodies and give birth to them when they are fully developed, have been on earth for perhaps seventy million years. Man-ape appeared less than ten million years ago. But human beings who were no longer apes emerged only about less than one million years ago.

Man or Homo sapiens is a mammal with a fully modern brain capable of thinking and reasoning. Man evolved on the earth about a hundred thousand years ago. Considering the time scale of history of life on earth, his era has just begun. Yet in this short time on the earth, he has learnt to control, to a great extent, his own surroundings and the lives of other living things.

Bodies of other animals are rigidly set to one special set of circumstances. In the long process of evolution, they have adapted many times to circumstances. But their changes were slow, taking hundreds of generations to complete. However, man by means of his brain and hands, can adapt quickly to changing conditions, not by altering slowly the shape and size of his body, but by making use of the things which surround him.

Man's ancestor was the ape. Though they are similar in some ways, they differ greatly too, in many important respects. When man descended to the ground from trees, he had to depend on his brain to escape from enemies. For living, hunting and working together, speech developed first and then the language. Language gave man the ability to share thoughts and feelings. This resulted in the formation of society and civilization.

Society may be defined as a group of people occupying a specific locality who are dependent on each other for survival, and who share a common culture. The economic systems and the family relationships of people show how they depend on each other. Members of a society are held together by a sense of identity. A society in which there are differences in the cultures of its people is called a pluralistic society.

All the living and the non-living entities that surround man is his environment. The early man took a great step forward when instead of depending entirely on wild plants for food he began to grow them himself. When he had sown seeds, he had to wait for them to sprout and grow. This took time and so he had to settle near his crops. Homes were built. Spinning and weaving, pottery and metal-beating developed.

By choosing the best seeds from the best plants and by watering the plants regularly, man gradually produced cultivated varieties. Later he found that he could pollinate and cross-pollinate his plants and produce better varieties. Wheat was probably the first grass to be cultivated, between ten and fifteen thousand years ago. It was much later that man understood the importance of fertilizers in cultivation.

The dog was probably the first animal to become friendly to man. Later he domesticated many animals for their flesh, skin, milk and wool. The silkworm has been reared for several thousand years now. Later, man tried his hand at making things artificially, copying nature, and succeeded. Inventions and discoveries continued at great speed.

Man, the Destroyer of the Environment: As the human population and the greed of man increased, the environment's degradation began. Earlier, man used to kill animals only for food and clothing. But soon, he started to kill for sport and pleasure. Thoughtless destruction of animals has always resulted in an increase in the number of their prey. Snakes are killed for various reasons, but then mice are multiplying fast when there are no snakes to keep their numbers down.

To cultivate crops, he destroyed large areas of wild forests, resulting in large-scale extinction of many valuable plants and animals. When man cuts down great forests for timber, disaster follows unless he plants new trees. In Syria, Palestine and North Africa, cities of the past have become deserts due to this reason. Loss of wild plants and animals is a permanent loss to mankind. We never know which plants and animals in the wild have useful properties. Who can say with certainty that chemical substances which can fight cancer or AIDS were not present in the plants or animals that have become extinct now?

Man, the Preserver of the Environment: Luckily, man is beginning to realize his responsibilities towards living things and to the earth as a whole. He understands that for every tree cut, another should be planted and that the soil must be enriched every time a crop is harvested. He knows that the wild plants and animals must be protected and that the water bodies must be kept clean.

How far he is putting these ideas, ideals and principles into positive action is what needs to be seen in the coming days. If man thinks that he is the lord of the creation and all things living and non-living have been created for his benefit, there is no escape for him from sure extinction. However, if he realizes that they are part of the same nature of which he is a part himself, that will be evolution of the human mind.

INDIVIDUAL'S RIGHTS AND RESPONSIBILITIES TOWARDS A CLEAN ENVIRONMENT

Rights go hand in hand with duties or responsibilities. Duty without rights is slavery; at the same time rights without duties is dogmatism. This principle is applicable, as in all other spheres, to environment also.

The problem is that the word 'right' is not always connected to courts, though it does look like a legal word. In law, 'rights' are always rights held by one person against another or others. They can also be held only by beings that are able to stand up and argue their case in a court of law. As this is not possible in case of animals, some people argue that animals cannot have rights. Similarly, how can a polluted river or an overgrazed pasture or a rainforest stand up and argue in court? Therefore, in the cases cited and the like, instead of talking about the rights of animals or ecosystems, we have to think of what our duties or responsibilities towards them are. Thus there are no separating duties from rights. They co-exist.

Some moral theorists insist that rights and responsibilities are correlative. In other words, if someone has a right, someone else has a responsibility to meet it. Put more strongly, no one can have a responsibility to anyone who does not have a return responsibility.

Individuals have duties or responsibilities, no doubt, towards other individuals and also towards their environment. They may fulfill it, they may not. They may think it is their own choice or wish. Moreover, in many cases, it is not clear to whom the responsibility belongs.

In such cases, it is better to term them moral responsibilities that are not legal. These had been clearly specified in ancient Indian scriptures. For example, while referring to the environment of life, *the Vedas consider the earth, the space, the sun and the solar system collectively.* All these are parts of nature and they all depend on each other. All natural resources and life forms (plants and animals) together are supporting man and so they must not be disturbed. Some examples:

- Pure *vaayu* (oxygen) is like elixir (*amrit*) and so it should be always available in its natural pure form.
- The earth and the atmosphere nourish us; if they are harmed, the result will be global disaster.
- We must take care of the earth, the atmosphere and the vegetation.
- We must not pollute water and destroy plants.
- A Vedic prayer to Mother Earth goes thus, "Whatever I dig out from you, O Earth! May that be regenerated fast; May we not damage thy habitat and heart".
- The famous 'Shanti mantra' or the Verse of Peace in Yajur Veda invokes peace (natural harmony) to the earth, sun, atmosphere, water, trees, medicinal plants, the entire vegetation, all bipeds (animals walking on two feet), quadrupeds (animals walking on four feet) etc.
- If we dig the earth for its treasures, we must fill it up. We should not harm the soft and sensitive parts of Mother Earth as that will lead to imbalance of the equilibrium. It has been reported that this verse was quoted by Late Smt. Indira Gandhi in an international conference on environmental pollution.

In addition to protecting all herbs, shrubs and trees, Vedic scriptures advised people to give special importance to Peepal, Tulsi, Neem etc.

- It is well known that Peepal absorbs a very high quantity of carbon dioxide from the air and releases an equal amount of oxygen.
- Tulsi releases oxygen for most of the day and night; it also has a high medicinal value.
- Neem is a powerful disinfectant acting against germs and insects. Nowadays, Neem is praised as the best biopesticide available in nature.
- The scriptures have given the uses of many other plant species and have instructed and also motivated people in different ways to plant and protect them. For example, medicinal plants (*vanaushaudhi*) and other vegetation (*vanaspathi*) of the forests were personified as Jungle Goddesses in the Vedas. Because of this, people cared for the forests, and protected them.

Not only the Vedas, other scriptures like Smritis and Puranas, *Kautilya's Arthashasthra* etc. give people very clear instructions about what to do and what not. Given below, are some examples:

- *Manusmriti* instructs people that only dry twigs should be used as fuel; those who cut green plants should be punished suitably.

- Varahamihira in *Brihatsamhita* (550A.D.) instructs us that we must plant trees on the banks of lakes and ponds to make them strong and attractive. He has also given a list of trees most suitable for this purpose.
- *Arthashasthra of Kautilya* tells us how to manage village tanks:
  1. Ponds, tanks etc. can be privately owned.
  2. The owner is free to sell or mortgage them.
  3. The ownership shall lapse if the tank has not been used for 5 years, except in cases of distress.
  4. Anyone who accepts a waterworks as a pledge, with the right to use them, shall maintain them in good condition.
  5. Owners may give water to others in return for a share of the profit made by them through fields, parks or gardens.
  6. In the absence of the owners, other individuals or the villagers as a whole shall maintain the waterworks.

It is amazing how ancient saints suggested simple methods of water conservation. For example, one instruction is: "Take out 5 handfuls of soil from a pond each time you enter it for a bath". Just imagine, if everyone follows this instruction dutifully, how well the depth of our water bodies would have been maintained! At present, due to siltation (the falling in of soil from the edge of a water body), our ponds and lakes are losing their capacity to hold water by 1 to 2% every year.

The *Puranas* state that:

- Those who plant trees will be blessed by good children.
- One who gets a well or pond dug for use by others will go to heaven.
- Dirtying rivers and ponds and their surroundings is a sin.

The custom of worshipping rivers and trees and animals and birds — Mother Nature as a whole — was very strong in the ancient days. Even in the 3rd or 4th millennia B.C., trees were worshipped by the Harappans.

Respect towards the Earth is strong among Indians because of the Vedic teachings. To make sure that everyone participated in the conservation of nature and natural resources, religious norms and social duties were set down by the scriptures. This is exactly what we mean by the 'community participation' or 'participatory management' professed by today's environmentalists.

All ancient Indian scriptures teach us to lead a nature - friendly life style. They encourage each individual to protect the environment. They motivate the citizens to such an extent in a positive manner that each individual is convinced that it is his responsibility to take care of all parts of the environment.

Many philosophers think that sense of duty has an emotional and sentimental basis. The 18th century German philosopher Immanuel Kant thought otherwise. According to him, duty is the result of a purely rational understanding of what must be and should be done. One understands what is right through reason. Understanding that something is right is also an understanding that it must be done. It is this that gives rise to a sense of duty.

We can see clearly how both these viewpoints have been beautifully combined in the Indian scriptures. The people used their reasoning power and understood that protection of the environment would be in their and everybody else's benefit; at the same time, the sense of duty they had was based on emotion and sentiment also because they had great respect and love for their seers.

Accountability is an important factor for determining responsibility. To say that someone is accountable for something gone wrong, first of all, that person's responsibility should have been specified. In such a situation, the violation of those duties makes the person answerable. In the case of environment, as stated earlier, the defendant (the polluted river or the overgrazed forest) cannot speak out; what punishment the accused should get is therefore decided by the community as a whole, as per the punishments prescribed by the law makers. This aspect has also been dealt with most admirably in the scriptures.

Kant has spoken at length on goodwill also. A person with goodwill is one who feels within that he has a certain set of obligations to others. He is different from another who will act according to his sense of duty, but only because the rewards and benefits it will bring to him. The former are at a higher level of thinking though the latter cannot be found guilty too. Both categories of people finally end up doing the same things. In ancient India, this was exactly what was happening. The environment was taken care of by all.

In contrast to this, *many people in the modern times believe that protecting the environment is the sole responsibility of the Government.* In many parts of India, we see the local people facing acute water scarcity; but they would not adopt any water conservation method themselves. They are waiting for the municipality to do it for them. The common open courtyards may be full of garbage; but the house owners do not think of a community compost pit or a biogas plant. They are waiting for the municipality to take the garbage away. It is because people think of their rights only. No one is bothered even to think about his duties.

There is the other side of the coin too. It is the duty of the government to provide its people with clean drinking water, with proper solid waste management systems. Is the Government fulfilling its duty?

In short, *everybody is after rights; nobody is for duties.*

Otherwise, how is it that we have National Human Rights Commissions in different countries and an International Human Rights Commissions too? Why don't we have National and International Human Duties Commissions?

If everyone does his duty, no one needs to worry about his rights. Everything will fall in place, and the world will be a better place to live in.

ENVIRONMENTAL ETHICS

Ethics is concerned with:

- What is morally good behaviour
- What makes a person good, and
- What principles we must use in decision making.

As one Greek philosopher said, the goal of studying ethics is not to know what is good, but to become good; otherwise there is no meaning in studying it.

Ethics involves a lot of reasoning; therefore it is based on logic. Yet, emotions influence the human mind more than reasoning. Therefore an easier way of motivating people to do anything is to create, encourage and enhance moral emotions in them. This point of view has influenced ethics also because ethical thinkers often argue that if we follow the right principles, that will lead to peace of mind, happiness and a more fulfilling life.

An ethical community comprises three elements — the family, the society and the country. These three factors together make an individual what he/she is. Finally, there are the universal ethical communities also. In the field of environment, there is definitely an increasing international consensus that there is an urgent need to protect the environment. The number of international treaties and agreements regarding various environmental issues is proof of this. All over the world, a general feeling has been expressed that we must respect the rights of the future generations. It is the duty of the present generation to protect the environment with all its biodiversity and its indigenous cultures for the generations to come.

The view that nature matters only where human welfare is concerned is wrong. This is a purely western thinking. Some westerners even believe that condemning cruelty to animals is good because it elevates the human race by stimulating compassion! According to this viewpoint, cruel practices towards animals are not wrong in themselves. They become wrong only if humans become less compassionate to other humans. However, as mentioned earlier, ancient scriptures of India show very clearly how the environment as a whole was treated as one entity in the country. We have always believed that man is part of nature, not that man is master of nature. Our principle has been 'Vasudhaiva kudumbakam' meaning the whole world is one family.

Yet *there is need to inculcate feeling of love towards nature in each and everyone on this earth.* The earlier this is done the better, because despite of the vast information we have on the dire consequences of human interference in the biosphere, many of our activities continue to harm the environment. We destroy forest covers, drain wetlands and pollute air, water and soil in various ways. The danger that human activities may lead to rising of sea levels and global climatic changes remains true. It is high time therefore that environmental education becomes part of the education process at all levels.

The different ethical issues related to environment are as follows:

S. No.	*Name of the country*	*GNP per person, 1998*
1	Japan	32,350
2	United States	29,240
3	Germany	26,570
4	Brazil	4,630
5	South Africa	3,310
6	Russia	2,260
7	China	750
8	Indonesia	640
9	India	440

Equity disparity among countries: There are vast differences between the economies of the developed countries and the developing countries (MDCs or more developed countries and LDCs or less developed countries, as they are referred to these days). Gross national product per person of some countries as per World Bank, World Development Indicators (2000) is given above: (Figures in dollars — value of 1 dollar was about rupees 35 in 1998).

The figures speak for themselves. Though there are poor people in the rich countries also, most of the poor in the world live in Africa and Asia. World Bank figures show that almost half the world's population survives on about 1 dollar per day.

More than 1 billion people do not have access to clean drinking water. The available water itself is polluted in many areas and this leads to many water-borne diseases. The number of people dying of cholera, typhoid, jaundice, dysentery and such other water-borne diseases is very high in the LDCs. Diseases that spread from garbage, malnutrition, infant mortality etc. are all more in these countries.

Equity disparity within countries: Within countries also, disparities are very clear. In general, in the rich countries, the difference in income between the rich and the poor is not as much as in the poor countries. *In the poor countries, the rural masses are affected most by environmental degradation.* The quantity of firewood collected by the poor from the forests is much less when compared to the huge quantities used by different wood based industries like timber, paper, plywood, rubber, package, etc. Thus the rich industrialists consume the forest produce the maximum while the rural poor take the blame for the deforestation and at the same time remain poor. Similarly, the poor fishermen are often blamed for over utilizing the marine resources. The fact is that the rich use trawlers go deep into the sea and collect fish in huge quantities while the poor fishermen using simple boats cannot penetrate far. It is thus the poor who are at the receiving end, always.

As such, *more and more people are migrating from the villages to towns and cities.* There is not enough of water or food or land for the migrants, aren't enough of roads or healthcare systems for them in the towns and cities. Yet, unable to endure the poverty in our villages, many people migrate to urban areas. The final result is that poverty spreads everywhere. If the rich in the rural and also the urban areas waste resources just because they can afford to waste, it is the poor all over who suffer.

Inequality in consumption of resources: There is no doubt that the *LDCs remain less developed because they have not been able to control their population growth* rate to the desired level. When the population is more, demand for everything increases — water, food, firewood, coal and petroleum — every type of natural resource. There is a big deficit in supply in many of these. Some LDCs do produce enough food and more, but then, these are exported to the MDCs; as a result, the starving poor are not benefited. Deforestation, desertification, pollution etc. all increase when population increases.

Even with all this, *consumption of natural resources is much more in the MDCs*. This is because the people in the rich countries demand more from nature. They use much greater quantities of energy resources like petroleum, electricity etc. Their usage of food and water resources is also more. Consumption of resources leads to industrialization and technology, which increase and consume still more resources. Similarly, machines and technology lead to surplus wealth, which is invested in more machines and more technology. In this way, resource consumption becomes a vicious circle in all rich countries. Above all, most of the rich countries are wasteful in nature.

USA has only about 4% of the world's population, but consumes about 25% of the world's resources. One reason for this high consumption is the high wastage.

The gender inequality: *Women who constitute 50% of the population do not get justice in any matter.* Women work for longer hours than men, for they have to do the household work and in addition work outside the home. They get up earlier and go to bed later than men. Women are mostly undernourished and often fall prey to anaemia. In areas of water scarcity, women have to walk long distances to bring water for household needs. Again, it is the women who go to the forest to collect firewood. Back home, they have to cook using fuel inefficient stoves. The fuel efficient stoves designed by educational institutions in the country have reached to only very few villages. In the field of education, women lag behind men everywhere. Whether the national average or the average of any state in India, truth remains that less women are literate than men. Because of this, they do not know their rights and also are not given any say in matters concerning society. They do not have the freedom even in matters related to themselves.

Environmental degradation affects women the most. Women feel the maximum brunt of water scarcity, deforestation, lack of good alternative energy sources etc. It has always been women whose health is not taken care of properly in society. On the whole, girls and women do not get the share due to them in society. Unless this state of affairs changes, society as a whole can never progress. And only in an enlightened society can there be any environmental awareness.

Destroying natural resources belonging to future generations: If the present rate of destruction of forests and water bodies and coral reefs continue, what will be left for the future generations? This is a question to ponder. We do not own the earth or its resources. We are only their custodians. We have to pass on these resources intact to our children. In fact, we have not worked hard to get these resources in the first place. These have been handed over by our forefathers. So now it becomes our duty to preserve them as best as possible for the benefit of future generations. Living in a clean environment is everyone's right. One generation cannot deny that right to another generation.

When forests are cut down, many animals have nowhere to go and they finally perish. The felling of even a single tree changes the environment. Many birds lose their nests, many squirrels, snakes etc. lose their shelter, many plants living in the shade of the tree die and countless microorganisms and small insects and worms in the soil are deprived of their food, as they eat the leaves fallen from the tree. In the absence of the tree, soil erosion, light intensity, temperature and wind velocity of the place increase, while fertility of the soil and atmospheric humidity decrease. Thus *cutting down of one tree is destroying one entire ecosystem.*

When the temperature of the sea water rises even slightly as a result of our activities, polyps that make the coral reefs die. When mercury and other such toxic chemicals from our factories are released into rivers, innumerable fish die. We drain wetlands, home to a variety of unique animals, only to make homes for ourselves. The list is truly endless. We owe it to our children that they also enjoy the benefits of the vast biodiversity on this planet. We have no right to destroy it.

Animal rights: While thinking or talking about animal rights, the first thing for us to remember is that we are also animals. We are one type of mammals; that is the truth. Agreed, we are more intelligent than all the other mammals, but that is all the more reason for us to understand that they, who depend on us in many ways, deserve our kindness. *Cruelty to animals is wrong ethically, just like cruelty to fellow human beings is wrong, ethically.* The animals are never cruel to us; they do not harm us for the sake of harming us. Lions and tigers kill their prey only when they are hungry. That is not wrong. Likewise human beings also catch fish or kill animals for food. That is part of the balancing system of nature. But men kill for sport too. This is cruelty. They cage birds; this is cruelty. We must respect the rights of animals, especially as they cannot talk. Finally, understanding that by protecting the animals, we are benefiting ourselves too is plain logic. In fact, what is good for nature is good for man as man is part of nature.

ETHICS AND MORALS

There are positive laws and natural laws in the world. *The positive laws are the laws that are enforced by the police and the judiciary*. But *to enforce the rights of the oppressed or the animals, a different kind of law is necessary. Such laws are called natural laws. These are based on ethics and morals*. These are the same to all, irrespective of countries, religion, caste, class, creed, age, sex etc.

If human beings have the same rights everywhere, it is because of ethical and moral standards. These are universal in application. Yet, in some countries, where the laws of the land are not being implemented truly, sections of people remain oppressed and there may not be any solace at all as the oppressors naturally have no ethical or moral feelings.

In fact, there is no practical difference between ethics and morals because ethics is the science of morals. It is concerned with human conduct and character. It is a system of morals or a set of rules of behaviour. Morals relate to character or conduct considered good or bad. Moral behaviour is always directed towards right. An individual with high morals is able to differentiate between right and wrong. Moral laws are supported by evidence of reason.

For an individual to feel ethical towards the planet earth and its resources, environmental education is a must from childhood. If a child grows up admiring the beauty of nature and its infinite variety, he will definitely be concerned about the environment throughout his life. This will lead to the making of successive generations concerned about nature and natural resources.

APPROACH OF SWAMI VIVEKANAND TOWARDS YOUTH, WOMEN AND SOCIAL DEVELOPMENT

Narendranath Dutta was born on 12th January 1863 in Bengal. Later he became a disciple of the great saint Ramakrishna Paramahansa. Soon he took the name Vivekananda and became a swami himself. He was an icon for his courage and fearlessness. He gave many speeches on the Hindu Philosophy. *He is also remembered for his positive outlook on the role of women in society and social problems* in general. *He used to inspire the youth to work hard for the country. Swami Vivekananda's birthday is celebrated even today, every 12th January, as National Youth Day.* He was against superstitions. He spoke against discrimination based on caste and religion.

Swami Vivekananda travelled throughout India. During these travels, he had stayed with kings as well as paupers. He saw for himself the disparity in riches and also how the lower castes were dominated by the upper castes. He understood that if India wanted to progress, a national awakening was necessary. He advised people to have faith and be holy and selfless.

In 1893, he addressed the World Parliament of Religions in Chicago, America. When he began his first lecture addressing the audience with the words, "Sisters and Brothers of America", there was a standing ovation for full two minutes! It was the first time that the Americans had heard such an affectionate way of address. He entered straight into their hearts with those words of his. In fact, that was the beginning of western interest in Hinduism.

After his return to India, he gave a series of lectures here. These lectures helped to raise the spirits of the demoralized Indians of those days to a great extent. *He encouraged people to serve the poor* and told them that this was indeed service of God. He named this type of service 'Daridra Narayana Seva'. He argued that if God resided in all, then there could not be any difference between different individuals, whichever caste he belonged to. However, there are many people who do not know this fundamental truth. So it is the duty of all who know this truth to teach the same to all those who do not, and help them come up in life.

Swami Vivekananda believed that religion must not mingle with the government. Social customs are usually formed with the permission of religious heads, but once these are formed, religion has nothing to do. The four basic castes have their own functions in society. All are essential for the well being of the society. Therefore, if any one dominates another, there will not be harmonious relations in society. Religion must not be used to change rules in the society. The rules would change by themselves if necessary and at the right time.

Here are some quotations from Vikenanda:

- "Aye, who ever saw money making the man? It is always man that makes money. The whole world has been made by the energy of man, by the power of enthusiasm, by the power of faith."
- "It is a tremendous error to feel helpless. Do not seek help from anyone. We are our own help. If we cannot help ourselves, there is none to help us."
- "All power is within you, you can do anything and everything. Believe in that, do not believe that you are weak."

No wonder then that many years after his death, Rabindranath Tagore said about him, "If you want to know India, study Vivekananda. In him, everything is positive and nothing negative."

Here are some more selected quotes of Vivekananda:

- When I asked God for strength
  He gave me difficult situations to face!!!
- When I asked God for wealth
  He showed me how to work hard!!!

The quotes speak volumes. *Swami Vivekananda believed in hard work.* His words are capable of motivating everyone, especially the youth. *He knew the role of science in bringing about development in society.* It was he who advised Jamshedji Tata to set up the Indian Institute of Science.

Swami Vivekananda always tried hard to instill confidence, enthusiasm and a sense of responsibility among the youth. *He tried to direct their energy towards nation building activities.* Youth must realize their inherent talent and proceed in the right direction; utilize energy and progress economically. They must utilize the available natural resources constructively, keeping away from antisocial activities. Youth must put the nation on the path to peace and progress.

According to Vivekananda, freedom is responsibility. *He focused on youth, women and the poor.* Youth must take the responsibility and try to wipe out poverty from the country. He told the youth, "Take the responsibility on your shoulders. Know that you are the creator of your own destiny."

Vivekananda provided a holistic vision of life. He went to the root of the problems. According to him first we must understand what is influencing us. Then we will know what is wrong.

He knew the correct meaning of education. He said, "Essence of education is concentration of mind, not collecting facts." He also believed that self-discipline and integrity are more important than bookish learning. A nation's future depends on youth, and nation building takes place through character building.

He had great love for mankind and sympathy for the poor and the downtrodden. He wanted them to fight against exploitation. He stood firm for women's rights. He was against the caste system. He believed that caste was the biggest dividing factor in the society. *Casteism was responsible for social divisiveness and poverty.* Brahmins are responsible for priest craft, untouchability and exclusive claim on spiritual scriptures. In India, the tyranny of the rich over the poor was also related to castes. Vivekananda wanted India to be raised by the spread of education. Priest craft must be removed and the poor must be fed. The upper castes are responsible for the upliftment of the lower classes.

Vivekananda studied the position of women in the past and in his own time. He said that *educated women would inculcate virtues in their children* and prepare the new generation who can distinguish these virtues. He stressed the importance of women in house keeping, religion, art, science and hygiene. He said that all girls must develop these skills.

He was also in favour of education of widows. It was indeed a bold step in those times. Vivekananda said that women must be put in a position where they are able to solve their problems in their own way. He believed that no one else ought to do this task for them because our women are quite capable as in many other parts of the world. *According to Vivekananda, upliftment of women and awakening of the masses are the first requisites before any real good can come to India.* "Arise! Awake! How much longer would you remain asleep?" His clarion call to the Indian masses is as much applicable today as it was more than a century ago.

APPROACH OF MAHATMA GANDHI TOWARDS YOUTH, WOMEN AND SOCIAL DEVELOPMENT

In the **Young India** Mahatma Gandhi wrote as far back as 1929, "This is my message to the youth....The youth must believe in the all-conquering power of love and truth." *Gandhiji wanted youth to do voluntary service to society.* He said that the youth must be involved in various social issues. They must have the spirit to serve the country.

Mahatma Gandhi believed that as India lives in her villages, the country cannot progress unless the village life becomes better. For this rural transformation, the youth must play a major role, according to him. He wrote that youth must be persuaded to make village life their goal. He agreed that there are many drawbacks in rural life, but all these can be corrected. *He wanted the youth to set up co-operatives in villages to better the conditions of the rural poor.*

There were more than 7,50,000 villages in India during Gandhiji's times. Urbanisation cannot support the rural population, as he wrote in the *Harijan* in 1934. Urbanisation is the death of villages. If the youth start living in the villages, the villagers will be served and the middle class job seekers will get honourable employment too.

Thus the *Mahatma believed that the youth could be agents of social transformation.* They must put their mind to rest and concentrate on energies. They must spread the message of peace and harmony in society.

According to Mahatma Gandhi, indicators of peace are:

- Elimination of wars
- Absence of conflicts between different classes, castes, religions and nations
- Love
- Rest
- Mental equilibrium
- Harmony
- Cooperation
- Unity, and,
- Happiness

In short, peace is must for social harmony. For peace, there must be proper utilization of natural resources in the society. *Youth must ensure that water, soil, forests, minerals, animals etc. should be utilized properly by uniform distribution and conservation.* The natural resources should not be exploited or destroyed.

Gandhiji exhorted youth to be patriotic. They must be loyal and affectionate to the country, national symbols etc. Patriotic feelings towards the country will encourage the youth to work selflessly for the country.

Society is a mixture of different religions, cultures and races. Gandhiji stressed upon the youth to bring about social changes in society. He especially extolled them to work for the abolishment of the following social evils:

- Sati
- Polygamy
- Untouchability
- Caste system
- Child marriage
- Dowry
- Prostitution
- Forced labour
- Sexual discrimination
- Religious misguidance, and,
- Exploitation

The Mahatma said that youth should believe in:

- Non-violence
- Equality
- Justice, and,
- Love

Violence destroys society. He also told the youth to cooperate with elders and children. Gandhiji put forward the concept of young entrepreneurs (young self-employed) for growth and equality in the country.

Mahatma Gandhi said that everyone is equal in the eyes of God, and so the youth should reduce the gap between the different sections of society. He appealed to the youth to eliminate all discrimination, oppression and exploitation and establish equality. Money in only some hands can cause political instability and social inequality. When there is equality, there is justice.

Gandhiji believed that men and women are equal and that they complement each other, though their functions are different in the society. None can be superior or inferior. He was the first Indian leader to *encourage women's participation in political and social work.* He believed that home life is entirely the field of women and so they should know the domestic chores. They should also have sufficient knowledge in the upbringing of the children. Women however must not be mere cooks or objects of pleasure to men. They are life companions of men. They are equal partners of life, sharing the joys and sorrows of men in all spheres. All our efforts to improve our country will be wasted, if these truths are not recognized.

The Mahatma extolled the youth to follow the principles of *Bhagavad Gita*, get internal happiness and utilize their energy positively. Truth was his main value. He believed in 'Sathyameva Jayathe'. He wanted the youth to practise truth in thoughts, words and actions. They must have the courage to say the truth, always.

Mahatma Gandhi said that basic education is everyone's right. Education gives the right values to people. Role models should come from the educated youth. They will build the nation.

For all this, social transformation is a must and the youth must be involved in:

- Community development
- Education, and
- Health.

He professed the dignity of labour to all. He said, "No labour is too mean for one who wants to earn an honest penny. If the sense of shame that wrongly attaches to physical labour could be got rid of, there is enough work to spare for young men and women of average intelligence." The aim should be a self-sufficient, non-exploitative society that has no social evils.

He had travelled throughout the length and breadth of the country. He saw that there were too many poor in India. He therefore suggested that *in addition to agriculture, people must own simple industries.* The spinning wheel can introduce successful cooperative societies. Handicrafts should not be displaced and machines should not replace labour. However, he did say that where necessary, machines must be introduced.

Gandhi knew, and truly so, that *India cannot progress unless her villages progress*. He therefore had given extremely detailed schemes for the upliftment of Indian villages. He had said, "We have to awaken villagers themselves and make them capable of tackling their own problems and forging ahead through their own strength." Gandhian approach to youth, women and social development is what we need now.

Similarly, Swami Vivekananda's views remain as fresh as ever. Even after almost sixty years of independence, if we have not progressed to the extent we should have, if there is poverty and inequality in our villages even now, it is because we have forgotten the wise words of Swami Vivekananda and Mahatma Gandhi. It is high time the energetic youth of India take up their responsibilities and lead the country from progress to progress.

THROW AWAY SOCIETY ETHICS

As science and technology advance, more and more finished products are being produced every year by the industry for the comfort and luxury of man. Part of the wastes produced during their manufacture may be blown away by the wind or carried away down the streams. Steps may not be taken to dispose them of safely. If the byproducts or side products are utilized in any way, throwing away of these can be avoided, which will also help in keeping the environment clean.

High level of urbanization has resulted in the mushrooming of towns and cities all over the world. While in the MDCs (more developed countries) the disparity between the rich and the poor is not very high, in the LDCs (less developed countries), this disparity is very high. More and more people believe that more money would improve their life. So people dream of a life of wealth. All their efforts are directed towards maximization of wealth. In this rat race, no one has the time to stop and think of the conditions of the poor. The poor are ignored. They remain poor without getting opportunities. Many a times, the poor become poorer because of the activities of the rich, like construction of skyscrapers in areas occupied by the poor. Unsuccessful in getting the promised compensation, they are compelled to move to still poorer localities. *The gap between the rich and the poor enlarges* further.

Most of the poor, especially in LDCs like India are uneducated. They have no idea of hygiene. They therefore dirty the environment by throwing away the unwanted things just outside their huts. *The slums in cities thus become breeding grounds of diseases*. The environment gets choked by used tins, plastics, rubber tyres etc.

In India, the educated are no better, in many cases. They also throw away plastic bags, plastic glasses, and many other unwanted things carelessly around, in a most unethical manner. The recent great flood in Mumbai was caused by the choking of drains by plastics. After a heavy rain for 18 hours on 26th July 2005, the city of Mumbai came to a complete standstill for the next 48 hours. People travelling in trains throw away the used plastic glasses and thermocol dishes just out of the windows. Nobody bothers to think of the effect of these on the environment. There are no proper waste management systems in trains or for that matter in most other places in India.

As people are getting wealthier, the throw away concept is also catching up. Earlier, people used to buy refills for their ball point pens. But nowadays, nobody bothers. As soon as the refill is over, the old pen is thrown away and a new pen bought. The concept of repair and reuse is almost nil now.

So what is happening to the environment? *Unwanted things are accumulating in the ecosystems.* Plastics are the main culprits. In India, we see cattle and buffaloes roaming around the streets. These animals eat the plastic bags in the absence of good green grass and develop tumors in their stomachs. In Lucknow, recently, cows have been operated to take out even ten or fifteen kilograms of plastics from their stomachs.

Plastics of today are the electronic wastes of tomorrow. There is no limit to the electronic equipment like televisions, computers, mobile phones, calculators etc. thrown away by people. Earlier, this was a problem for the MDCs alone. But now this is a serious problem for LDCs like India also.

With growing computerization, the quantity of the e-waste is increasing day by day. For example, over 85 million ***cell phones*** are in use now, in 2006. This number is expected to go up to 275 million or more by 2010. Now 7 Indians out of 1000 own a *computer*. By 2008, this is going to double. Along with these, are the other electronic equipment like

calculators, music i-pods* etc. *Televisions are more in number than the land line phones now. The IT industry is growing in India, and such changes are bound to happen. All this is undoubtedly good for the economic boost of India. *The question is how we are going to dispose of this scrap.* There is no safe method yet. *The old computers of the USA and Europe all land in Africa, India or China.* There, they are manually dismantled, releasing the toxic materials like lead, mercury, cadmium etc., all highly toxic, along with copper. The LDCs are becoming the waste bins of the MDCs.

Most of the industrialized nations are thus high-waste or throwaway societies. They try to sustain their economic growth by increased input of raw materials and also energy. Utilization of fossil and other forms of energy is also maximum in developed countries. *The wastes finally end up in low waste societies. This is absolutely against all ethics.*

It is however a welcome change that *'recycling' is catching up* everywhere. Many companies are also eager to become '*zero waste companies*' by recycling or reusing its waste materials in various suitable ways. These business houses recycle their by-products and do not send them to landfills, as was being done earlier. New revenues are generated also in the process. This has led to cost reduction and increased profits. Sanyo, Canon, Toyota etc. have reported that they have achieved zero waste production. Such activities will help in reducing the number of throwaway societies in general.

SUSTAINABLE EARTH SOCIETY ETHICS

Many aspects of nature can be classified as sustainable and non-sustainable. Resources, economy, development, society etc. can all be sustainable or otherwise. *Sustainable development has the following characteristics*:

- It is an overall socio-economic growth of the society.
- It is done through rational exploitation and optimal utilization of resources.
- Ecofriendly technologies are adopted for the process.
- Environmental education is necessary for sustainable development.
- It involves better organization.
- It leads to sustainable economy.
- Environmental quality and ecological balance are maintained.
- It ensures continued yield of resources to the present and future generations.

This means that the final goals of sustainable development in a society are two:

1. To provide a strategy (action plan) that will improve the quality of human life, and,
2. To provide a strategy that ensures the vitality and diversity of the earth.

Sustainable development must be beneficial for the people and also for the environment as a whole. The quality of the environmental resources must not decrease and at the same time the society should develop. If development is sustainable, then the society will automatically become sustainable. *A society where sustainable development is practised in all fields is called a sustainable society.*

Some guidelines (principles) for a sustainable earth society are given below:

- Improve quality of human life.
- Respect each other for harmonious coexistence and community life.
- Conserve biodiversity.
- Minimise utilization of non-renewable natural resources.
- Ensure that renewable resources remain sustainable.
- Change personal attitudes towards resources and environment.
- Maintain earth's carrying capacity.
- Motivate every individual to take care of the environment.
- Link development with conservation.
- Establish global treaties.

Due to the wide disparities in the economies of the developed and the developing countries, it is difficult to form policies for establishing sustainable earth societies. Developing countries like India account for a very large population; in fact every 6th world citizen is an Indian. The developed countries are pointing fingers at such countries (mainly India and China) and accusing them of depleting natural resources much faster than they can regenerate by natural processes. On the other hand, the industrialized west is utilizing a major portion of the earth's resources, in spite of having low population levels. *Accusing each other* in this manner *is not going to solve the problems. The rich and the poor nations together should formulate policies for establishing a sustainable earth society.* Every individual will then follow the ethical principles involved in such a society.

ETHICAL GUIDELINES

Ethics had been strong in all societies in the earlier times. All over the world, people led simple lives. In those days, natural resources were aplenty. It was believed that these will never end. Gradually, population increased greatly in some countries like India and China. Industrial revolution that spread in the west increased people's desire for luxury and comforts. Everywhere in the world, man's needs increased as his greed grew. This led to a depletion of the natural resources. The situation remains the same today. There is no limit to man's desires and wants. If this tendency is not checked and if we do not take care to replace the resources we are using, the shortages are sure to grow to dangerous proportions.

In India, nature used to be worshipped in the earlier days. The wisdom of the sages and other knowledgeable persons had prevailed upon the common masses. Thus many trees like Peepal, Bargad, Neem, Kadam etc. and small plants like Tulsi used to be worshipped. In tribal areas, there were festivals related to trees like Karam. They also protected Mahua trees which provided edible flowers and oil from seeds and alcohol.

Many animals were also considered sacred. Many of them have bees depicted as vehicles of gods and goddesses or associated to them in different ways in the Hindu mythology. For example, eagle, bull, peacock, swan, mouse, tiger and owl are respected as vehicles of Vishnu, Shiva, Karthik, Saraswathi, Ganesh, Durga and Lakshmi respectively. Animals like cow, snake and monkeys are associated with Krishna, Shiva and Rama respectively. Hanuman was in the form of a monkey while Jambavant was in the form if a bear. Lord Vishnu's earlier incarnations were all in the form of different animals like fish, tortoise, boar etc. Such associations used to motivate people to love and respect these animals and birds.

There were unwritten rules against polluting water bodies. Many rivers in India are considered holy. Some well known examples are Ganga, Yamuna, Krishna, Cauvery, Godavari, Narmada, Tapti, Gomati, Sarayu, Son, Netravati, Gandaki, Sindhu etc. Mountains like Himalaya, Vindhya, Satpura etc.; hills like Nilgiri, Nallamala, Mahadeo, Parasnath, Rajmahal etc. are all venerated by the people. Forests are thought to be governed by Van Devta's, i.e. forest gods and goddesses. In this way, *plants, animals, forests, mountains, rivers, in fact, all parts of nature were loved and respected in India.*

There was the practice of setting up sacred groves in many parts of India. For example, in Kerala, these are called 'Kavus'. These are dedicated to different gods and goddesses. Kavus dedicated to snakes are also common. *A sacred grove is a patch of dense vegetation including a small shrine and a pond or stream.* The vegetation is so dense and undisturbed and pollution free that many rare species of plants and animals live here. The ponds and streams never dry in the cool and humid environment and conserve precious water. The 'kavus' are owned by families who take every care to maintain them. Big sized sacred groves are maintained by a few families together or the whole community. The belief that gods lived in the groves helps the owners to look after them well. The non-believers, among Hindus themselves and also other religions, have often contested to the existence of the sacred groves. But it is a good sign that environmentalists have come to their rescue. Now the government has also decided to protect them.

Mahatma Gandhi and many other great men of India preached and practised 'Plain living and high thinking'. His Principle of Enoughness is very popular: "The earth provides enough for everyone's needs but not everyone's greed." *Simple living will automatically reduce overexploitation of natural resources.* Many great men of the past had motivated people to protect the plants and animals. Ethically, Swami Vivekananda had said, "Not injuring any living being is a virtue; injuring any being is a sin." Though this is not possible fully as many plants and animals are eaten, the words can be followed to avoid cruelty to them and also not to kill or injure for sport.

It was environmental ethics that motivated Rabindranath Tagore to set up Shanti Niketan. This is a beautiful concept where lessons are imparted to students in the open, under the trees, in natural surroundings. This makes the students nature lovers.

There can be many ethical guidelines to prevent the earth from further degradation. If these guidelines are followed, environment on the earth will improve greatly:

For conservation of vegetation:

- Plant trees wherever possible.
- Prevent cutting of trees.
- Know the medicinal values of various plants and put them to use.
- Pass on the information you have to others so that others will take interest in protecting medicinal plants.
- Do not overwater your plants; they may die due to waterlogging.
- While going out of station for holidays, place a long hose pipe with holes at intervals along the row of plants in your garden and keep the tap partially open. This is drip irrigation for homes. The

plants will remain alive; at the same time, no water will be wasted.

- Paper is made from wood; reduction in use of paper amounts to cutting of lesser number of trees. Therefore, use recycled paper if possible.
- Do not crumple paper, only tear it into pieces before throwing, because crumpling breaks the fibres and reduces the quality of recycled paper.
- Do not waste paper; use both sides.
- Reuse gift papers.
- During functions, give potted plants in place of bouquets to guests.
- If your compound walls are covered by moss, do not scrape them off unless you want to whitewash them for some purpose; mosses look green and beautiful; they also absorb carbon dioxide from air and release oxygen into air during photosynthesis, like all other green plants.
- Cultivate love for nature in the young children at home; tell them all about the usefulness of each part of nature; tell them stories concerned with nature.

For conservation of animals:

- Take good care of pet animals and birds.
- Do not throw stones or disturb small animals like lizards, chameleons, birds etc.
- Do not tease animals in the zoos.
- Arrange for water containers outside your homes for wandering animals, if possible.
- Attract birds to your gardens by providing water for them.
- Break leftover food into small pieces and throw them to birds.
- Instead of throwing banana skins etc. in the dust bins, give them to cows that are roaming in your locality.
- Do not catch or kill butterflies, moths etc. They along with many other insects help in pollination.
- Do not kill earthworms; they improve soil quality.
- Do not catch or kill dragon flies; they eat other harmful insects. Remember that dragon flies are gradually becoming extinct.
- Do not use products made from the skin of snakes, crocodiles, leopards etc. These are endangered animals.
- Do not use ivory products for decorating homes.
- Do not bring small wild animals home for rearing them; most probably they will die; moreover, it is against the law.

For water conservation:

- Conserve water by practising rooftop rainwater harvesting.
- Do not waste water at home.
- Report about leaking pipes in public places to the authorities.
- Do not direct sewage water on to roads, rivers etc.
- Do not keep the tap on while brushing or shaving.
- Do not throw garbage into water bodies.
- For potted plants, reuse the water used for washing vegetables and fruits.
- Water the plants in the morning to reduce evaporation from the soil.
- Use eco-friendly cleaning solutions in place of chemicals.
- Do not cement your courtyard; maintain a lawn instead. The green grass allows water to filter down the soil, it looks beautiful too.

For waste management:

- Use dustbins and do not throw garbage on the streets.
- Keep parks and gardens clean.
- Do not throw used plastics here and there.
- Use paper or cloth bags in place of plastic bags for shopping.
- Make compost manure at home from waste materials like skin of vegetables and fruits, and other biodegradable waste materials.
- Grow plants like banana, ol, kachu etc. near outlets of water from bathrooms, kitchens etc.
- Segregate solid wastes at home so that all degradable wastes go into the compost pit while only the nondegradable wastes like plastics are deposited in the municipality bins.
- Do not burn plastic wastes at home; the fumes are also harmful.

For soil conservation:

- Do not leave open areas in your garden without any vegetation.
- Provide a cover to the soil by dead leaves and twigs, cut grass etc.
- Raise the edges of the garden space so that no water flows out, carrying precious soil away. Let the water sink in place.
- Use organic manure like cowdung, green manure, compost, vermicompost etc. in place of chemical fertilizers to maintain soil fertility.
- Use biopesticides like neem oilcake in place of chemical pesticides.

- Get involved in community programmes like tree plantation on vacant lands nearby.

For Conservation of energy:

- Reduce electricity consumption by all possible methods like:
  a) Switching off fans and lights, television, computer etc. while not in use.
  b) Using tube lights in place of ordinary bulbs.
  c) Using energy saver bulbs.
  d) Turning off electric heaters in the kitchen slightly before the food is cooked as the heat that lingers on will complete the cooking.
  e) Placing study tables near windows to use sunlight in daytime .
  f) Not keeping hot food in refrigerators as that will consume more electricity.
  g) Getting electric metres checked etc.
- While using gas stoves in the kitchen, remember that:
  a) Avoid reheating of food; this is bad for health also
  b) After the curry starts boiling, bring it sim
  c) Soak pulses beforehand so that they will cook faster.
  d) Use pressure cookers as much as possible.
  e) Take out eggs or other cooled items from the fridge beforehand so that extra gas does not burn to bring them to the room temperature first.
  f) Use flat bottomed vessels for cooking on flat hot plates, but while cooking on gas, use round bottomed vessels to utilize the energy to the maximum.
  g) Cook items covered whenever possible.
- Use alternative energy sources like:
  a) Biogas
  b) Solar energy
- Avoid waste of petrol by:
  a) Walking or cycling as much as possible
  b) Arranging for car pools instead of each person using his own car. If a few neighbours have to go in the same direction for work or college at the same time, car pooling system is ideal.
  c) Trying to use public transport like trains and buses.
  d) Servicing vehicles regularly and getting emission checks done.

e) Not keeping the car switched on while parked.

f) Stopping fully at traffic signals if the light is red.

The list can go on and on. All the practices mentioned here and more are easy to follow. The only thing needed is determination.

The ethical guidelines can be summarized as follows:

1. We should understand that we are part of nature and behave accordingly.
2. We should try our best to conserve all the resources that nature has gifted us with and not waste any resource.
3. Where we have to take out resources, we must see to it that they are replaced
4. We should not degrade the earth's physical, chemical or biological capital.
5. We must do Environment Impact Assessment (EIA) studies before starting any project.
6. To find out whether our projects will be harmful for the future generations, we can do a GIA study also (Grandchild Impact Analysis) — this is a recent concept.
7. We should honour the right of all beings to live on the earth.
8. We should try hard to preserve the present biodiversity in nature
9. We should not allow certain human communities also, like forest dwellers, to perish.
10. We must be kind to animals.
11. We must try to lead simple lives and motivate others to do so.
12. We can and must think globally, but we must act locally to solve different problems.

Those who follow these ethical guidelines are True Earth Citizens.

GROWTH OF CIVILIZATION AND SOCIAL STRUCTURES

Human beings belong to the species *Homo sapiens*. We have been on the earth, as per fossil and anthropological studies, for the past 40,000 years or so, while our planet, as per estimates, has been in existence for the past 4.6 billion years. So man can be said to be a newcomer.

In many ways, man is inferior to other animals. He does not have glossy fur or glowing feathers. He does not possess the strength of the bull or the speed of the deer. He cannot swim under water like a fish, nor can he fly in the air like a bird. Yet he is the head of the animal family as a whole. He is at the top of the ladder of evolution. He can adapt to any situation, using his hands and brain, unlike the animals.

One of the most important events in modern science was the publication of Charles Darwin's "On the Origin of Species" in 1859. His theory inspired intense research, both in biology and in anthropology.

The anthropological evolutionists believed that society and culture developed in a particular order. These stages, they say, can be reconstructed. They started to study in detail the science of culture on which human behaviour and development are based. They believed that similar isolated societies had developed in different parts of the world, though there were no contacts among them.

As far back as in the 19th century, three basic stages of social evolution were recognised:

1) Savagery
2) Barbarism, and,
3) Civilization

The use of bows and arrows and the gathering method of living, according to them, is savagery. Domestication of animals took place in the stage of barbarism. Civilization as we understand today evolved much after these two stages.

Like a person, a civilization is more than a sum of its parts. If we describe a man's features, his life history, where he lives and how, elaborate about his family, friends and foes, define his ethics and politics — still that is not the man himself. What is left will be what makes him himself, and different from other men. Same is the case with a civilization. Even when we have described its people, arts, religion, villages, towns and cities, caste and class, economics and politics, science and technology, methods of defence, ethics and morals — in fact everything one can think of, even then the inner style of civilization will be left out.

In the nineteenth century, the evolutionists spoke of social and cultural evolution in society. They tried to formulate universal laws of human behaviour and cultural evolution which would explain the similarities and differences among various cultures in the world.

The evolution of religion in different cultures was also traced in the nineteenth century. *Growth of civilization based on religion is supposed to have taken place in 3 stages.*

The 1st stage was *animism* where people believed in spirits and ghosts.

The 2nd stage was *polytheism*, i.e. belief in many gods.

The 3rd stage was *monotheism*, i.e. belief in one god. In this stage, people stick more to ethics and morals as religion becomes more associated to concepts of what is right and what is wrong.

According to this view, these stages were progressive and it showed that man had become more and more rational.

Some other anthropologists gave more stress to social evolution in civilizations compared to religious evolution. Accordingly, *there was a social evolution from polygamy to monogamy* and said society became more specialized and complex as evolution progressed.

By the end of the nineteenth century, many anthropologists started voicing objections to the theories of the evolutionists. They objected to the method of study (without fieldwork) and also to the fact that westerners were considered civilized by them in contrast to others who were pictured as savages. This was wrong.

The belief that no society was more civilized than any other became strong. *Ethnography became a new branch of studies. It describes a society based on the cultures and characters as per the viewpoint of the natives.* This paved the way for the recognition of the various cultures of the world. This was long overdue. In this method of research, collection of data became very important.

Meanwhile, some anthropologists argued that all cultures began in one or more specific areas. From these areas, they spread all over the world. This *belief that different cultures had a common origin is called diffusionism.*

The extreme diffusionists believed that there was only one point of origin. Some believed that the origin of all cultures was Egypt. But this view was wrong, as it would mean that Egypt had a primitive culture, and the western cultures were more evolved.

Most of the diffusionists however believed that there were many *centres of origin of civilization and social structures.* These were *called culture circles.* It was because the culture circles overlapped that cultures also overlapped.

Later, some other diffusionists modified the idea of culture circles. They put forward the concept of culture areas. A*reas of origin of civilizations and social structures related to geographical areas are called culture areas.* From this concept, the age-area-hypothesis evolved. *As per the age-area-hypothesis, as cultural characters spread out in all directions from the point of origin, the age of these traits can be determined from the geographical distribution. Older cultures cover wider areas.*

However, this hypothesis was also rejected later as culture cannot be explained only on the effect of the environment. This led to more research and the formation of *Functionalism. Functionalists believe that there is no need to trace the history of civilizations based on speculations; instead, present civilizations must be studied without trying to compare them with the past ones.*

Functionalists believe that each custom and belief of a society has a specific function, which helps to maintain the structure of that society. They compare present day-societies in terms of similarities and differences in structure and function both.

People all over the world have biological as well as psychological needs. There are other needs also, like instrumental needs (law and education), integrative needs (religion and art) etc. The ultimate function of any society is to fulfill these needs. Suppose science and technology are not able to provide answers to some questions, people may turn to religion, magic etc. for answers.

By the twentieth century, another trend developed, called *neo-evolutionism*. Many anthropologists from America and Europe were neo-evolutionists. According to them, all *civilizations consist of 3 essential components — techno-economic, social and ideological.* The economic aspect includes environment also, according to him, as society will naturally adapt to the prevailing environment. *This is called the cultural materialist approach.*

All modern anthropologists search for the general laws that determine the way civilizations and social structures form, grow and adapt to changes in the environment.

***Social groups*:** Social groupings are important to humans because they give identity and support to the members of the group.

***Marriage and Family*:** Men and women are interdependent. Parent-child bonds are also strong in human beings. This gave rise to families. Family provides for economic co-operation and also a setting for child bearing. To control sexual activity, the system of marriage developed. Different types of marriages and family patterns came into existence in different parts of the world. Establishment of families helped to solve many problems which individuals could not solve independently, by themselves.

***Societies*:** Some problems could not be solved by families. In matters of defence, resource distribution for works of large scales etc., as the challenges were too great, societies were formed. In general, extended (joint) families evolved in agricultural societies and nuclear families in industrialized societies.

***Social organizations*:** Kinship and marriage were sufficient to deal with most organizational problems. But other groupings also came into being in society. The most important of such social organizations are age grouping, common interest associations and stratifications.

***Age grouping*:** Children grow together. Childhood gives way to youth, then middle age and finally old age. Each group has a different status in society.

***Common interest associations*:** Men and women form associations in society for various goals like friendship, recreation, governance, defence, economic interests etc. Men's associations are more noticeable.

***Social stratification*:** A stratified society is divided into two or more groups. These are ranked high or low. This resulted in inequality. Divisions based on class, castes etc. are examples. Social stratification is seen in all societies. Though it has many disadvantages, it has become deep rooted in societies all over the world and is practically impossible to change.

HUMAN SETTLEMENT PATTERNS AND THEIR IMPACT ON THE ENVIRONMENT

After air, water and food, in that order, what human beings need is shelter. People build houses and develop settlements to protect themselves from harsh weather and to enjoy their social life.

A settlement is a collection of houses and their inhabitants, other necessary buildings along with the surroundings. In a settlement, people live in harmony with the environment.

***Origin of settlements*:** The earliest human beings wandered from place to place in search of food and water. In those times, they depended entirely on wild plants for fruits and roots. Later on they killed animals also. That was a nomadic existence (life of wanderers). When the early man understood how new plants sprout from seeds and those plants again produce seeds, they started to grow plants for grains on their own. This took a few months. During that period, man had to settle near the crops instead of wandering from place to place. For that, man learnt to make huts from bamboos and dry leaves and other building materials. These were the earliest settlements of human beings.

***Sites of settlements*:** The early men must have searched for availability of water, and enough plants and animals to decide where to settle. Where the vegetation grew well, the soil was naturally good. Such places were forests or areas near forests. The early settlements were all temporary or unstable. Later permanent or stable settlements came up.

***Unstable settlements*:** Temporary settlements are set up by nomads (wanderers). These account for about 2 to 3% of the settlements in the world. People who practise shifting cultivation and people who are hunters live in unstable settlements. For example, some tribes of the North Eastern India like Jhumias' live by cultivating land. They first clear forest land and plant their crops there. When the soil becomes infertile, they leave that settlement and move to some other place where they clear new parts of forests for cultivation. This is called shifting cultivation. Tribes who do this in the North East have no other alternative as the terrain there is full of hills and slopes only. Slopes lose fertility very fast due to soil erosion. However, shifting cultivation is bad for the environment as more and more land becomes infertile as the tribes keep on shifting.

Unstable settlements become a necessity in areas with extreme weather conditions. Gujjars and Bakarwals of Jammu Kashmir live in the valleys in winters which are comparatively warmer. They live grazing their goats and sheep in the pastures. In summer, when temperatures rise in the valleys, they move up the mountains, along with their animals.

Stable settlements: *Stable settlements can be rural or urban. Settlements develop in the areas between villages and cities also. This is the concept of rural-urban continuum.*

Stable Rural settlements: *Any settlement in which most of the people are engaged in agriculture, forestry, mining or fishery is a rural settlement.* Most of the settlements in the world are rural and stable.

Rural settlements are broadly of three types:

1.Compact or nucleated settlements

2.Semi-compact or semi-nucleated settlements, and,

3.Dispersed settlements

Compact or Nucleated settlements: A compact rural settlement is based on agriculture. It includes farms, grazing fields, religious structures and other buildings. Croplands are not more than about 3 km away from the houses and market places are in the centre.

In the beginning, people needed each other's help for clearing forests, tilling the land, to protect themselves from wild animals etc. This must have been the reason for the development of compact settlements.

In India separate localities may be allocated to upper and lower castes in compact settlements. It is high time social reformation takes place in India so that such wrong practices are corrected. The upper castes must realize that all are equal in the eyes of God and the lower castes must gather courage to fight such injustices.

Other than farming communities, fishing communities also may live in compact settlements. This is because mutual co-operation is essential for making and maintaining fishing boats. Hunting communities of Africa also form compact settlements for the sake of safety from wild animals.

Compact settlements develop on the plains or near rivers or other water bodies. If the resources are not much, the settlements are small. In the deserts of Rajastan, such a settlement has only a maximum of 10 houses. But in African jungles, where resources are aplenty, very large compact settlements are seen.

***Semi-compact or semi-nucleated settlements*:** This is a phase between the compact and the dispersed settlements. When the population in a compact settlement increases, houses are built in the surrounding open spaces. This is how semi-compact settlements are formed. In the hilly areas of M.P. and the north eastern states, such settlements can be seen.

***Dispersed settlements*:** In areas where extreme environmental situations are present, like very dense forests, very hot deserts, land devoid of proper nutrients etc., dispersed settlements develop. In such areas, the farmers will have to be always present near their farms. In parts of Rajasthan, Jammu & Kashmir and the north-east, such settlements have developed.

***Patterns of rural settlements*:** *The relationship between one house or building with another is called settlement pattern.*

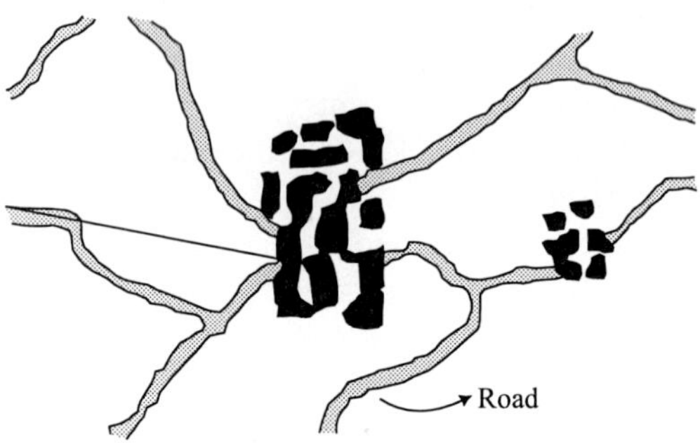

1

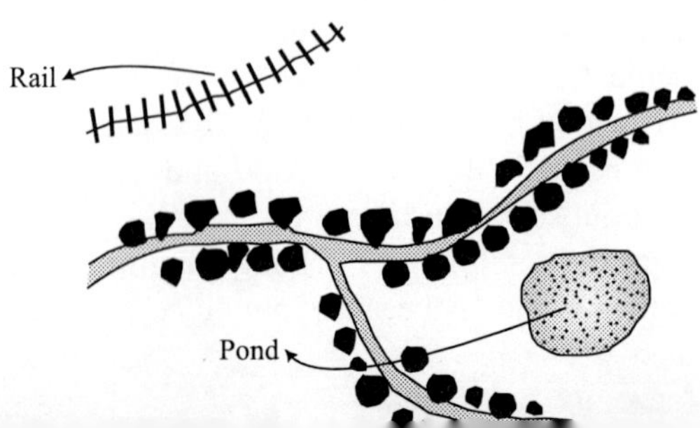

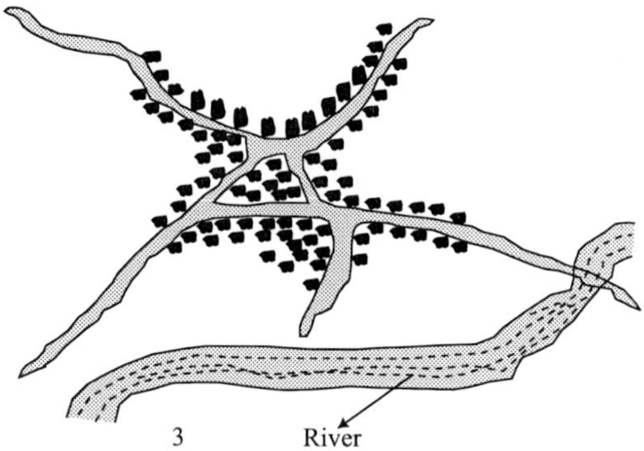

3

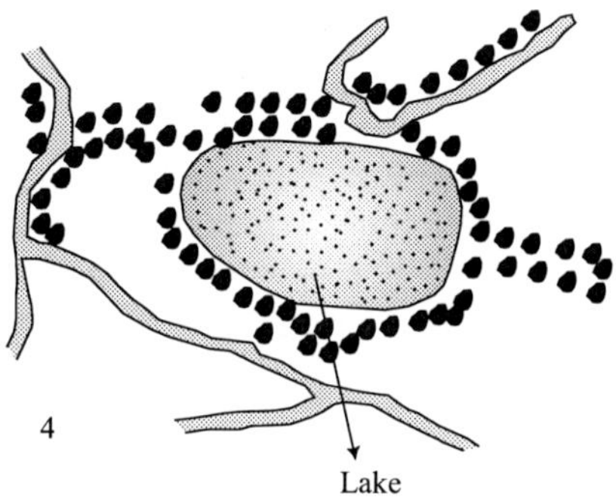

4

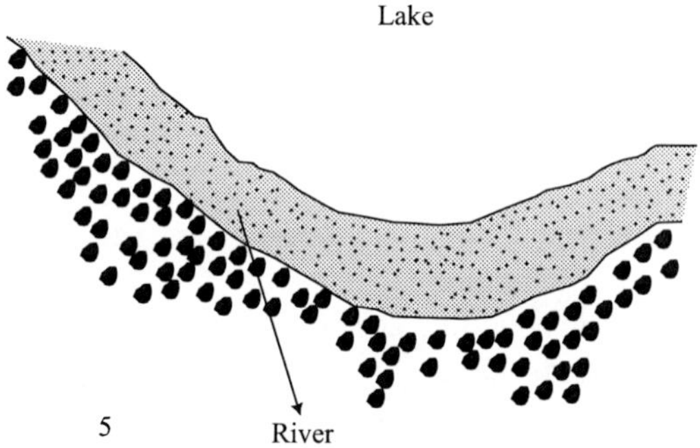

5

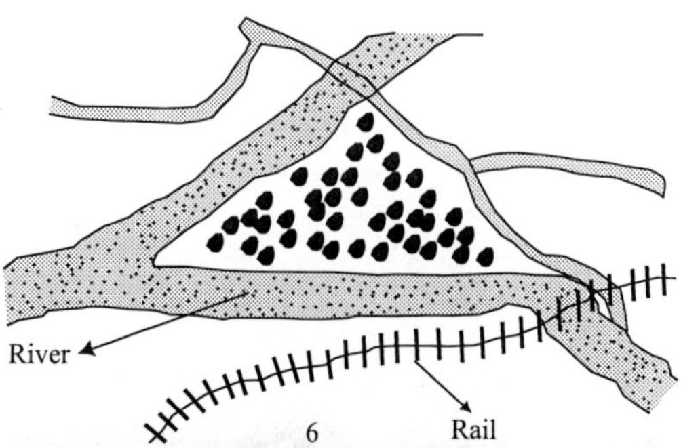

6

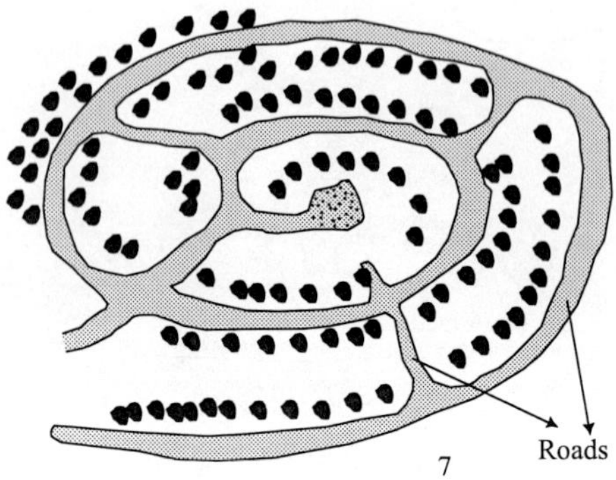

Settlement Patterns

1. Rectangular 2. Linear 3. Star like 4. Circular
5. Semicircular 6. Triangular 7. Nebular

- *Rectangular pattern*: Most of the rural settlements are rectangular in shape. They develop in productive alluvial soils.
- *Linear pattern*: Here, houses are built on either side of a road, river, railway line etc. These are also common.
- *Star-like pattern*: If many roads meet at one point or points near each other, houses come up around the roads and form a star-like settlement.
- *Circular pattern*: These develop around water bodies like lakes, large ponds etc. Fishing from these freshwater bodies is the occupation of these people.
- *Semicircular pattern*: These are also settlements of fishermen, salt workers etc. They build their houses parallel to rivers or the sea coast in a semicircle.
- *Triangular pattern*: In places where two rivers meet, settlements are triangular in shape.
- *Nebular pattern*: Where roads form a spiral, houses come up between the roads. This pattern looks like a nebula.

Planned rural settlements: These are seen in many countries including India. They have many advantages:

a) Only crops giving good yield are planted.
b) Infrastructure like water, electricity, roads etc. are available.
c) Social forestry is done here.

They have some disadvantages also, like:

a) It takes time for the benefits to show. Some settlers get impatient and leave.

b) When population increases, some families have to move out.
c) If monoculture is done, soil loses fertility. (Crop rotation must be done to avoid this.)

Rural-Urban continuum: *The land between the rural and the urban areas is known as rural-urban continuum. It shows characters of both rural and urban communities.*

Urban settlements: *If the people in a settlement are engaged in secondary, tertiary or quaternary activities, it is called an urban settlement.* (Primary activity is agriculture.)

These have developed along:

- Trade routes
- Rivers
- Seacoasts
- Mountain crossings
- Harbours
- Estuaries
- Sites of resources like minerals, precious stones, energy resources like coal and petroleum etc.
- Spots of scenic beauty
- Spots of religious importance etc.

The word 'Urban' means related to towns and cities. In the case of cities and mega cities, the definition is based on the number of people living there. These numbers are different in different countries.

Cities: In our country, an urban area with more than one lakh people is considered a city. Cities have many business houses and many houses. It has a specific administration.

Towns: Urban areas with population of less than one lakh are towns.

Metropolis or Metropolitan area: This has many urbanized areas.

Mega city or mega polis: A city with a population of 8 million or more is a mega city. Examples are Tokyo, London, Paris, Mumbai, Kolkata, Delhi etc.

Suburbs: Areas outside the boundaries of towns and cities are called suburbs.

Impact on the environment: Both rural and urban settlements have an impact on the environment.

Impact of rural settlements on the environment:

- Deforestation due to cutting down or burning of forests for cultivation.
- Soil erosion due to deforestation.
- Desertification due to deforestation and overgrazing.

- Salinification (soil becoming saline) due to excess watering of crops.
- Depletion of groundwater due to overuse and lack of conservation.
- Pollution of air, water and soil due to excessive use of fertilizers and pesticides.
- Pollution of air due to smoky stoves using coal (*koyla chulhas*), burning dried cow dung cakes (*goytas* or *uplas*) etc.
- Formation of garbage heaps in the absence of proper solid waste management.
- Spreading of diseases due to lack of hygiene and lack of education etc.

Impact of urban settlements on the environment:

- Air pollution due to large number of automobiles
- Air, water and soil pollution due to fumes and solid as well as liquid wastes from industries.
- Fumes of sulphur dioxide and nitrogen oxides from industries and sulphur dioxde from automobiles released into the air cause acid rain.
- Large amounts of carbon dioxide released into the air from industries and automobiles lead to greenhouse effect, ozone depletion and global warming.
- Effluents from industries containing many chemicals, some of them toxic, like the heavy metals cadmium, mercury etc. are released into rivers without treatment. These kill fish and other aquatic animals in large numbers.
- Noise pollution due to vehicles and all other types of noises associated with city life.
- Soil erosion along the unpaved footpaths on the sides of the roads.
- Shortage of water due to increased population and also absence of water conservation schemes.
- Accumulating garbage on roadsides in the absence of properly functioning municipalities.
- Diseases spread by houseflies, mosquitoes and also due to air and water pollution, adulterated food items etc.

HUNTING AND GATHERING SOCIETY

For many thousands of years, the primitive men lived as hunters and gatherers. They collected edible fruits, berries, roots, stems, nuts, seeds, leaves etc. from the forests where they lived. Initially, they ate the remnants of meat left by other animals of prey. Later, they made tools for hunting by sharpening stones, sticks and animal bones.

One of man's earliest relationships with animals was that of hunting. Soon after man learnt to hunt, he also learnt how to wear skins of killed animals to protect himself from cold. Thus animals became very important to man as they provided him with food and also clothing.

As hunters, men had to follow the animals and learn their nature and behaviour. He learnt to fear some animals and admire some other animals. He realized that some animals were very fast, some were strong and some others were intelligent. In this way, he also noticed that some were slow and stupid and that he could hunt them easily.

Dogs might have followed men hoping to pick up some meat or bones. They would have fed on the leftover meat which men left in the jungle after taking the best parts away. Gradually, dogs became men's friends and barked at his enemies — whether other men or animals. Seeing this, men also must have started rewarding them with food and being kind to them. The hunters started to make use of the dogs' keen sense of smell in hunting wild animals. This was the beginning of taming of animals.

Some hunters learnt how to catch fish. They became fishermen. They also killed other aquatic animals. They made their settlements near water bodies, while hunters and gatherers settled in or near forests. When hunters came to know that meat burnt in forest fires was tastier, they tried hard to make fire. Soon, they were successful in making fire by rubbing a stone against another. The early hunters used to try to maintain the fire for as long as possible by feeding it with dry leaves and twigs. The possession of fire was a matter of great pride for the different hunter-gatherer groups.

The earliest hunter gatherers used to live in caves. Later, they formed small settlements in groups of about fifty. They worked together for collecting food and other necessary things like bamboos and dry leaves for making huts. But they were nomads, shifting settlements when food became scarce in their original area.

Soon, they found out that some herbs could be used as medicines. They also learnt how to predict weather, where to go for which plants in which seasons, where to find water etc. Sometimes, they followed birds and animals going in search of honey, found the beehives and took honey out of the hives.

There is evidence that the hunter gatherers used to try to keep their numbers low so that there was no difficulty in times of shortage of food. However, the methods practised by the early men were barbaric. Children and old people used to be killed by the young and strong in the settlements. Later, herbs were found out which could be used as contraceptives and also for abortion. They also understood that as long as young mothers breast-fed their babies, chances of getting pregnant were very low.

Due to lack of medicines and proper childcare, infant mortality was very high those days. Moreover, life expectancy was also low, as low as 30 to 40, again due to diseases and the rough and dangerous wildlife in the jungles. Therefore, the growth rate of hunter gatherer populations was very low.

The impact of hunting and gathering societies on environment: The activities of the hunter gatherers had a harmful effect on the environment. As they developed better and better tools for hunting, they killed more animals than they needed for food. They also planned group activities like while some chased the animals out of the jungles, others who waited outside killed them. Quite frequently, the scared animals used to run, causing a stampede in which many other animals were killed unnecessarily.

After the invention of fire, hunter gatherers became more confident of their capabilities. Often, the fires they made spread and killed animals. It is thought that the extinction of some animals in the stone-age was caused by the activities of the hunting and gathering societies.

However, *people were very few in number and the resources in abundance; therefore, the impact of the early men on the environment was not much*. The impact was only minor and local. They used limited resources. The negative effects were soon rectified by nature in natural processes. As the hunter gatherer societies lived on the earth for tens of thousands of years, it can be assumed that they had not harmed the environment to any great extent. In fact, *it is not that the hunter gatherers lived only in the distant past. Even today, many such tribes are scattered all over the world.*

Hunting-gathering societies in the present world: *Pygmies of Central Africa are the simplest and the most primitive people*. They are less than 5 feet tall and are mostly naked. Seen in countries like Congo, Cameroon, Rwanda etc, they collect roots, fruits, leaves, frogs, lizards and other animals for food. They do not know farming, animal husbandry etc. They have no permanent settlements, social rules, warfare, religion, or any knowledge of ceramics, weaving, basket making etc. Other tribes in Africa are Twa, Tswa etc. who live by fishing and hunting. There were 200,000 Pygmies in Africa in 2001.

Examples of some other hunting gathering societies of the present day:

- Semangs and Sakais of Malayasia.
- San bushmen in the Kalahari desert of Angola, Botswana and Namibia in South Africa.
- Blackfoots and other Red Indian tribes in parts of America.
- Eskimos near the North Pole, living in four countries — United States, Canada, Russia and Greenland.
- Aborigines of Australia.
- Andamanese of Andaman Nicobar Islands of India.
- Gujjars and Bakarwals of Jammu and Kashmir in India etc.

AGRICULTURAL SOCIETY

From being hunters and gatherers, man changed his lifestyle when he learnt to till the soil. The first harvest must have been gathered 10,000 to 15,000 years ago. In Egypt and Iraq, land was certainly tilled in 6,000 B.C.

The *Agricultural Revolution* had begun in different parts of the world at approximately the same time. In the beginning, agriculture was used as a supplement to hunting and gathering. Soon however, it became a main practice. Reasons for a shift from hunting gathering to cultivation must have been different in different parts of the world. Climate changes might have been one reason.

Plant cultivation must have developed in different areas of the world. In the tropical forests of South East Asia, North East Africa etc, cultivation became popular. The method used was slash-and-burn. *During slash-and-burn, people cleared a small forest area by cutting down the trees and bigger plants and by burning the ground vegetation. The ash obtained was mixed with the soil to make it fertile.*

Water is very much essential for cultivation. In the early days, men could choose locality of his fields as they wished. But as the number of people in a tribe increased, the fertile land became rare. In such situations, the strong tribes drove the weaker tribes towards the deserts. In some cases, where rivers or streams were nearby, the early men even dug canals with their simple tools and diverted water towards their fields.

Agricultural societies practised shifting cultivation in the earlier days. *When a plot became infertile due to continued cultivation, it was discarded. The whole settlement moved to another area with fertile soil. This is called shifting cultivation.* This method of cultivation is of course bad for the soil, but in the earlier days, land was available in plenty and people were very few. So there was ample time for the degraded land to regain its fertility by natural processes. Later, the same plot could be used again for the same crop. However the early growers knew that degraded land could be used for tree crops, medicinal plants, firewood, etc. and they tried this. In 20 to 30 years, the land became fully fertile again. *Cultivation where the land is not degraded is sustainable cultivation.*

The early cultivators grew only how much food they wanted for themselves; this practice is called subsistence farming; meaning 'to plough or till'. In the early days, the tools used for farming were only sharpened stones and sticks and the like. The first ploughs were probably hooked pieces of wood. Such ploughs helped only to make the surface of the ground rough. This was one reason why large plots could not be cultivated. It was about 7,000 years ago that metal was discovered and used to make simple tools. With the advent of metallic ploughs with wooden handles, farming became easier. Soon, animals were used to pull the ploughs. And men guided the animals. Now areas where soil was crisscrossed with dense roots of plants could also be tilled and made into croplands. Men planted a mixture of crops in the cleared fields. *Food crops and tree crops were planted as a mixture; this is the earliest known agro-forestry.*

The shift from hunting gathering society to agriculture society had many effects on the life-style of people and the nature of society itself. Some of these are:

- More food produced could support more people. Earlier, life depended fully on hunted or gathered food and so death rates were high. Now death rate fell and thus the population increased.
- When men began using animals to pull their ploughs, food production increased. Now more food could be produced with less human labour.
- More and more areas were put to agriculture as man learnt to build canals and irrigate their fields.
- Hunter gatherers could not reserve food for times of scarcity. Whatever they collected they had to consume immediately on a daily basis. As agriculture became the method to get food, seeds, rhizomes etc. could be stored and consumed later.
- As fewer people were needed to produce food for the whole community, others left farms and took to other professions like pottery etc. Gradually, towns were set up and villages became different from towns. This was the beginning of urbanization.
- When wild animals entered cultivated fields, they had to be driven away or killed. Thus many animals became man's enemies. Birds,

rats etc. which tried to share the crops also became man's enemies. He tried to find out ways to protect his crops from all of them.

- Ownership of land became a family affair. As a result, fights between families for land, canals etc. became common.

The impact of agricultural societies on environment: As the number of people increased due to availability of food in times of need, more and more land was cultivated and more and more wood was cut for clearing forests and making huts. The resulting deforestation led to soil erosion and loss of fertility of soils. Forests were converted to grasslands or even deserts. Many wild plants and animals became endangered or even extinct. With the discovery of metals, mining began. This led to more degradation of land. Towns and cities were established. Accumulation of garbage and neglect of waste water led to diseases. Thus the whole face of the world changed to a great extent by the activities of agricultural societies.

Some agricultural societies of the present times: Agriculture has become the main occupation of many in the present world. As such, it is not possible to designate all those who have farms as agricultural societies. However, *there are some agricultural societies even today in different parts of the world*. Some examples:

- The Oceanians are agriculturists. Oceania is a group of scattered islands in the Pacific Ocean. Often, Australia and New Zealand are also included in Oceania. Agriculture is the main occupation of the Oceanians. They cultivate coffee, cocoa and coconuts.
- Yorubas of Africa, mainly in Nigeria and parts of Congo, cultivate maize, millets and banana. They also export cocoa.
- Bhils are agricultural tribes of India. They are spread in the four states, namely Gujarat, Madhya Pradesh, Maharashtra and Rajastan. They are the 3rd largest tribe in India, after Santhals and Gonds. In 1991, there were 25 lakh Bhils in India. The heroic character Eklavya of *Mahabharata* was a Bhil. They cultivate cereals, vegetables and fodder crops; at the same time, they also hunt and gather food. So theirs is a mixed society.
- Gonds seen in parts of Orissa, Chhattisgarh, Madhya Pradesh, Maharashtra and Andhra Pradesh are also a mixed society, practising agriculture, hunting and gathering.
- The Nagas of Nagaland practise *jhum* cultivation and shifting cultivation. Cutting and burning the forest vegetation and using the ash as fertlizer in the same place is called *jhum* cultivation. The crops planted are always mixed. The Nagas make excellent handicraft items with bamboos etc. also.

INDUSTRIAL SOCIETY

Industrial Revolution was a great step forward in the history of mankind. It began in England in the middle of the 1700s. Gradually it spread to the US in the 1800s. It led to great social and cultural changes. Industrialisation is regarded by many as one part of modernisation. The other parts, according to them, are technological development, agricultural development and urbanization.

Industrial revolution started in England when the forests there had been depleted to a great extent by the original dwellers. There was shortage of fuelwood. It was then that they found the coal reserves under the ground. Immediately, they started using coal as fuel. This meant a shift from the renewable wood to non-renewable coal as source of energy.

When coal was used to boil water, the steam engine was discovered. Many types of machines were soon invented one after the other. The discovery of petroleum pushed industrialization further ahead. Machines run on petrol and natural gas became common. Manufacturing goods in large quantities became easy. This affected handicrafts sector. Number of factories increased and industrial towns and cities were established. The urban people depended on the industries for livelihood. Soon, rural people started coming to the towns and cities in search of work.

Many towns were based on coal mining. The whole urban area became covered with black soot. Safety measures were not followed in the mines and many lives were lost. Lung diseases increased. Quantities of dust and smoke in the air increased. Agricultural machineries run on fossil fuels were also invented. When machines replaced men in the farms, more men migrated to the cities.

After the First World War, more advanced machines were invented. New techniques came up for mass production of many articles of daily use. This is what we see in today's industrialized nations like USA, Canada, Japan and many countries of Western Europe. Advanced industrial societies give many benefits to the workforce. Many needs of men are taken care of by the companies, like water supply, electricity, medical care, entertainment and leave travel costs, education, old age pension for the retired employees etc. This is what attracts more and more rural people to the urban areas.

Impact of industrial societies on environment: *Pollution of air, water and soil is the most important effect of industrialized societies on environment* (see Unit 5 for details on industrial pollution). Other than this, the inter-relationships among people has declined due to urbanization which is a result of industrialization. In general, we see that the rural masses know each other more and help each other better in times of need. The urban people are mostly too preoccupied with their own lives that they have not much time to spare for thinking about the welfare of the community as a whole. This makes people more self-centric. People come to believe that they are different from other beings of nature. They consider themselves superior to other forms. This leads to neglect of nature and even destruction of nature without a thought for the future. People get farther and farther removed from nature, which makes them bitter in nature. Many diseases are in general due to stresses and tensions of various kinds.

Agriculture versus Industry: Whether agricultural development is better or whether industrial development is better for a country is a topic of debate. The debate became stronger after the Second World War. It was argued by some that in the developing countries, agriculture is the main occupation. As populations in these countries are growing fast, they need more and more food and other agricultural products. So the developing countries must pay more attention to agriculture. When they produce more, they can even export the extra to the developed countries and earn foreign exchange which can be used for other purposes. The developed countries on the other hand need not pursue agriculture much as they can import agricultural produce from developing countries.

This argument is opposed by many. According to them, developing countries also should go in for industrialization as use of modern technology will boost up the agricultural produce. It was industrialization that made the developed countries prosperous. So, all countries must follow the same pattern for development. The drawbacks of industrialization are mainly different types of pollution and the cost involved in preventing and controlling pollution.

Whatever method is adopted for advancement, it is important that development should be uniform. Balanced growth is always better than unbalanced growth. The revenue from leading sectors must be used for the betterment of the lagging sectors. Decision on whether to adopt the agricultural society model or the industrial society model must be taken at a local level. Only this will ensure all round growth in any country.

KNOWLEDGE SOCIETY

Man, undoubtedly, is the most intelligent of all creation. He, with his more efficient hands and brain, has proved himself superior to all other creatures on the earth. He was a hunter gatherer first, just like other animals.

But man's observing power was more than all of them. He observed how seeds sprouted in nature to give rise to new plants; he observed how the seeds were formed on the mature plants. He utilized this knowledge to cultivate the plants of his choice. *It is believed that wheat was the first cultivated crop.* He had understood that wheat grains have a tough coat which needs to be broken to reach the soft interior. For this, he used two stones of suitable shapes. This, we can say, was the first grinding device.

Later man found out that he could self-pollinate and cross-pollinate his plants. By this method, he produced better quality plants. For example, the date palm bears male and female flowers on separate trees. The ancient Babylonians used to cut branches bearing male flowers and shake them over female flowers. This method of cross-pollination made sure that good dates were formed.

In this way, gradually, *as man observed what was happening in nature and how it was happening, he learnt more and more things*. He successfully copied many feats of nature. In later years, he built boats which were shaped like fish and aeroplanes which were streamlined like birds. This studying nature and copying it will never end. In the present time, when man has advanced so much with all the discoveries and inventions, he is capable all the more of copying nature with perfection, using the vast knowledge he has gained from his past experiences.

As population is increasing steadily, man needs to have more food, more of everything. For this, he is trying to increase the efficiency of photosynthesis (making of food by green plants) by various means. He is also trying to increase the area of cultivation by reclaiming land degraded due to different reasons. To multiply plants in thousands, tissue culture is being adopted. To bring various useful genes in one combination, while plant and animal breeding were practised earlier, the same is being attempted by genetic engineering and biotechnology at present. Genetically modified (GM) foods and organic farming are becoming popular rapidly.

That fossil fuels will not last long has led to search for alternate energy sources. As a result, various such sources have already been found out. For example, *solar energy is being used for various purposes now*. Flat plate collectors using solar energy for heating water for domestic purposes have been in use for many years now. From the 1980s onwards, solar panels have been in great demand in the United States. Solar thermal electric power generation has also become common. Here, electricity is produced by using sunlight to boil water to steam, which is then used to drive an electric generator. More recent is the device which converts solar energy directly into electricity. These are called photovoltaic cells. Solar power towers collect sunlight from many reflectors to achieve high temperature. The reflectors are steered by computers to ensure that they face the sun throughout the day.

Even as photovoltaic electricity is catching up, *use of Hydrogen as a non-polluting fuel for automobiles has been found out* and made practical. Companies in Germany and Japan have already come up with such models. It is hoped that they will become popular in a short span.

The *energy of the ocean currents can be used to rotate giant turbines under water to generate electricity. Geothermal energy and tidal energy are also alternate energy sources, just like wind energy. Nuclear energy is made use of in many countries to generate electricity.* As the term 'nuclear' is connected with nuclear weapons and safe nuclear waste disposal, people in general are scared of nuclear energy. Yet it definitely has some advantages and so must be considered a future option. In a nuclear power reactor, atoms of Uranium are split to produce huge quantities of heat energy. The turbine and generator are as in the coal burning power plant. (See Alternative energy sources in Unit 6)

Research work is going on in tens of thousands of laboratories all over the world in all possible fields to improve the life of man. In addition to food materials, all the articles human beings want to fulfill all types of necessities and also luxuries are being made available to them one after the other. Coming to think of it, there is no end to it.

With all this progress made, it is sad to see that *we are nowhere near eradicating hunger completely.* When the present millennium began, it was estimated that 1.1 billion people were undernourished and underweight. This, after a near tripling of food output in 50 years before 2000. In India, the number of poor people has decreased, but in Africa, it has increased. Yet, *population explosion* is the main reason for this sorry state of affairs. Therefore, it is high time that with all his knowledge, man understands this simple truth — that if the world has to be hunger free, the number of mouths to be fed must decrease.

It is also high time that we understand the harmful effects of robbing the nature of natural resources continuously. We must take to sustainable methods of managing the environment. Another important truth that we must realize is that there is no end to materialistic desires. *The simpler we live, the lesser will be our needs* and the lesser will be the burden on environment because of us. We have industries to manufacture the products we need, but if we are releasing all the waste into the nearby river without treating the effluent, how can we say that we are a knowledge society? If we are cutting down forests and killing endangered species, how can we say that we are a knowledge society? So the first thing that we must know is that as part nature, we must protect nature. If we don't, we will perish. The day we realize this is the day when we can pride ourselves that we are a true knowledge society.

ENVIRONMENTAL MOVEMENTS AND PEOPLE'S PARTICIPATION

Europeans first went to North America in the 15th century. They saw that the resources were unlimited, and they began conquering the wilderness. This went on, with enormous wastage of resources. It was between 1832 and 1870 that some individuals became scared at the rate of loss and decided to spread environmental awareness. *By late 1800s a clear conservation movement began in the United States.*

In contrast to this, *the first environmental movement was organized in India in 1730, more than a century before such a movement originated in the U.S.* It was fought against indiscriminate tree felling in Jodhpur, Rajastan. Amrita Devi, a village woman, collected 362 other women from her village and marched to the forest nearby. They declared that the forest trees could be cut down only over their dead bodies. And each woman clung to a tree and stood motionless when the contractors came. All the 362 women were axed to death by the tree fellers. This heart rending incident compelled the then king of Jodhpur to ban tree felling in that forest.

People's participation is a must in all environmental movements. Pollution of air, water and soil, and mindless utilization of natural resources have raised environmental degradation to dangerous heights across the globe. One of the best methods to reverse this trend is to organize movements emphasizing importance of environment.

Participation of local people is essential for any such movement to be successful. While ensuring people's participation, the following points must be kept in mind:

- As the people in charge of the programmes will be mostly outsiders, they will not have answers to all the questions. They should listen patiently to the local people.
- If there are any conflicts between individuals or groups of individuals in the locality, these conflicts should be removed so that people will interact a better-way.

- Some financial benefits must be there for the local people. Then only they will participate wholeheartedly.
- The decision makers should find out what issues are more important in the area from the local people themselves.
- The decision makers should always remember that their sole objective is to solve local environmental problems.

Some case studies

1. Gxalingenwa forest in South Africa covers an area of 616 ha and has great biodiversity. Here, outsiders are involved in illegal trapping of wild animals and trade of valued bird species, some of them endangered. Two environmentalists John Robertson and Michael J, Lawes (2005) surveyed the attitudes and opinions of the local people. The study showed that 77% of the local people wanted participatory forest management (PFM) in the area. They also wanted to share the benefits from the forest. They believed that the policies of the Government are obstructing their rights to the forest. Thus if participation includes sharing of profits, the users will ensure that the environment is not degraded. (The same must be true of the tribals of Jharkhand and other parts of India.)
2. Environmental conditions of a slum in Chandigarh were improved greatly by an awareness campaign involving the local people by Swami, Thakur, Gupta and Bhatia (2004). They involved mothers, childcare workers, medical students, teachers and school students in the programme and taught 40,000 slum dwellers the importance of hygiene.
3. Recently, in Plachimada in Kerala, villagers arose in protest against the powerful multinational Coco-Cola when they realized that the company was drawing an enormous quantity of ground water. "Drinking water is our birthright" read one placard.

SCOPE AND OBJECTIVES

Environment is everyone's concern. By the wrongful activities of human beings, nature has already been partly destroyed. Before the situation worsens further and the damage becomes permanent, it is our duty to preserve nature at least as it is now so that future generations will not accuse us of negligence.

Realizing the gravity of the situation, environmental movements have sprung up all over the world. *Environmental movements are most essential to create awareness among the general public and make them understand the importance of the environment.*

Environmental movements have become common in many parts of the world, including India. The more educated the people are and the more aware they are of the environment related problems in the country, especially in their localities, the more will be the number of environmental movements. These can be for protection of any part of the environment or to fight pollution.

In the developed world, most people are educated, books and newspapers play an important role in spreading information on environmental issues. For example, in 1962, the book 'Silent Spring' written by Rachel Carson described in detail the pollution of air, water and wildlife from pesticides such as DDT. Her book was well received and it helped to increase the knowledge of people regarding the harmful effects of pesticides. From the 1960s onwards, many environmentalists have been writing articles on environment. These have inspired the formation of many environment movements all over the world. *The real scope of such movements is that*:

- They help to solve environmental problems.
- They educate the people and create environmental awareness.
- They inspire more and more individuals and groups of individuals to start environmental movements.
- They force governments to formulate necessary laws and to amend existing inefficient laws.
- They instill a sense of responsibility in the people towards the proper upkeep of the environment.
- They increase the sense of hygiene in people which leads to prevention of many diseases etc.

The objectives of environmental movements: The objectives of environmental movements vary according to the place and time.

- Against big dams.
- Against the setting up of polluting factories in residential areas.
- For creating awareness about the dangers of deforestation.
- For teaching the people how to manage domestic solid wastes in the most beneficial manner.
- For teaching the poor people hygiene.
- For conserving water by rainwater harvesting methods.
- To demonstrate against discharge of toxic chemicals from industries into rivers.

The list will be quite long. In this way, any issue related to environment is picked up by the environmentally conscious local people and a movement is born. Then they garner support and put their schemes into action. It is essential that the results of the activities of the movement reach the grass root level for a positive change to take place in the environment.

There have been many environment movements in India. The oldest of them is the Bishnoi Movement. It is believed to have been started by a sage called Sambaji 400 years ago in Rajasthan. People worship certain trees and the black bucks. The Bishnois worship and protect these plants and animals even today very sincerely.

Then there was the *Baliyapal Movement.* This was against the taking over of the thickly populated village of Baliyapal village's fertile land for testing missiles.

More famous Movements in India are the *Chipko Movement, Silent Valley Movement, Narmada Bachao Andolan (NBA) etc.*

WELL KNOWN ENVIRONMENTAL MOVEMENTS IN INDIA

***The Chipko Movement*:** The first movement against indiscriminate tree felling was perhaps started in India in 1730, under the leadership of village woman Amrita Devi. Though at the cost of the lives of 363 courageous rural women, the movement was successful in that the King of Jodhpur banned illegal tree felling. However, nothing much happened as per records after that for more than two hundred years.

In 1972, the 'Chipko Movement' was initiated in Uttar Pradesh. Again, women were the pioneers. 'Chipko' means 'hug'. Timber contractors were felling trees in large numbers in the Uttarakhand hills. Under the leadership of Sunderlal Bahuguna, the Movement started in the Chamoli District of Garhwal region in U.P. The local women embraced trees when the contractors came to fell the trees. As a result, the contractors had to leave.

Chipko's plan is to planting the 5 F's, i.e. trees for

1. Food
2. Fodder
3. Fuel
4. Fire
5. Fertilizer

Every tree is a sentry to protect us from erosion, flood, draught etc. The most important product of trees is not timber as we think, but soil, water and oxygen. The movement soon became famous all over the world. Environmentalists from France, Germany, Sweden, Switzerland etc. came to Inida to visit Bahuguna to learn about the unique Chipko Movement. On April 29th 1983, a group of school students assembled at the Union Square Park in New York and hugged a big tree to observe the Chipko Day.

Chipko Movement spread along the hill regions of India and reached Karnataka in 1983. On September 6th 1983, early in the morning, a group of villagers gathered near a pond and walked three kilometers towards the hills of Uttar Kannada. As they neared the foot of the hills, as expected, they saw men with axes about to cut down trees. The people, including women and children, ran towards them and each one stood hugging a tree. That was the beginning of the Appiko Movement in Karnataka. (Appiko means 'hug' in Kannada.)

It started without any propaganda, without media to cover the event or photographers to capture the moment in frame. They had no leader. They had only heard that somewhere in North India, some people had saved trees by hugging them and offering themselves to the axe men. It was only after the incident that Bahuguna visited the area.

The Chipko Movement of India was one of the four recipients of the 1987 Right Livelihood Awards totaling $100,000. The Right Livelihood Awards Foundation introduced the awards in 1980. Every year, it honours and supports "those working on practicable and replicable solutions to the real problems facing us today". In a citation, the Foundation said, "Chipko is honoured for its dedication to the conservation, restoration and ecologically responsible use of India's natural resources."

Tehri Dam: Tehri Dam is built on River Bhagirathi, south of its meeting point with River Bhilangana. The dam is 260.5 metres high — more than three times the height of Qutab Minar. It is the highest rock fill dam in the world. It is more than 1 km wide in the riverbed. It can generate 2,000 MW of electricity. The dam can fully irrigate 2,70,000 ha and complement irrigation on another 6,00,000 ha of agricultural land.

The engineers who built it claim that the dam can withstand earthquakes of up to 7.2 on Richter scale. This magnitude is the same as that of the earthquake which rocked Kashmir in 2005. The engineers are thus extremely proud of their achievement.

The Central Himalayan region is a seismically active (earthquake prone) region. According to seismologists, an earthquake of more than 8.0 is expected to shake the whole of Tehri region before the end of this century. God forbid, but if this prediction comes true, the dam is sure to burst.

Even otherwise, whenever a dam is built and huge quantities of water are stored in a reservoir, the tremendous pressure of the water itself has the potential to cause shifting of landmasses underground, resulting in earthquakes. If the dam is in an earthquake prone zone, this possibility is enhanced greatly.

Another danger is landslides. If a major landslide occurs and a huge landmass slides into the river upstream, the resulting wave will be so big that it will flow over the dam and flood all the areas downstream.

During the construction of the Tehri Dam, as in the case of other big dams, the engineering work was given all the importance. True, it is a major work and must be of excellent quality; but what about human lives? What about nature itself? In building the Tehri Dam, the social and environmental issues have been neglected. Bahuguna says, "This is a dam built on our tears". The reservoir will submerge 40 villages totally and 80 villages partly. This is in addition to the Old Tehri town. A total of 18,000 families had been displaced. But even though land acquisition had begun in 1979, only 13,000 local people have been rehabilitated till now.

Pratapnagar tehsil of the Tehri-Garhwal District is in between the two rivers, Bhagirathi and Bhilangana. The bridges connecting this tehsil with other places have been submerged. So the people have to take a much longer route to go anywhere. People of Lambagaon have to travel 80 to 100 kms to reach the District Head Quarters, in place of the 25 to 30 earlier. It is most difficult for pregnant women and other sick persons to go to the doctor along this long route. The cost of things and the transport cost have shot up steeply after the submergence of the bridge.

In general, in addition to submergence of good land and lack of proper rehabilitation packages for the displaced local people, mostly tribals and other rural poor, destruction to valuable biodiversity is a major concern when big dams are built. In the case of Tehri, these harmful effects of the dam do raise concerns in addition to the possibility of earthquakes.

***Narmada Dam*:** River Narmada originates from a plateau in the district of Shahdol in the state of Madhya Pradesh. Its name is Amar Kantaka Plateau. The river flows westwards for about 1,300 km and empties in the Arabian Sea. It is the longest and largest west flowing river of India. Its annual water flow is 41 billion cubic metres. Between the Vindhya and Satapura mountain ranges, River Narmada drains almost 10 million hectares of land. Thus the basin of Narmada is very vast.

The Government of Madhya Pradesh decided to put to action a major plan called the Narmada Basin Development Programme. Accordingly, the plan was to build 35 large dams on River Narmada and its tributaries.

450 medium dams and several thousand minor structures were also planned. The total cost estimated was Rs. 25,000 crores.

The benefits projected by the Government of Madhya Pradesh were:

1. Irrigation of many million hectares of parched land, making it cultivable or increasing the present yield as more than 1 crop could be harvested on the completion of the dams
2. Thousands of industries in the industrial belt of the state would get water supply from the reservoirs
3. Several thousand megawatts of electric power would be made available to the citizens of the state from the hydroelectric projects that would come up on the dams

However, *environmentalists think that the harmful effects of the dams far outweigh the benefits*. They point out that:

1. The basin of Narmada is a dense forest; building major dams along the river would need clearing of vast tracts of the forest; this is suicidal for the unique biodiversity of these lush green forests and it will also cause loss of rich soil, floods and draughts. Even climatic changes are possible if large areas of the dense forests are cleared.
2. More than 1 million people would be displaced. From past experience with other projects in other parts of the country, it is clear that not much would be done for these poor people, mostly simple minded tribals. The displaced are usually promised compensatory land, but this is seldom given. Even if land is allotted, after much delay and misery, it is uncultivable, in some far off rocky terrain or wasteland without any provision for irrigation. Uprooting of whole families result in many social problems also like employment of adults and education of children.
3. More than 1000 villages would be submerged totally. All structures in these villages including houses, schools, hospitals, religious places etc. will all go under water, causing immense loss of money and mental agony. The monetary compensation is very late in coming or does not come at all.

Because of all this, many environmental groups protested the building of big dams on Narmada. However, most of the major, medium and minor dams were built, disregarding the protests by the environmental groups and the affected villagers. *The latest controversy is over the Sardar Sarovar Project (SSP)*. Activist Medha Patekar, who is the leader of the Narmada

Bachao Andolan (NBA), asks why this project is essential when 35 big dams have already been built on Narmada, in addition to many small ones. Literary figure Arundhati Roy also joined Medha Patkar in the struggle. Recently, Hindi film actor Amir Khan has joined the movement too.

Building of big dams has been found to be counterproductive all over the world. This was proved beyond doubt at the World Commission of Dams (WCD), an independent body. The consensus was that the price of environmental damage and social disruption of local communities are not worth the benefits of providing water and power.

In fact, Medha Patkar says that the solution to the water problem lies in conserving water in as many places as possible so that runoff to the rivers is reduced. This will irrigate land where necessary and also prevent floods and soil erosion. The construction cost will be much cheaper and there will be no displacement of indigenous people. The case went to the court.

However, the Supreme Court's order on the SSP was a blow to the anti-dam environmental activists. Though some judges were with them but they were in a minority. The majority judgment was that construction of the dam would continue up to 90 metres and after that further permission would be given only after the Government met with all requirements. Medha Patkar is still fighting.

***Silent Valley*:** Silent Valley is a valley of virgin (uncut) forests surrounded by Nilgiri forests to the north and Attapadi forests to the east. It is a place in the Palghat district of the state of Kerala. The total area of the Silent Valley is 8950 hectares. All the forests together occupy 40,000 hectares.

Virgin (primitive) forests are also called old growth forests, not been seriously disturbed for several thousands of years. 60% of the world's tropical forests are primitive. Silent Valley of the Western Ghats is one such forest. A large variety of wild life species live in old growth forests. They also have large numbers of standing dead trees and fallen logs as the forests are never disturbed. Great varieties of species live on this dead wood and return the minerals to the soil.

As the primitive forests are very dense and mostly inaccessible, many species of plants, animals and microbes that live here have not been studied yet. We never know which of them may be useful to man. One or more of them, singly or in various combinations in varying proportions, may prove to be a drug for cancer or AIDS. There may be precious genes in some of them which scientists are searching everywhere for use in their genetic engineering studies. Whatever the case may be, it is a crime to destroy forests, which are vast gene pools of unique organisms.

Therefore, when the State Government of Kerala decided to construct a dam in the Silent Valley for generation of hydro power, there was a great furore. It was in 1976 that the Kerala Government put forward this plan of generating 120 megawatts of electricity by implementing the Silent Valley Project. The Government said that the dam would store 270 million cubic feet of water in a large reservoir, which would be 700 hectares in size. The estimated cost was Rs. 25 crores in 1979.

The educated people of Kerala are very much environmentally aware. They began protests immediately under the banner of the NGO, Kerala Sasta Sahitya Parishad (KSSP). The Parishad was supported by students, teachers and the general public of Kerala. A variety of protest methods were adopted.

It was a long struggle. Under the guidance of KSSP, people petitioned the Government to stop the project. Cycle and jeep rallies and hartals were organised, audio- visual films were prepared and articles were written in newspapers to create public awareness. In later stages, people of other states also came forward to help.

The Silent Valley Movement soon attracted the attention of environmentalists and environmental organizations in many foreign countries. Many, like IUCN (International Union for Conservation of Nature and Natural Resources) supported KSSP strongly. Finally, Prime Minister Indira Gandhi intervened and the Govt. of Kerala had to drop the project in 1984, by when the estimated cost had escalated to Rs. 51 crores.

This is the success story of how an NGO, supported by an enlightened public, converted an Environmental Movement into a mass movement and forced a state government to change its decision.

ROLE OF TRIBAL PEOPLE, WOMEN AND NGOs IN ENVIRONMENTAL PROTECTION

Role of Tribal people: As stated earlier, *hunter gatherers were the first tribal people in the world.* Many such tribes exist even today in different parts of the world. The hunter gatherers do not practice agriculture or animal husbandry. So they live in places close to the source of available water and food. So they move about a great deal. When food supply in one place gets over, they move to new places. This is of course bad for the forests where they stay, but the hunter gatherers live in small groups and the effect they can bring about in vast jungles is not much. Utilization is never more than regeneration. Hunting gathering tribals' requirements were limited to the forest produce. The relationship between the tribals and the environment is therefore very friendly. Even when they hunt animals, the environment is not degraded as the requirement is limited.

Some other tribes took to pastoralism. Pastoralists are people who make a living by raising domesticated animals like cattle, sheep, horses and camels. Pastoralism is an adaptation to dense grasslands, mountains, deserts, etc. where agriculture and horticulture are not possible. Pastoralists are nomadic, moving far and wide in accordance with the weather conditions. One example is the Bakhtiaris of Iran. They also do not destroy the environment as they are people who tend sheep and goats and move from Iran to Iraq and back when in search of grass in different seasons. However, they do clear forest areas by burning to set up temporary homes for their animals and themselves. Still due to the small numbers of animals and tribals, the damage to the environment is well within limits.

When the early man learnt to cultivate land, the agricultural tribals came into being. People became food producers about 10,000 years ago. When cultivation of food crops started, different river valley civilizations came into existence. Population increased. Modern tools were made which were used to clear forests by fire for jhumming cultivation. Shifting cultivation was also started when the original plots became infertile. Thus destruction of ecosystems start in the nature. Even today, jhumming is done in many parts of the North-East India and other countries of South and South-East Asia.

Some diverted from agriculture to horticulture. A good example of a horticultural tribe is the Gururumba of New Guinea. The family that keeps its garden neatest has the most prestige in the community. So the gardens are always well maintained. They use different soils for different crops. They also rotate crops. They do not use fertilizers. The crops are harvested at different times of the year. Altogether, the agricultural practices of the Gururumbas are environment friendly.

A tribe can be defined as a group of people in a primitive or barbarous stage of development acknowledging the authority of a chief and usually regarding themselves as having a common ancestry. They have a common dialect and occupy a common territory. In India, they are also called Vanavasis (forest dwellers) or Adivasis (primitive tribes).

Tribes are an integral part of the population of India. They are part of nature like others. The difference is that the tribal people know that they are part of nature, while others do not realise this truth. In such a situation, it is not possible that they will destroy nature in any way. *Next to Africa, India has the largest tribal population in the world.*

Geographically, the tribes of India can be divided into three categories:

1. South Indian Tribes
2. North Indian tribes
3. Tribes of Central and Chotanagpur areas

In South India, there are many tribes. *Malapandarams are a very small tribe living in the Central Travancore Hills of Kerala.* They live in caves or very simple sheds. They are mainly gatherers, though at times they hunt with the help of tame dogs. They love forests and will never even imagine of destroying them. Other tribes in Kerala are hunters, fishermen or agriculturists. *Tribes of Tamil Nadu are also many, the most important being Todas.* They are a purely pastoral people, living in the forests of Nilgiris. *Of the Dravidian tribes, Gonds are the largest.*

Santhals are a very large tribe scattered all over Chota Nagpur and Santhal Paraganas. Santhals and Oraons depend on cultivation using ploughs. *Santhals, Oraons, Mundas and Hos are common in Jharkhand. Bhils are seen in M.P., Gujarat, Rajasthan,* etc. *Khasis and Garos are tribal people of Assam.* Many more tribes live all over India.

The tribal people live mostly in forests or nearby villages. They are a treasure house in the matter of knowledge in medicinal plants. In the forests of Uttarkhand, there is a special species of grass. It is famous all over the world for being inexhaustible. It is said to grow at night to compensate the loss happened due to grazing during daytime.

The Gonds are the most civilized and advanced aboriginal tribes of India. They are an agricultural tribe. They maintain the fertility of the soil by applying cow dung and all the garbage generated in their households to the soil. They always settle in elevated places so that there is a proper drainage. They also hunt animals. Their fields are fenced for protection. Oraons plant groves of fruit trees outside their village.

Kukis of Assam are a migratory race. They live in dense bamboo forests. They raise only one crop and then shift to another part of the forest where bamboos grow thickly. They cut down the bamboos and start cultivating. This is very bad for the soil. Garos also cultivate land for two years and then abandon the land. Khasis, a tribe living in Assam, burn down whole forests for cultivation.

In general, damage to our forests is brought about mainly by illegal tree feeling by unscrupulous contractors and poaching of wild animals and birds by smugglers. These activities take place in many cases with the knowledge and permission of forest officers. Industries also use a large quantity of timber for different practices. There are no laws that they must plant trees to make up for the loss. *The role of tribal people in damaging our forests is minimal.* They do not bring about large scale destruction as they depend on forests for their livelihood. These days, some tribal people have started selling forest produce illegally to urban traders, but this does not amount to much, though this tendency should be nipped in the bud.

In India, even after independence, the ancestral land and forest rights of our Adivasis were not recognized. This has been a historical injustice done for these past 59 years. Fortunately, Scheduled Tribes (Recognition of Forests Rights) Bill was drafted by the Ministry of Tribal Affairs in 2005. This has been supported by all political parties, tribal MPS cutting across party lines, tribal rights activists, anthropologists, social scientists and historians. Let us hope that this bill will be passed in Parliament. The state forest management has been proved to be ineffective in managing our forests. It is high time the forest rights are handed over to the tribal people, for they are sure to conserve the forests in a much better and sustainable manner.

Role of women: Throughout the developing world, women are often the most intimately connected to the use and care of the environment. This is partly because they are the most affected by changes in the environment. For example, in South and South East Asia, 60% of women are affected by HIV/AIDS while the percentage of affected men is less. Similarly, effect of toxic chemicals on health is maximal in women and children.

Women make half the population of the world. Yet they are a neglected lot. Though they work for two thirds of all the hours worked in the world, they receive only one-tenth of the world's income. They own only 0.01% of world's property. Women in the poor countries die more than men due to malnutrition and diseases related to pregnancy, which could have been prevented.

One of the main reasons for these discrepancies is the lack of education in women. Among the illiterate in the world, which was 950 million in 2001, more than two thirds were women. In many developing countries, literacy rates of women are less than half of those for men.

The UN Fourth World Conference on Women was held in Beijing, China in 1995. This Conference recognized that women have an important role to play in sustainable development — in both sustainable production and sustainable consumption.

Women in general have a better idea of management of natural resources. To help women, especially in rural areas, for increased participation in sustainable development, it is advisable to provide them with:

- Training in natural resource management
- Training in productive resource management
- Opportunities to take part in development programmes
- Land
- Credit
- Cooperatives, etc.

UNEP hosted the 1st Women's assembly, which was aptly named as 'Women As the Voice for the Environment (WAVE) from 11th to 13th October 2004 in Nairobi. 140 prominent women environmentalists from 60 countries including 7 Environmental Ministers and other high level representatives took part in the proceedings. They discussed the crucial role women should play in conservation of natural resources and sustainable development. It is women who are most vulnerable to all sorts of environment related problems. Role of women in environment management is always ignored. Their knowledge is undervalued. They have fewer rights and a lower status compared to men.

One main reason for environment related problems, as most other problems, in the less developed countries is the population explosion. To reduce birth rates, it is most essential to educate and empower women. Such women will have fewer and healthier children and will be healthier themselves. This will help the rural women to take more active part in constructive activities.

Through management and proper use of natural resources, women sustain families and communities. They play a major part in promoting sustainable development as:

- Consumers
- Producers
- Caretakers
- Educators of families, and
- Preservers of nature for the present and the future generations.

Therefore, women must have an important say in policy matters regarding the environmental problems facing a country. They must be part of the decision making group working in the areas of conservation, protection and rehabilitation.

Women are natural leaders in:

- Promoting environmental ethics
- Reducing resource pollution
- Recycling resources to minimize waste
- Reducing excess consumption of resources, etc.

It was in the Tehri Garhwal District, where the Chipko Movement led by Gandhian Sunderlal Bahuguna began, another Movement originated - the Beej Bachao Andolan (Save Seeds Movement or SSM). Here also, like in the case of the Chipko Movement, village women took the initiative. *The rural women took part most enthusiastically in saving traditional seeds and motivating farmers to practice organic farming.*

In farming villages, though men plough the fields, it is women who do most of the other work in the fields. Men bring new hybrid variety seeds as they get subsidies for such seeds. Men say green revolution seeds are better as they give a higher yield and this will increase profit. But the women know that these will ruin the land, and so they do not want to go in for short term gains. They prefer traditional seeds and it is well known today that these are best for long term sustainable development.

Some activists of the Save Seeds Movement participated in the famous Kashmir to Kohima foot march organized by Bahuguna. Forest Protection Committees were organized and watchmen (actually, watchwomen) were appointed for protection of many forests. This resulted in remarkable regeneration of forests in Jardhar, Kuri, etc.

Women of a village in Orissa decided to take the responsibility of protecting forests on their own shoulders when the male members who were doing the same began to face serious threats from the timber mafia. On October 26, 1999, 200 people with 70 carts were seen entering the forest for illegal tree felling. The men ran to the forest department but received no help from them. The women then gathered outside the forest, divided themselves into two groups and stood on either side of the path to the forest. When the men came out, the women attacked them with spades and sharp weapons. The men were scared of fighting the women due to social reasons, and ran away! The women then sent for the forest officials. The felled timber was sold by the villagers. The money was deposited in the village fund. This is a real woman power.

In hilly villages, women fetch fuel wood, fodder and water for their households. So it is the women who suffer most from loss of forests. They will have to walk longer distances with heavy loads on their heads. They may also have to go to dangerous slopes to collect fuel wood and fodder. This was a major reason why the Chipko Andolan was supported more enthusiastically by women.

It is women who do all types of work at home which need water, like cooking, washing of clothes and utensils, etc. Therefore *scarcity of water affects the women most.* The result is that they will use water very sparingly, as they know the real value of water.

Women nature writers play an important role in environment protection. Alice Hamilton (1869 to 1970) of the US was the first to write in detail about industries polluting the environment by toxic waste materials. She continued writing for the safety of workers in such industries. Rachel Carson (1907 to 1964), also of the US, was a nature lover. Her book 'The Sea Around Us' published in 1952 was a best seller. She also spread awareness in effect of pesticides among the common people. In India, Vandana Shiva, winner of many awards, published 11 books and 74 articles on research and analysis on environmental themes, mainly conservation of biodiversity.

Role of NGOs: Non-Governmental Organisations (NGOs) have become common in all countries in the world. This is because the Governments on their own have not been able to solve all the problems facing their peoples anywhere in the world.

NGOs play an important role in promoting:

- Nature conservation
- Pollution control
- Rural development
- Literacy
- Family welfare
- Social forestry
- Environmental awareness, etc.

There are many NGOs worldwide. Some of them are international, some national and some local. All are non-profit organizations. NGOs try to bring about changes in attitude in governments and business houses also, in addition to the people. NGOs are spreading very fast everywhere. In 1998, a study by Johns Hopkins University found that this non-profit sector employs 19 million people in 22 countries.

Formation of NGOs is a comparatively recent phenomenon. More than half number of all the NGOs in Europe were founded only in the last 10 years or so. 70% of those in the US are less than 30 years old. The total number of international NGOs was just 176 in 1909. The number increased to 23,000 by 1998.

The strength of NGOs has been proved time and again. The signing of a treaty to ban landmines can be taken as one example. An international NGO, called the International campaign to Ban Landmines (ICBL) was formed in 1993. The aim was to promote a treaty to ban landmines. The NGO never had an office; nor did they have any permanent staff. Only with the help of e-mail and the internet, ICBL contacted more than 1,000 organizations in 60 countries and requested them to get governments to ban landmines. By 1999, 131 nations had signed a treaty which bans the production, stocking and use of landmines.

The method adopted by ICBL was simple. They did not project use of landmines as a military practice. Instead, they publicized the fact that 80% of the people who died or got injured were civilians. They also involved the injured in the campaign. By doing this, they gained sympathy from the public and the governments. Thus the whole issue was humanized. The result was extremely good.

The most well known international NGO is Greenpeace. This is a non-profit organisation with a presence in 40 countries. It organises public campaigns for almost everything related to environment. It is based in Amsterdam. In India, its office is situated in Bangalore. *Unilever Mercury thermometer production plant* in Kodaikanal, Tamilnadu, used to import mercury from the US and re-export thermometers back to the US. Lax safety standards at the plant left workers with mercury poisoning. Large mercury contaminated wastes were dumped in the surrounding areas by the company in complete disregard of the local communities. Local environment groups and Greenpeace exposed this in May 2001. The plant was closed, but the waste was left where it was. Cleanup has started recently. The Greenpeace campaigner in India says that several thousands tons of toxic waste contaminated soil are still to be identified. The beginning of the cleanup is a big victory for the campaigners in India.

This is a classic case of chemical industries of selling plants that are too dirty or inefficient for the US or Europe to countries such as India where environmental and safety standards are lower. Thus these industries continue to make profits from pollution in poorer countries. It is high time we Indians become aware of such nefarious designs. For this also, creating environmental awareness is of paramount importance for the third world countries. The latest example is the sending of an *asbestos laden ship from France to India.* The ship is said to carry many more tons of asbestos in excess of the declared quantity. Greenpeace has protested the deal, saying that the workers who do the ship breaking in Gujarat will fall victims to slow asbestos poisoning. (Asbestos is a carcinogen, i.e. a cancer causing chemical.)

The role of NGOs in creating environmental awareness and promoting environmental education is most significant. The Bombay Natural History Society (BNHS) began as a small society of 6 members in 1883. *BNHS is India's oldest research oriented NGO, working in wildlife research.* It publishes many magazines. Dr. Salim Ali, a greate ornithologist, was a scientist in BNHS. The BNHS has helped Governement of India in formulating many laws on wildlife. Though started for promoting wild life, *World Wide Fund for Nature - India (WWF-I) with Head Office in New Delhi works on many environmental issues.*

Kerala Sastra Sahitya Parishad (KSSP) is one of the pioneering NGOs in India. KSSP initiated the Silent Valley Movement and successfully stalled a hydroelectric project in the Silent Valley, one of the best biodiversity spots in the world. It was a long struggle.

Some other examples in India are:

- Khonoma is a village in Nagaland where the warrior tribe Angami lives. A lot of timber was being stolen from the forests. Khonoma is a tourist village. So the local Tourism Development Board is conducting an Environment Impact Assessment (EIA) study to know the impact of tourism on the environment. In this study, the NGO, 'EQUATIONS' helped the Board a lot.
- Satara Tukum is a small tribal village in Maharashtra. Till 1970, the forests in the village were quite healthy. But later, degradation set in slowly and very soon habitat degradation and loss of wildlife were noticed. It was the Nagpur based NGO, Vidharbha Nature Conservation Society, that formed a Nature Club with the village youth.
- The NGO 'Vasundhara' in Orissa helps in convening monthly meetings of women from nearby villages to discuss matters related to forest protection.
- Magsaysay award winner Rajendra Singh did a lot for water conservation in Rajasthan. He got the Ramon Magsaysay award for his work in 2001. He became famous as the Water man of Rajasthan. He was involved in the watershed management projects

in Alwar District. Rajendra Singh did not call any engineers to solve the water related problems of Rajasthan. With the help of the NGO named Tarun Bharat Sangh (TBS) that he had founded, he taught the local people how to conserve water by constructing ponds, deepening rivers that have gone dry, etc.

- Bangladesh has successfully banned the use of plastic bags, completely in Dhaka and partially in other cities. This has been made possible by the combined efforts of the Ministry of Environment and various NGOs and environmental activists.

In this way, *there are thousands of NGOs working for conservation of nature, prevention of pollution and other environment related issues.* The combined efforts put in by them are making a great difference in the global environment today.

QUESTIONS

A] Very short answer type questions:

a) Multiple choice type - Tick the right answer:

1. Ozone depletion in the stratosphere leads to:
   (a) Forest fires (b) Global warming
   (c) Water pollution (d) Smoke pollution
2. Soil erosion in deserts takes place because of:
   (a) Rain (b) Glaciers
   (c) Wind (d) Landslides
3. Greenhouse effect results in:
   (a) Reduction in temperature
   (b) Reduction in light intensity
   (c) Increase in temperature
   (d) Increase in wind velocity
4. Sheet erosion is caused by:
   (a) Fast moving Glaciers (b) Earthquakes
   (c) Heavy rain (d) Strong wind
5. Greenhouse effect is the result of an increase in the concentration of:
   (a) CO_2 (b) O_2
   (c) CO (d) CH_4
6. A row of tall trees planted in deserts to prevent soil erosion by wind is called:
   (a) Wind vane (b) Wind mill
   (c) Wind break (d) None of these

7. Planting food crops and tree crops as a mixture is called:
   (a) Social forestry (b) Afforestation
   (c) Reforestation (d) Agroforestry
8. 'On the Origin of Species' was written by:
   (a) Charles Darwin (b) George Mendel
   (c) Carolus Linnaeus (d) George Bernard Shaw
9. An urban area with more than 1 lakh population is a:
   (a) Town (b) Municipality
   (c) Small town (d) City
10. Examples of mega cities are:
    (a) London (b) Paris
    (c) Mumbai (d) All of the above
11. Simplest and most primitive tribals are:
    (a) Bhils (b) Pygmies
    (c) Nagas (d) Eskimos
12. Arthasastra was written by:
    (a) Aryabhatta (b) Kalidas
    (c) Kautilya (d) Varahamihira
13. Swami Vivekananda's birthday on 12th January is celebrated as:
    (a) National Youth Day (b) World Environment Day
    (c) World Population Day (d) National Integration Day
14. Bhils are present in:
    (a) Gujarat (b) Madhya Pradesh
    (c) Maharashtra (d) All of the above
15. The oldest Environment Movement in India is:
    (a) Silent Valley Movement
    (b) Narmada Bachao Andolan
    (c) Bishnoi Movement
    (d) Chipko Movement
16. Chipko Movement is associated with:
    (a) Medha Patkar (b) Arundhati Devi
    (c) Sunderlal Bahuguna (d) Salim Ali
17. Tehri Dam is built on River:
    (a) Yamuna (b) Sindhu
    (c) Brahmaputra (d) Bhagirathi
18. Shanti Niketan was set up based on environmental ethics by:
    (a) Jawaharlal Nehru (b) Mahatma Gandhi
    (c) Rabindranath Tagore (d) Vidyasagar

19. River Narmada originates from:
    (a) Deccan Plateau (b) Amar Kantaka Plateau
    (c) Chotanagpur Plateau (d) None of the above
20. Silent Valley Movement was fought under the guidance of the NGO:
    (a) Greenpeace (b) BNHS
    (c) NBA (d) KSSP

b) Define the following:
1. Man
2. Society
3. Environment
4. Ethics
5. Sustainable society
6. Earth citizens
7. Ethnography
8. Culture areas
9. Sacred groves
10. Settlement
11. Tribe
12. Age-area hypothesis
13. Rural settlement
14. Urban settlement
15. Rural-urban continuum

c) Name the three elements that make an ethical community.

d) Enlist the ethical issues related to environment.

e) Explain briefly, in 2 or 3 sentences:
1. Sustainable cultivation
2. Agroforestry
3. Evolution of man
4. Subsistence farming
5. Society

f) Name any five hunting gathering societies of the present world.

g) Name any five agricultural societies in the present times.

B] Short answer type questions:

Write short notes on:
1. Man's role as a destroyer and preserver of environment
2. Teachings of ancient Indian scriptures on environment
3. Role of accountability in determining responsibility
4. Equity disparity
5. Inequality in consumption of resources
6. Gender inequality
7. Natural resources and future generations
8. Sacred groves
9. Animal rights
10. Zero waste companies

11. Diffusionism and culture cycles
12. Functionalism and society
13. Origin of settlements
14. Rural settlements
15. Urban settlements
16. Impacts of settlements on environment
17. Wild life Protection Act 1972
18. Agricultural societies and their impacts on environment
19. Agricultural societies of the present times
20. Industrial revolution and its effects
21. NGO's

C] Long answer type questions:

a) Write long notes on any two:

1. (*i*) Characteristics of sustainable development
   (*ii*) Comparison between stable and unstable settlements
   (*iii*) Hunting gathering societies and their impact on environment
2. (*i*) Views of early anthropologists on evolution of civilization and society
   (*ii*) Guidelines for sustainable earth society
   (*iii*) Throw away society ethics
3. (*i*) Sustainable society ethics
   (*ii*) Knowledge Society
   (*iii*) People's participation in environmental movements
   (*iv*) Scope and objectives of environmental movements:
4. (*i*) Chipko Movement
   (*ii*) Narmada Dam
   (*iii*) Tehri dam
   (*iv*) Silent Valley

b) What are individual's rights and responsibilities towards clean environment?

c) What do you understand by Environmental Ethics? Explain.

d) What were Swami Vivekananda's views on youth, women and social development?

e) What was Gandhiji's approach towards youth, women and social development?

f) Describe in detail the ethical guidelines which prevent degradation of environment.

g) Trace the growth of civilization and social structures in the world.
h) Give an account of human settlements.
i) With the help of suitable diagrams, describe settlement patterns.
j) Discuss the role of tribal people in environmental protection.
k) What is the role of women in protecting the environment?
l) How do NGOs help in environmental protection? Explain with examples.

Unit 3

Social Issues and the Environment

It is said, "Man is a social animal". This is a true statement. We do not have normal existence separated from society. Therefore all types of social issues are important to us, our well being. Many of these issues are closely connected to the environment in which we live. Thus any study of Environmental Science will be complete only if the relation between environment and social issues are well understood. Some of the most important of these issues are included in this unit. These are:

- Sustainable development
- Conservation
- Global climate changes
- Acid rain
- Ozone depletion, and
- Various environment related Acts

UNSUSTAINABLE TO SUSTAINABLE DEVELOPMENT

The belief that economic development is the only desirable development is wrong. This is clear from the fact that if the economically developed countries do not pay attention to sustainable development, the environment will become impossible to live in. Only economic improvement is not enough; people must have a clean environment to live in. The more developed countries (MDCs) realized this fact by the early 1970s. They understood that many of the problems facing them were caused by indifference to environmental issues. Many of these problems are created by unsustainable developmental processes like:

- Cutting of trees in the forest without replacing them
- Monoculture (planting the same crop year after year in the same area)
- Excess use of artificial fertilizers
- Excess use of toxic pesticides, fungicides, herbicides, etc.

- Overgrazing
- Continuous use of hybrid seeds at the cost of traditional seeds
- Overexploitation of ground water without recharging it
- Over irrigating croplands
- Overexploitation of fish and other aquatic animals
- Draining marshes for constructing buildings
- Construction of major dams
- Poaching wild animals from forests
- Illegal trading of wild animals and plants
- Mining
- Shifting cultivation
- Jhum cultivation
- Sand mining from river beds
- Polluting air, water and soil by different means, etc.

From this it is clear that *any development during which natural resources get depleted faster than their renewal is unsustainable development.* This leads to environment degradation, global climate changes, increase in the occurrence of various diseases, etc. To become more industrialized and as a result, more economically superior, the MDCs like America, Canada and Western Europe used up a large quantity of their natural resources and exploited the less developed countries (LDCs) in Africa and South and South East Asia too. The result was that the former became richer while the latter became poorer.

The same happened in different areas of the LDCs. While some areas showed great economic advancement, some others remained poor and undeveloped. This type of discrepancy, inequality in growth, led to great discontentment and ill will.

To prevent all these from happening, it is essential to follow sustainable development and discard unsustainable development. *Sustainable development can be defined as the development that fulfills the needs of the present generation without harming in any way the needs of the future generations.* Such a development will also remove all types of discriminations between:

- Continents
- Countries
- Religions
- Castes
- Classes
- Races

- Genders
- Age groups, etc.

All will be equal, as it should be. However, this is practically impossible. Yet we can and should try to reduce the existing gaps between the extremes in the various categories. This is definitely possible by following the principles of sustainable development.

We have to pay a price for development. Thus though industrialization, urbanization, etc. are all beneficial to mankind, they are not without pitfalls. Social and psychological impacts set aside, developmental activities upset the environment as a whole. The entire biosphere is affected by these. Disturbances in any one part of the biosphere are conveyed to the other parts and cause grave consequences, most importantly, depletion of natural resources. This shows that development should not be at the cost of natural resources. We must develop, but do not disturb nature, of which we are essentially a part. This is what, in short, meant by sustainable development. It is all about conserving our natural resources along with developmental processes.

India has only 2.4% of the world's land area, but has 16% of the world's population. It is therefore obvious that there is a heavy pressure in India on the natural resources including land. This is so in spite of the fact that an average Indian consumes a much lower amount of resources than an average person in the developed world.

It has been realized over the past few years that technology alone cannot solve our problems. We must combine the best of old practices with the benefits that modern science and technology offer. It is meaningless to imitate the Western World. In fact, they themselves have started realizing the follies they have committed. They have become more conscious about the environment now. What we must learn from their experience is that we should not try to be a developed country while ignoring our environment. It is true that right to develop is a fundamental human right. Yet, we must never follow unsustainable patterns of production and consumption. *True development is environmentally sound sustainable development.* We must give utmost importance to the ecosystem balance, in other words, nature itself, while withdrawing resources from the nature, processing them and using the products. For this, a holistic approach to development is a must.

It was in this spirit of sustainable development that India participated actively in the *Earth Summit held at Rio de Janeiro, Brazil, in 1992*. All participating countries realized that protecting and restoring the biosphere to its original state is the responsibility of all countries. The Third World countries felt that since most of the environmental damage has been done by the developed countries, they should bear the cost of damage control. It was calculated that $625 billion would be needed to correct the damage done till then. The developed world, led by U.S.A. talked of common responsibility, but it is a fact that the developing world cannot afford to pay an equal share. The words 'differential responsibilities' were included in the document. This was a great achievement for the developing nations.

The four crucial documents adopted by the Earth Summit were:

1. Rio Declaration
2. Conventions for climate change and biodiversity
3. Conventions for forestry, and
4. Agenda-21, i.e. the Action Plan.

Conservation of nature has been ingrained in the minds of Indians for thousands of years now. For example, the earliest reference to forest conservation is seen in the classical Indian epic, 'Mahabharata'. We have worshipped nature as Mother Earth for years. We have given divine status to the Sun, the Moon and the planets, to mountains and rivers and to various plants and animals and birds. The aura of divinity that we have conferred on them prevents us from causing them harm. As a result of this, the ecosystems here were always in balance in the earlier years. Destruction, in the form of deforestation, started at first during the British rule. But it is most unfortunate that now our own greedy contractors and poachers are plundering our own forest wealth. It is high time to stop all this.

Afforestation, Reforestation and Forest Conservation must be done on a very large scale in India. *Social forestry* is a very good method. It has two aims.

1. The use of public and common land to produce firewood, fodder and small timber to the local poor and also thereby achieve soil and water conservation.
2. To relieve the pressure on forest conservation.

The programme involves villages and school children. In India, the area under social forestry has increased substantially in the past few years. Van Mahotsav and other such local programmes and the work of NGOs have helped in this.

Agroforestry is a modification of social forestry. Here, trees are planted in suitable locations on agricultural land. This practice along with integrated animal farming has been an ancient practice in India. Now the activity has been given the name of Agroforestry and is being popularized.

Deforestation must be checked by all means. For this, a plan of action can be followed.

1. Land degraded due to monoculture must be converted to forest.
2. Special species of trees should be planted on saline, alkaline and water logged soils.
3. Trees must be planted on road sides, near railway lines, canal banks, farm boundaries, around villages, school yards, public buildings, market squares, etc.
4. Fuel efficient stoves must be developed and alternative energy sources like solar heat and biogas must be popularized.

The main natural resources are minerals, soil, water, forests, etc. During developmental processes in a country, degradation and depletion of these natural resources will of course take place due to consumption, but these must be immediately replaced. As far as possible, more should be given back to nature than the amount of natural resource consumed, because we have already caused immense harm to our environment due to our activities over the past centuries. Thus it has become an added responsibility of each citizen on the globe not only to protect nature, but also compensate for past mistakes. This is real sustainable development and real planning for the future.

Some of the most important minerals like petroleum and coal are used as fuel for power. These minerals and natural gases are all fossil fuels which form a major chunk of the fuel energy used all over the world. These resources were evolved millions of years ago. Since the beginning of the machine age, industrial societies have become increasingly dependent on fossil fuels. This must cause our concern for two main reasons.

1. *Fossil fuel resource (coal and petroleum) is limited in amount.* These were produced by solar energy in the distant past and when they are finished, they will not be replenished. Fossils are still being formed, but at a very, very negligible rate compared to the rate at which we are consuming them. We first started consuming fossil fuels at an appreciable rate only about 150 years ago. On a global scale, we still have some coal to consume for a few more centuries, but natural gas and oil will be in short supply in a few decades.

   In the United States, the situation is worse than the global average because they are depleting their resources at a faster rate than other fossil fuel rich countries. In fact, the entire stock of fossil fuel available has been held in storage under the earth's surface for more than a hundred million years; and now it is being completely exploited only in a few centuries.

2. *Unexpected environmental consequences result from the extensive use of fossil fuels* for everything from cooking in our kitchens to powering our automobiles. When we burn coal, natural gas or oil to obtain energy, gaseous compounds are formed and dumped into the atmosphere. This is causing problems. For many years, it was felt that the emitted gases were not significant as the earth's atmosphere is very vast. But now with increasing world population and industrialization, this is no longer true. The atmospheric pollution is causing health problems and even death. Carbon dioxide emissions are threatening to produce climatic changes over the entire globe.

It is thus high time we find solutions to these problems of resource depletion and environmental pollution so that we can achieve our aim of sustainable development everywhere.

The solution is that we should *turn to alternate energy sources* as soon as possible and as much as possible. The main forms of alternative sources of energy are solar and wind energy, biogas from municipal solid waste and agro waste etc. These are far less polluting than conventional sources. They also help in managing the solid wastes in a profitable way. Preventing pollution is one main aspect of sustainable development.

We must *improve our agricultural practices* so that precious land is not degraded. For this, crop rotation, use of organic manure and biopesticides, drip irrigation, use of traditional seeds, etc. must be encouraged.

Mining sand from river beds should be stopped as it leads to frequent floods. To prevent floods, it is best to have water conserving structures in various places so that not much water flows into the rivers in the first place.

Urbanization is taking place at a very fast pace in many parts of India. This is a part of development, no doubt, but *unplanned development is harmful*. For example, rice fields conserve precious water and raise the ground water levels. Permission should not be given to level them for making buildings.

Rainwater Harvesting was done in ancient times in many countries, including India by different methods. It is high time we followed these traditional methods and also implemented some new methods along with. By practicing rooftop rainwater harvesting, we are solving our own water problem. At the same time, we are contributing towards solving a national problem of gigantic proportions. We are also taking part in the global effort towards attaining environment friendly sustainable development in a local way. Localities make a nation and nations make the world. If development goes hand in hand with maintaining the balance of natural resources in all countries, the whole world will show sustainable development.

Educating the common populace about environment is most essential for popularizing the concept of sustainable development, especially in the developing (less developed) countries like India.

CONCEPT OF CONSERVATION

'Conservation' in Latin means 'Guard together' (*con* = together; *servare* = guard). Conservation does not mean hoarding or even rationed use of resources. *Conservation is a constant effort to increase resources, while at the same time utilizing them for our needs.*

We have to *conserve the following resources* for ourselves and also the future generations:

- Forests
- Water bodies
- Soil or land
- Grazing pastures or ranges
- Energy resources
- Germplasm
- Biodiversity
- Fish and other aquatic animals, etc.

Forest Management

The forest management includes:

- Forest conservation (conserving forests in the present state, or prevention of degradation and destruction of forests)
- Afforestation (planting trees in new areas), and
- Reforestation (planting trees in areas where there were trees earlier, but have been destroyed).

Rising population, developmental activities, industrialization, pollution and illegal tree felling have decreased the forest cover in India to a great extent. However, it is heartening to see that of late, efforts are being made to improve the situation.

Silviculture is a cultivation of trees and forest regeneration. The basic feature of all forest conservation measures is the *sustained (continued) yield concept.* Using forest produce for making different articles of human use and at the same time ensuring forest regeneration is an example for sustained yield concept. The articles of use may be small like toothpicks, match boxes or pencils or huge like carts, telephone poles, boats or ships.

Conservation of Water Bodies

Both freshwater and marine water bodies are important to us and also the aquatic plants and animals in various ways. Freshwater bodies are misused more and more by the growing population. Their conservation means managing them properly while at the same time maintaining their high productivity levels. For this, knowledge of the following aspects is necessary:

- Drainage pattern (pattern of water flow from the highest point of the area to the water body)
- Runoff (the water flowing into the water body along slopes)
- Mineral receipts (the types and quantities of minerals carried by runoff water)
- Aquatic life (Plants, animals and microorganisms living in the water body)
- Carrying capacity of pollutants (the maximum amount of pollutants the water body can carry without causing harmful effects), etc.

Dams are the largest structures for water conservation. However, it is better to construct many small dams in suitable places than to construct 1 or 2 big dams. All over the world, this fact has been recognized now, though India is yet to wake up to the truth that major dams lead to flooding of large areas, deforestation of vast tracts, loss of precious biodiversity, problems of rehabilitation of the displaced rural poor, etc.

Interlinking of rivers has been suggested by some in India. But this will be an environmental disaster. Also, it is extremely costly. It will cost lakhs or crores of rupees. Different rivers flow at different levels and it will be difficult to join them. Conflicts between States and also between India with Nepal and Bangladesh will increase. It will also lead to salination, water pollution, sedimentation and water logging. Moreover, it will be dangerous to mix waters of different qualities and temperatures as many aquatic organisms will perish then. Another major problem will be displacement of local people. However, interlinking at local levels can be done without grave consequences.

Thus *conservation methods adopted must be meaningful*. These are:

- Water harvesting in several places
- Recycling and reuse of water
- Stopping wastage of water in rural and urban areas
- Adopting drip irrigation in croplands, etc.

Conservation of Land or Soil

Soil supports all terrestrial (land) ecosystems. A good soil consists of 40% inorganic nutrients, 10% organic matter, 25% water and 25% air. It is essential to maintain this proportion in any cropland. One inch of top soil takes two hundred to one thousand years to form; therefore it is a crime to destroy the soil cover of the earth. Land degradation takes place due to different reasons. There is an urgent need to correct faulty practices. *The different means of soil conservation are*:

- Preventing soil erosion by various methods
- Improving soil fertility by different methods
- Planting trees and grass wherever possible
- Controlled grazing
- Adopting proper agricultural practices
- Preventing soil pollution
- Stopping over cropping
- Neutralisation of acidity, alkalinity and salinity by chemical and other means, etc.

Conservation of Ranges

This is also called range management. Grazing lands for cows, goats, sheep, buffaloes, etc. should be maintained well. Though heavy grazing is bad for the pastures, moderate grazing is good. This is because the grasses grow better and produce more branches when the tops are pruned.

Ranges are qualitatively classified into:

- Excellent
- Good
- Fair, and
- Poor

Common methods used for range management are:

- Stock level policy — Every range has an ideal carrying capacity, i.e. the maximum number of grazing animals it can support. In some years when grass is less, some animals may die. To avoid this, dried grass is stocked in good years and used in times of need. Also, it is better to keep animals at 60 to 70% carrying capacity.
- Reseeding — When ranges get damaged, reseeding can be done.
- Pest control — To protect grasses from diseases, pest control must be practised.
- Water holes — Animals graze more in area where water is nearby. Thus, such areas parts are overgrazed and other areas remain under grazed. To avoid this, water holes must be dug for animals in different areas, if the ranges are large.

Conservation of Energy

As the fossil fuels that are commonly burnt for energy are limited in the world, we have to depend on alternate energy sources like:

- Solar energy
- Wind energy
- Hydroelectric power
- Tidal energy
- Geothermal energy
- Nuclear energy
- Biomass from municipal waste and agro waste
- Hydrogen power, etc.

We will have to shift to these energy resources sooner than later. The added advantage is that they are all non-polluting. (See Energy resources in Unit 8)

Conservation of Germplasm

It is very essential to maintain the equilibrium of the ecosystems in nature. By this, we are conserving valuable germplasm. *The reproductive cells of any organism together make its germplasm. It is the germplasm that later gives rise to the whole plant or animal.* It is through germplasm that reproduction of species and continuation of generations take place. So conserving germplasm is of utmost importance. Germplasm is stored in the form of mostly seeds in gene banks or seed banks. This way, no particular gene combination is lost forever. If germplasm is not conserved, many species of the present day will gradually become extinct.

Conservation of Biodiversity

The existence of a large variety of living organisms in any locality is called biodiversity. It is conserved by:

- Protecting natural habitats
- Limited exploitation of species
- Maintaining species in protected areas like national parks, sanctuaries, etc.
- Establishment of biosphere reserves for plants and animals
- Strict implementation of laws
- Restricting export of rare species
- Educating people for environmental protection, etc.

There are many Non-Governmental Organisations (NGOs) in India like Bombay Natural History Society, Wildlife Preservation Society of India, etc. which are very active in conserving biodiversity and also spreading awareness regarding this among the masses.

To conserve biodiversity, following two methods are used.

1. *In situ conservation* - This is conservation of the plants and animals in their own habitats. For this purpose, a Protected Areas Network has been made in India. This includes National parks, Wildlife Sanctuaries, Nature Reserves, Biosphere Reserves, etc.
2. *Ex situ conservation* - This is conservation outside the original habitats of the organisms. Conservation done in zoos, aquaria, botanical and zoological gardens, etc. are examples of *ex situ* conservation.

Conservation of Fish and other Aquatic Animals

Fish is an important renewable resource in freshwater and also sea water. There are 30,000 types of fish in the world. Japan is a major fishing nation.

Fish catch in the world has been increasing steadily. Conservation of fish has become necessary as the catch is exceeding levels of natural regeneration. As a result of over fishing, fish like Halibut, Salmon, etc. are gradually becoming endangered. To prevent over fishing, the following should be done:

- Catching young fish should be banned. For this, sizes of the holes in the net can be fixed.
- Offenders who pollute fresh water and sea water should be punished.
- Pisciculture (rearing fish) should be encouraged.

CLIMATE CHANGE

Climate of any place is influenced by light, temperature, humidity, wind, etc. Major changes in climate usually do not take place in nature. They do occur because of volcanoes, rivers changing courses, spontaneous (natural) forest fires, etc. When the climate changes in a place, many other changes result, like changes in the species composition, occurrence of floods and droughts, etc.

In the earlier days, the climate never changed as a result of human activities. But nowadays, this has become common. One of the most important changes in climate is increase in the atmospheric temperature and its effects. This is called global warming. Other changes are — more rainfall leading to floods in some places while less than normal rainfall leading to drought in other places, increased occurrence of storms, earthquakes, cyclones, tsunamis, etc.

Climatic changes can seriously affect availability of plant and animal food, drinking water, forest produce, etc. They can also bring about changes in the disease causing germs and make them stronger and of more varieties. This can lead to epidemics of known diseases and also newer types of diseases in human beings, animals and plants.

Increased occurrence of diseases coupled with floods, droughts, etc. can lead to migration of whole populations from one place to another on the earth. This will cause problems of settlements of refugees and their livelihood.

The climate of a place is the average of its weather over a long period. Rainfall, sunshine, wind, humidity and temperature are the important aspects of climate. Over the last 150 to 200 years, climatic changes have been taking place too rapidly. Many species of plants and animals are not able to adapt themselves to these changed climates. Many have become endangered or even extinct.

Greenhouse gases in the atmosphere have increased to a great extent. Carbon dioxide is the most important greenhouse gas in the atmosphere, mainly because of:

- Fossil fuel combustion in industries, and
- Increased use of automobiles

Methane is another greenhouse gas, which comes from:

- Domesticated animals like cows, goats, pigs, buffaloes, camels, horses and sheep.
- Water logged paddy fields, where anaerobic bacteria live in abundance

- Waste dumps and land fills
- Leaking pipelines, etc.

Nitrous oxides are emitted from fields fertilized extensively by nitrogenous fertilizers.

The natural causes for global climatic changes are continental drifts, volcanoes, ocean currents, etc. We cannot do anything to stop these. However, we can do much in the case of the human causes. Climate changes that started with industrialization have been continuing ever since step by step.

- Large scale use of fossil fuels had started during the period of the Industrial Revolution in the 19th century.
- The industries created jobs for people.
- To get these jobs, there was migration from the villages to the towns and cities.
- More and more land came to be used for housing.
- Natural resources were used extensively for construction, transportation, etc.
- Wastes increased tremendously.
- Number of vehicles increased many times.
- New industries came up, which again polluted air, water and soil.

The process has thus become a vicious cycle. It is going on and on. Along with the urban areas, rural areas changed too. These were mainly based on agriculture.

- To feed the increasing population, it became necessary to create more croplands.
- For this, more and more forests were cleared.
- This resulted in erosion, floods and droughts.
- To increase yield, artificial fertilizers were used liberally, emitting greenhouse gases like nitrogen oxides into the air.
- Dams were built for irrigation; this led to further deforestation and flash floods.
- The increased number of power plants released large quantities of greenhouse gases into the air.
- More irrigation led to more salination of soil.

In this way, villages have also contributed to large scale atmospheric changes leading to climate changes.

Our industries, vehicles, and croplands all pollute the air and increase greenhouse gases in the atmosphere. When this happens, climate is bound to change. This is a global problem. Thus, all countries in the world must get together to solve this problem. As the global climate change is mainly because of our activities, we can rectify the problem by various methods:

1. Cut down the use of fossil fuels
2. Reduce consumerism
3. Stop deforestation
4. Use more environment friendly methods of agriculture
5. Reduce demand for energy
6. Use alternative sources of energy
7. Preserve wetlands
8. Tax the polluting industries to compensate for cost of pollution control
9. Use non-polluting substitutes for raw materials in polluting industries
10. Create public awareness about climate changes and its effects, etc.

In short, climatic changes due to human interference in the working of nature do pose a serious threat to the present generation. The future generations will have to face these problems on a larger scale if we do not take any action at this moment.

GLOBAL WARMING

Our environment provides us with a variety of natural resources. The atmosphere is one of them. It forms a protective shell over the earth. Under normal conditions with a 0.03% by volume of carbon dioxide (CO_2), the temperature at the surface of the earth is maintained by the energy balance between the sun rays that strike the earth and the heat that is radiated back into the atmosphere. However, when there is an increase in CO_2 concentration (the major sources of which are industries, automobiles and burning of fossil fuels), it prevents heat from being re-radiated out. *The thick layer of CO_2 in the atmosphere functions like the gas panels of a greenhouse, allowing the sunlight to pass through, but preventing the heat from re-radiating out. This is called the greenhouse effect.*

The major sources of increased concentrations of carbon dioxide in the atmosphere are industries, automobiles and burning of fossil fuels. Most heat is absorbed by CO_2 and water vapour in the atmosphere, which adds to the heat already present. The net result is the heating up of the earth's atmosphere. This is called *global warming.*

Some amount of CO_2 in the atmosphere is absorbed by the oceans. But as industrialization accelerated and the number of automobiles increased, rate of CO_2 release has become more than the rate of absorption by the oceans.

All gases which contribute to the greenhouse effect are known as Greenhouse Gases or GHGs. These are:

- CO_2 (Carbon dioxide)
- N_2O (Nitrous oxide)
- CH_4 (methane)
- CFCs (Chloro Fluoro Carbons)
- SO_2 (Sulphur dioxide)
- HFCs (Hydro Fluoro Carbons), and
- About 25 other gases

These are discharged into the atmosphere mainly by industries and partly by other human activities, leading to an increase in surface and atmospheric temperature. The pattern of this rise has been studied by different agencies. Though the results have not been the same, all agree that there has been a gradual increase in global temperature.

The overall increase in temperature over the past one hundred years has been estimated to be about 0.5 to 0.7° C. According to one study, average air temperature started increasing in the Northern Hemisphere first because of rapid industrialization in the developed countries, while the increase started in the Southern Hemisphere comparatively late.

In India, during 1997-1998, a rise in temperature of 2° C from the normal was recorded in the Indian Ocean. This had resulted in the death of 70% of the corals in the Andaman Nicobar and Lakshadweep islands.

There have been predictions that if the present amount of carbon dioxide is doubled, the temperature of the earth's surface will increase by 2.9° C. In such a situation, continued global warming may have grave consequences like melting of the polar ice caps leading to a rise in sea level resulting in flooding of low lying cities like London, Dhaka, Kolkata, etc.

Realising the seriousness of the issue, *UNEP (United Nations Environment Programme) chose the slogan "Global Warming: Global Warning" to alert the people* on World Environment Day, June 5, 1989.

Soon, the UNCED (United Nations Conference on Environment and Development), commonly called the *Earth Summit, was held at Rio de Janeiro, capital of Brazil, in 1992.* It was a historic assembly of world leaders from 170 countries. This marked the setting of a global agenda in international relations and economic development.

The Kyoto Protocol was in fact a follow up of the Earth Summit. This treaty is called so because it was negotiated in Kyoto, Japan, *in 1997*. As of April 2006, a total of 163 countries have signed the agreement. These countries are bound to reduce their emissions of greenhouse gases.

Advocates of the Kyoto Protocol claim that reducing these emissions is crucially important; carbon dioxide, they believe, is causing the earth's atmosphere to heat up. Most prominent among advocates of Kyoto have been the European Union and many environmentalist organizations. The United Nations and some individual nations have also issued reports favoring the Kyoto Protocol.

India, China and Brazil are the countries having the greatest potential for the new Clean Development Mechanism (CDM) proposed by the protocol. There are many opportunities in the non-conventional (alternate) energy sector and other sectors based on increasing energy efficiency, utilisation of wastes, etc. To cite one example, using fly ash to make cement can be a CDM project. For each ton of fly ash used in this manner, the amount of CO_2 is released into the atmosphere is reduced by one ton.

If countries do not heed to the warnings by scientists, global warming is sure to bring about drastic climatic changes on the earth. Rise in temperature will not be the only result, rainfall patterns are sure to change, leading to more floods and droughts, more cyclones and storms. The recent erratic rainfall in India, according to some, is an effect of global warming. Therefore, if the countries that signed the Kyoto Protocol do act as promised, and more countries sign the Protocol, the rate of global warming at present will surely fall.

A counterargument has been put forward by some. According to these arguments, global warming is not a very serious problem; moreover, nature has the capacity to maintain its equilibrium. But not many believe in this theory. Majority believe that global warming is a reality and it will have serious consequences if not controlled by the present generation.

ACID RAIN

When fossil fuels like coal, petrol, natural gas, etc. are burned, large quantities of carbon dioxide, sulphur dioxide and nitrogen oxides are formed and are released into the atmosphere. Exhausts from automobiles also contain some of these gases. These gases dissolve in the water droplets in the atmosphere and form acids like sulphuric acid, nitric acid, etc. They fall down on earth along with rain. This is called *acid rain*. Sometimes, they remain in the atmosphere in the form of clouds or fogs.

Acid rain is more common in the developed countries because they burn more fuel for their industries and the number of automobiles there is much higher than in other countries. They also have more power stations. In fact, burning of fossil fuels for power generation contributes to almost

60-70% of the sulphur dioxide released into the atmosphere in the world annually.

The oxides may travel for very long distances in the air. So other countries are at a risk too. Moreover, during these travels, the oxides may be converted to more hazardous chemicals which may remain in the atmosphere.

Acid rains result in many problems. These are:

- Formation of acidic soils
- Acidification of fresh water bodies like ponds, lakes, streams, etc. by direct rain and also through runoff from land
- Reduction in yield of different crops
- Reduction in number of aquatic plants and animals, mainly fish
- Human health problems
- Corrosion of buildings, bridges, monuments, fences, statues, etc.
- Leaching (moving downwards) of nutrients from the top layers of the soil
- Death of many nitrogen fixing bacteria and blue-green algae
- Loss of fertility of soils due to washing away of nutrients
- Damage to leaves leading to reduced photosynthesis (making of food by green cells)
- Destruction of wax covering of trees leading to more attacks by pests; such trees can not tolerate cold too
- Toxic chemicals like Cadmium and Mercury dissolve in the acid rain and are washed down from the acidic soils to water bodies
- Lowering of pH of water leads to death of many planktons (microscopic organisms) which are the food for insects, small fish, etc. This affects the whole food chain
- Damage to monuments like the Taj Mahal
- Fish caught from acidic waters not fit for human use
- Acid rain of nitric acid leads to excess nitrogen levels in the soil. This can help some plants to grow better at the expense of others, thus disturbing the pattern of vegetation
- Aluminium ions are released from the soil, which damage roots of trees
- Aluminium ions from the soil when washed into water bodies kill fish by increasing the production of mucus which blocks their gills
- A very interesting chain effect of acid rain has been suggested by some Dutch scientists. Of late, in Dutch forests, there is a reduction in the numbers of song birds. The chain effect being cited is:

1. Acid rain leaches out calcium from soil
2. As a result, the population of snails falls
3. Song birds now get fewer snails to eat; so they eat other food
4. Due to calcium deficiency, the shells of the song birds' eggs are thinner and porous
5. Chances of the eggs breaking and also drying fast increases
6. Most of the chicks that are hatched have poorly developed bones; such defective birds die faster

Just see how balanced the whole ecosystem is in nature. What right do we have to disturb this perfect balance?

Solutions to the problem of acid rain:

1. Use less fossil fuels
2. Use more environment friendly fuels
3. Use natural gas in place of coal and petrol
4. Use coal with lesser sulphur content
5. Do research on safe automobiles. Cars running on Hydrogen fuel have already been found out in Germany and Japan. These will hit the market shortly for sale. Still, more research is required as these cars are costly.
6. Reduce the number of automobiles used. For this,
   a) adopt car pooling system instead of each person driving a car
   b) use bicycles in place of scooters and motor cycles whenever possible
   c) walk short distances; this is excellent for health too
7. Install scrubbers in industries to prevent the harmful gases from entering into the atmosphere
8. To neutralize the acidity of soils damaged by acid rain, add gypsum (Calcium sulphate) to the soil

Acid rain is not a very serious problem in LDCs (Less Developed Countries) as of now. However, it is assumed that acid rain will be more in these countries in the near future as they are following the path of development adopted by the Western countries. At the same time, the West has realized its folly and is trying hard to correct the mistake. Therefore, instead of blindly copying the West, we must adopt more eco-friendly methods of living.

OZONE LAYER DEPLETION

The atmosphere which forms a protective covering around our earth comprises many layers. *The lowermost layer of the atmosphere in which man and other organisms live is called the troposphere.* It is about 20 km thick. It is a mixture of several gases. Above this is the stratosphere, about 30 km thick. In this layer, ozone is formed by a photochemical reaction. Solar energy splits the oxygen molecule to form atomic oxygen. When this combines with molecular oxygen, ozone is formed. It is also formed through chemical reactions involving pollutants like sulphur dioxide, nitrogen oxides, etc. with ultraviolet radiations. Above the stratosphere is the mesosphere, about 40 km thick. Ozone layer is present in the lower part of the mesosphere also. Thermosphere follows mesosphere for about 400 km and this is followed by the exosphere or outer space, which extends up to about 32,200 km from the earth.

The ozone layer in the stratosphere protects us from the harmful effects of the ultraviolet rays radiated by the sun. Its depletion by human activities may have serious implications. For examples, pollutants like CFCs (Chloro Fluoro Carbons) emitted by air conditioners, refrigerators, etc. entering the stratosphere remain there for years, react with ozone and deplete it, forming what is called the 'ozone hole' in the ozone layer. The thinning of the ozone layer leads to many serious disturbances in the natural phenomena.

Effects of thinning of ozone layer on the environment:

1. Temperature changes
2. Rainfall failure
3. Increase in the filtration of UV radiation to the lower layer of the atmosphere leading to:
    - Decrease in the chlorophyll content of plants
    - Increase in harmful mutations in all organisms
    - Reduction in fish productivity
    - Increase in skin cancer among humans, etc.

A thick ozone umbrella is thus essential for the harmony and balance of our ecosystems.

Ozone depletion was the first global environmental problem recgonised by the world community in mid 1980's. In 1987, an international conference was held at Montreal on substances that deplete the ozone layer. Came to be known as the Montreal Protocol, it was signed by 175 parties. The countries agreed that they would phase out CFCs, HFCs (hydro fluoro carbons), halons, methyl bromide, etc. which are ozone depleting chemicals. Under the treaty, countries were required to gradually stop production of such chemicals. Developing countries were given more time. CFCs are supposed to be phased out completely by 2010. Similar phase out time tables exist for the other ozone-depleting substances covered under the treaty. Scientists believe that if all countries comply fully with the Protocol, the ozone shield will gradually begin to heal within the next few years, with a full recovery to pre-1980 levels expected by about 2050.

Most countries have met their obligations to the Montreal Protocol in a timely fashion. By 1997, global CFC production was down 85% from its 1986 level as the treaty began to encourage development of alternatives to these harmful chemicals and the illegal CFC trade violations in the US and Europe declined.

Some reasons for illegal trade in CFC:

1. Regional refrigerant management programmes were poorly organized
2. Alternatives were perceived to be too costly
3. Capacity to detect CFC release was inadequate
4. Officials were untrained in identifying the harmful chemicals, etc.

Anyway, enforcement in the west is improving. Moreover, the World Bank and several other international donors have stepped in with millions of dollars as funding to countries which lagged behind, like Russia, Azerbaijan, Bulgaria etc. Most of the developing countries are also well on their track to meeting their commitments.

The good news is that the ozone layer has started healing. The World Meteorological Organization and the United Nations Environment Programme announced this good news in 2002. In August 2006, two UN agencies confirmed the news. The main reason for the healing is that shorter lived harmful gases like methyl chloroform and methyl bromide have been removed. The report will be published any time this year, in 2007. As per this report, by 2050, large areas over Europe, North America, Asia, Southern Africa, Latin America and Asia will have their ozone layers repaired. Yet the time is not ripe for any celebration. For, unexpected delays are also occurring in the process of repair. This is because some types of CFCs are still in use. Also, the substitutes for CFCs that cause some depletion will be used more in future.

There is thus no relaxation. We just can not rest assured that all is well. In 2005, the Ministry of Environment and Forests, Govt. of India, had found that many companies are still using ozone unfriendly substances in the country. The technologies used here are different from those being used in developed countries. Awareness creation among the erring companies is therefore necessary in our country. *As individuals, what we* *all can do is plant and protect more trees in our surroundings. This helps to heal the ozone hole faster, as plants use large quantities of CO_2 for photosynthesis, thus reducing its level in the atmosphere.*

ENVIRONMENTAL LAWS

All countries have various laws for environment protection. Even in old times, laws had been formulated by Governments to protect the environment. In India also, several Acts were in force even before independence. Some examples are Indian Fisheries Act (1897), Bengal Smoke Nuisance Act (1905), The Motor Vehicles Act (1935), etc.

However, environment became a big issue only after the United Nations Conference on Human Environment was held at Stockholm, Sweden, in 1972. In this conference, Prime Minister Indira Gandhi had spoken strongly and effectively for India and other developing countries. It was after the Stockholm Conference that the world woke up to the seriousness of environmental issues.

The Environment (Protection) Act of 1986

Union Carbide, an American company, had set up a chemical plant in Bhopal. On the midnight of December 3, 1984, forty tons of highly toxic methyl isocyanate (MIC) stored in the company leaked into the atmosphere and spread throughout the city. Though according to the Govt. of India, death toll on that night was 2500, according to environmental activists, 8000 people died on that night itself. Another 8000 people had died later due to toxic effects. In addition to this, 200,000 people were injured due to the gas leak. Thus the Bhopal Gas Tragedy was the worst industrial accident in history.

The exact number of deaths and injuries happened that night will never be known, as there was total confusion after the tragedy. However, so many American lawyers came to Bhopal immediately. Seeing this and to speed up the payment of compensation to the relatives of the dead and the affected, Govt. of India passed the Bhopal Act in March 1985.

It was with the Bhopal Act in mind that the Govt. of India passed the Environment (Protection) Act in 1986. This is called EPA 1986 for short. The Act was passed under Article 253 of the Constitution.

The purpose of the Act:

1. To implement the decisions of the United Nations Conference on the Human Environment of 1972

2. To protect and improve the human environment
3. To prevent hazards to human beings, other living animals, plants and property
4. To provide a framework for the Central Government to co-ordinate activities of the Central and all State authorities established under previous laws.
5. To give wide powers to the executive to enable bureaucrats to frame necessary rules and regulations.

The scope of the Act: The Act was passed to provide protection to the environment and to improve the environment. The act includes 26 sections in 4 chapters. The Act is applicable all over India.

The scope of EPA 1986 is quite broad. Section 2 of the Act defines the following terms thus:

- "Environment" includes water, air and land and the inter-relationship that exists among and between water, air and land, and human beings and other living creatures, plants, microorganisms and property.
- "Environmental pollutant" means any solid, liquid or gaseous substance present in such concentration as may be, or tend to be, injurious to environment.
- "Environmental pollution" means the presence in the environment of any environmental pollutant.
- "Hazardous substance" means any substance or preparation, which, by reason of its chemical or physicochemical properties or handling, is liable to cause harm to human beings, other living creatures, plants, microorganisms, property or environment.

As per Section 3 (1) of the Act, the Central Government can "take all such measures as it deems necessary for the purpose of protecting and improving the quality of the environment and preventing, controlling and abating environmental pollution."

Based on the Act, the Centre can:

- Set new national standards for the quality of the environment (ambient standards)
- Set standards for controlling emissions and effluent discharges
- Regulate industrial locations
- Prescribe procedures for managing hazardous substances
- Establish safe guards for preventing accidents, and
- Collect and disseminate information regarding environmental pollution

As per Section 23 of the Act, the *Central Government can delegate its functions and powers to:*

- Any officer
- State Government, or
- Any other authority

According to Section 24 of the Environment (Protection) Act, the provisions and the orders based on the Act will over ride any other law. Under Sections 6 and 25, the Centre has the power to make more rules.

Chapter 3 of the act details on environmental pollution. Almost all the subjects listed in Section 3 are about industrial pollution.

All this shows how powerful the Act is. It also shows how vast the powers of the Central Government and through it, the state Governments are in controlling pollution and protecting the environment. The Centre has rightly used the powers the Act has given it in passing several rules and regulations regarding pollution and toxic substances. For example, the Centre has extended protection to ecologically sensitive areas like the sea coast. It has also introduced a system of labeling products with 'Ecomark'.

The details of the Act in different areas are:

1. *Pollution control*: Section 7 of EPA prohibits discharge or emission of pollutants in excess of the standards set. There are 3 types of standards.

- Source standards — The polluter must reduce at source the emission and discharge of pollutants.
- Product standards — The new manufactured products must follow pollution norms.
- Ambient standards — These are standards set for maximum pollutants in the air so that the concerned people live healthily.

In Schedule 1, 89 industries have been listed. These industries must follow the standards fixed for each of them regarding disposal of wastes. In the case of other industries not listed in Schedule 1, general standards must be followed. The General standards are also called Minimum National Standards. These have to be met by every industry.

The Pollution Control Boards can set strict standards where necessary. Their duties are:

- To advise the industries to set up waste water treatment plants.
- To encourage industries to follow the 3 R's, i.e. Reduce, Reuse and Recycle wastes.
- To motivate industries to set up biogas plants and use other alternate energy sources.

- To see the effluent carrying capacity of the medium into which it will be discharged, etc.

New motor vehicles must meet emission and noise limits. Air conditioners, refrigerators, air coolers and some types of construction equipments manufactured after December 31, 1993, must meet the noise levels set for them.

Levels of ambient air quality have been fixed, for various air pollutants and noise, for industrial, residential and commercial areas. Some zones like areas near hospitals must be 'Silent Zones'.

2. *Hazardous substance regulation*: Many rules have been formulated based on EPA. According to these rules,

- A person generating hazardous waste materials must do proper handling, storage and disposal of these wastes.
- Hazardous industries must identify major accident hazards and submit a safety report to the authorities stating the preventive measures being taken.
- An importer of hazardous chemicals must give complete information on product safety to the authorities. It must transport the chemicals according to the rule set.
- Biomedical wastes must be disposed properly by hospitals and clinics.
- Manufacture, import, export, storage and use of hazardous microorganisms and genetically engineered cells have to get proper license for such activities.

3. *Environment Impact Assessment*: An assessment of the possible effects of an industry, construction work, etc. on environment is called Environment Impact assessment (EIA). It was in 1994 that our Government decided to make EIA reports compulsory for industries. Only after the Impact Assessment Agency clears the project the industry can be started. In some cases, further review by experts will be necessary before the clearance can be given. The Agency is the Ministry of Environment and Forests.

4. *Coastal Regulations and Protection of Special Areas*: Special regulations have been made in ecologically sensitive areas. The Coastal Zone Regulations were issued on February 1997. As per these regulations, developmental activities and tourism can not be done within 500 metres from the sea shore. This rule is applicable for the entire coast of India. Setting up of new industries cannot be done within this range. Commercial activities are restricted. Building activity is regulated according to the level of urbanization and the ecological sensitivity of the coastal region.

Some horticultural belts and the Himalayan foothills in some places, parts of the Aravalli range, etc. are also ecologically sensitive areas. These regions face special environmental risks. All such areas are also covered by this rule.

5. *The Ecomark scheme*: Manufacturers of consumer goods can use the label 'Ecomark' if the products follow the rules for environment friendly products set by the Government. This scheme was announced in 1991. Its aim is to encourage manufacturers to make environment friendly products. It will also help consumers to make choices between environment friendly and other products. However, till 2003, no product carrying Ecomark had reached the market.

Another rule brought recently, in 1999, prohibits packing of food stuff in recycled plastic bags. They must be virgin bags, or they must be at least 20 microns thick. This ensures that thin bags made of plastic recycled over and over again cannot be used to pack food.

Violations and Penalties under the Act: The EPA was the first Act to give the Government sweeping powers for environmental protection. It can issue direct orders to:

- Close, prohibit or regulate any industry or operation or process
- Stop or regulate the supply of water, electricity or any other service

Officers of the Central Government assigned with the specific duty can enter the industry for

- Examination
- Testing of equipment
- Taking samples of air, water, soil or any other substance from any place for analysis

Discharging of pollutants in excess of the standard permitted is unlawful. Persons who break EPA must prevent pollution and report the discharge to the authorities.

Penalties are as follows:

- For each failure to follow the rules, a prison term of up to 5 years or a fine of Rs. 100,000/- or both
- For every day of continued violation of the law, fine will be Rs. 5,000/-
- If violation is for more than 1 year, prison term to be up to 7 years

Just an outline of EPA is given here. It is an elaborate Act, with many sections and sub-sections. It is essential to create awareness regarding this Act among the public. It must be followed by one and all. Punishments should be given to all offenders.

Air (Prevention and Control of Pollution) Act of 1981

The Air (Prevention and Control of Pollution) Act was passed by the parliament under Article 253 of the constitution. This was done to implement the decisions taken at the United Nations Conference held at Stockholm in 1972.

The Act displays the 'detrimental effects (of air pollution) on the health of the people as also on animal life, vegetation and property.'

The main provisions of the act are:

- Providing for prevention, control and abatement of pollution
- Providing for the establishment of Central and State Pollution Control Boards
- Give powers to the Boards to implement the Act

The Act is similar to the Water Act which was passed in 1974. The Central Pollution Control Board (CPCB) and the State Pollution Control Board (SPCB) were already having certain powers under the Water Act. The Air Act expanded the authority of CPCB and SPCBs to include air pollution control.

All sources of air pollution come under this Act, such as:

- Industries
- Power plants, and
- Vehicles

As per the Air Act, the *pollution control boards can check the emission of all types of air pollutants like*:

- Carbon dioxide
- Sulphur dioxide
- Carbon monoxide
- Nitrogen oxides
- Lead
- Smoke
- Volatile organic compounds, etc.

All industries operating within such areas must get a permit from the state Board. Under the Air Act, all industries operating within such areas must get a permit from the State Boards.

The states must prescribe emission standards for different types of industries and automobiles after consultation with the Central Board. While doing so, the states must note the ambient air quality standards of the industries.

Before 1987, the Air Act was not very strict. It used to give mild penalties to the law breakers. In 1987, there was an amendment which made the authorities more powerful and the penalties stiffer. Now the PCBs can stop water supply or electricity supply to a defaulting industrial unit or even close it down. The PCBs can also apply to the court directly to restrict the industries to keep emissions or effluents under the permitted limit. Also, noise pollution was included in the Air Act. This was a good step as it is air that gets polluted with noise and so noise pollution is definitely one type of air pollution.

The Air Act of 1981, as amended in 1987, became more powerful in certain areas:

- The amendment gave permission to state governments to designate certain areas as pollution control areas. All industries operating within such areas must get a permit from the State Board. Under the Air Act, all industries operating within such areas must get a permit from the State Boards.
- Magistrates were given power to restrain air polluters from discharging emissions
- Citizens can file suits; they can also ask the PCBs to give them the necessary emission data so that they can file suits.
- The State Boards were given the power to notify standards independently

Violations and Penalties: Anyone disobeying any order or breaking the Air Act can be put in jail for 3 months or can be fined for Rs. 10,000/- for each day of disobeying. Imprisonment and fine both can also be given to the industries, based on the seriousness of the offence.

Air Act Cases: There have been many Air Act cases in India. Many are pending in different courts too. Some famous examples are:

- Dwarka Cement Works case
- Stone quarrying case of Dehradun
- The Taj Trapezium (The Taj Mahal) case
- Motor Vehicles Case, etc.

Water (Prevention and Control of Pollution) Act of 1977

Discussions went on between the Centre and the states for more than ten years before the Water (Prevention and Control of Pollution) Act was passed in 1977. It was passed under Article 252 (1) of the Constitution. This Article empowers the parliament to enact laws on subjects under state control if the state legislatures of two or more than two states demand such laws. In this case, twelve states wanted the Act and so it was passed in the parliament.

Since 1982, the Central Board has been attached to the Department of Environment, Forests and Wildlife of Central Government. *The Water Act covers different types of water bodies like*:

- Streams
- Inland waters
- Underground waters
- Sea water
- Tidal waters
- Wells
- Sewers, etc.

The Act provides for a permit system to prevent and control water pollution. The Act prohibits discharging polluting matter into streams, wells and sewers or on land in excess of the standards established by the State Boards. A person must obtain permission from the State Board before starting any industry, operation or process, any treatment or disposal system or any extension or addition to any such system.

As per the Water Act, the CPCB (Central Pollution Control Board) can:

- Advise the Central Government on water pollution issues
- Fund investigation and research related to water pollution
- Develop an all inclusive plan for control of water pollution

Important features of the Water Act are as follows:

- The Water Act suggests that State Governments can make water pollution laws.
- The Act gave power to the State Boards to regulate pollution
- The State Boards will encourage, conduct and participate in research and investigations on water pollution.
- They will develop reliable and economical methods of treatment of sewage and effluents from industry and other sources.
- The Central Board will coordinate the activities of the State Boards.

Powers of the State Boards with regard to the Water Act:

- The State Boards can establish and enforce standards of effluents discharged into water bodies by factories.
- They can control sewage discharges by industries
- State Boards can approve, reject or modify applications submitted by factories in any matter concerned with effluent or sewage discharge

- To minimize water pollution, they can advise State Governments which sites would be best suited for starting new industries
- The State Boards can give conditional permission to industries, setting restrictions on location, construction and use of outlets and nature and composition of the discharges
- They can start necessary construction, if even after 30 days, the defaulter has not started it, and charge the amount on the defaulter.
- The State Boards can enter and inspect the industries or operations when necessary

Violations and Penalties: Before the amendment in 1988, the Water Act was not very powerful. Enforcement was done by the following methods prior to 1988:

1. Boards could start criminal proceedings against the erring industries
2. Applications could be submitted to magistrates to restrain polluters.

After the amendment, State Pollution Control Boards have more power. They can:

1. Withdraw power supply of the industry at fault
2. Withdraw its water supply
3. Close a defaulting industrial plant
4. Put severe punishment on the defaulters

Water Act Cases: The first and second cases mentioned were against the Central Board for Prevention and Control of Water Pollution, while the third was versus State of Rajasthan.

- Pondicherry papers Ltd. Case
- Delhi Bottling Co. Pvt. Ltd. Case
- Agarwal Textile Industries Case

The Water (Prevention and Control of Pollution) Cess Act was passed in 1977 to help meet the expenses of the Central and State Boards. Cess means tax. As per this Act, a differential taxing system is followed — higher rates for water are to be paid by the defaulting units. This revenue is used to implement the Water Act.

Wildlife Protection Act of 1972

Based on the resolutions passed by eleven states under Article 252 (1), Parliament enacted the *Wildlife (Protection) Act in 1972*. It *has the following features*:

- Wildlife advisory boards can be set up and wildlife wardens and other staff can be appointed to implement the act.
- Schedule 1 animals and birds cannot be hunted at all.
- For hunting animals and birds of Schedules 2, 3 and 4, special permission is needed.
- State governments can declare any area of ecological importance as a sanctuary or a national park.
- In both these, public entry is restricted.
- No wildlife or habitat can be destroyed here.
- Cattle and other animals can be grazed in sanctuaries, but this cannot be done in national parks.
- Regulates trade in wild animals, animal products and trophies
- Imposes penalties for violating the Act

Species have been listed in different Schedules by the Act. These are as follows:

Endangered Species	Schedule 1
Species requiring special protection	Schedule 2
Big game	Schedule 3
Small game	Schedule 4
Some species classified as vermin	Schedule 5

Bringing any harm to the endangered species is prohibited throughout our country. There are regulations for hunting other species. Species belonging to Schedule 5 can be hunted without any permission. It is the responsibility of the wildlife wardens and staff to see that the Act is followed by all.

The Act was amended in 1982. As per this amendment, wild animals can be captured and taken to other places for scientific management of their populations. For example, *ex situ* conservation of wild animals is done in zoos (See Unit 8). This beneficial procedure should not suffer due to the Act. That is why such an amendment was brought in 1982 for the Wild Life (Protection) Act of 1972.

In 1991, the Act was amended again. In this amendment, three major additions were made:

1. Special chapters were included to protect specific plants and animals.
2. Functioning of zoos was also regulated.
3. The needs of tribal people and forest dwellers were recognized and suitable changes were made keeping their welfare in mind.

The Wildlife Act adopts two policies for protection of wildlife. Accordingly,

- Some species (the endangered species) are protected everywhere.
- All species are protected in special areas like Sanctuaries and National parks.

The Wildlife Act was passed in parliament under Article 252, after eleven state legislatures passed the necessary resolutions. The Forty-Second Constitutional Amendment shifted the subject of wildlife to the Concurrent List. By this, the parliament got the power to bring laws on wildlife more easily.

Violations and Penalties:

- Any violation of the Act attracts imprisonment and fines.
- Citizens, after giving a notice of 60 days, can prosecute the Government.

There are some *pitfalls in the Act.* For example,

- The Tehri Dam will destroy 12 rare and endangered species. Is this hunting? If so, can the construction of the dam be stopped?
- For hunting animals and illegal wildlife trade, punishment is very mild in Jammu and Kashmir, while it is severe in all other states. Are such discriminations right?
- Why is there a provision for supplying certificates of ownership to those who possess skins, horns, etc. of rare animals?

Some cases:

- The Sariska Case — Tarun Bharat Sangh, an NGO wanted to know how the Government of Rajasthan could allow mining in a protected forest.
- Jamnagar Marine Park case — Gujarat Navodaya Mandal challenged the permit given to Reliance Petroleum Ltd. by the Chief Wildlife Warden for laying a pipeline through a sanctuary
- The most recent cases are against some celebrities for hunting endangered animals.

The Forest (Conservation) Act of 1980

The Indian Forest Act of 1927 used to be followed by the forest departments of individual states for many years. When India got independence, forests were placed under the State List of the Constitution. At the time of independence, forest cover in India was 25%. This went on decreasing and touched an all time low of 10% according to some environmentalists in the 1970s. However, due to large scale reforestation programmes, there has been some improvement. The forest cover in India stood at 18% in 1997. Even this is much less than the stipulated percentage of 30 to 33% for ecological stability.

The harmful effects of felling trees indiscriminately were becoming very prominent in India. At the same time, specialists in different fields debated how to balance the different demands of the nation as forests had to be tapped for various purposes. It was then that the Government of India passed the Forest (Conservation) Act in 1980. It was later amended in 1988. A revised National Forest Policy was also adopted in 1988.

Under the Forest Conservation Act, an Advisory Committee has been constituted. This committee will advise the Centre on funding and the various other issues. *As per the amended Act, states have to get approval of the Central Government for:*

- De-reserving a Reserved Forest
- Using forest land for non-forest purposes like mining, building dams for irrigation, hydroelectric power, etc.
- Assigning forest land to a person or corporation
- For clearing forest land for reforestation, etc.

At the same time, the *State Governments have the power to*:

- Use the forest for forestry purposes.
- Construct water holes, trenches, check posts, fences, pipelines, etc.
- Stop any undesirable activity in the forest as soon as observed.

The Forest (conservation) Act of 1980 was passed by the Centre in alarm. Until then, an average of 1 million hectares of forest land was being cleared for various purposes in India. After the Forest (Conservation) Act was passed, deforestation rates did come down to some extent. But this was not enough. Therefore *some amendments were made in the Act again in 1992. Accordingly,*

- Cultivation of crops like tea, coffee, spices, rubber, etc. are non-forest activities.
- Cultivation of oil yielding plants, medicinal plants and fruit trees need prior permission. No permission is needed if the crop is a native species.
- Taking stones and boulders from forests is a non-forestry activity.
- Tusser culture in the forests is a tribal activity to make silk from insects. It is the livelihood of tribals. So this is allowed. But local host plants must be used. Cultivating mulberry bushes in the forest for rearing silk worms (sericulture) is not allowed.

- Any proposal for any non-forest activity must provide a cost-benefit analysis first. An EIA study (Environment Impact Assessment study) must also be done to know the possible effects of the proposed activity on the surrounding environment as a whole.

Violations and Penalties:

The Act prohibits the cutting of trees without the prior approval of the Central Government. Forest land cannot be used for non-forest purposes. Disobeying the Act will lead to an imprisonment of 15 days.

Some Cases:

- The Dehradun Valley litigation — This was the first case where the Supreme Court had to balance ecological integrity against industrial needs. Thousands of acres of Mussorie Hill Range of the Himalayas were blasted by dynamite for limestone mining.
- ARC Cement case — This company was operating a cement factory in the Dehradun valley violating the Forest Act. The company was ordered to shift to some other area.

ISSUES INVOLVED IN ENFORCEMENT OF ENVIRONMENTAL LEGISLATION

It took long years for man to know that he was degrading his own environment; it took even many years for him to realize that there should be environmental legislations to prevent environmental degradation. Now that legislations are there, it is our duty to obey the rules and regulations. It is the duty of the Government to see that we obey rules and regulations, and if we do not, then punish us.

Bringing up issues in parliament, pass them and make them Act is all good. Yet enforcement of legislations on everyone, irrespective of class, creed and every other divisive factor in society, is as important as enacting them in the parliament.

For enforcement of legislation, an effective agency will be needed. Such an agency has three important functions.

1. Collect all necessary data
2. Process the collected data in detail, and
3. Pass on the data and the analysis of the data to the law enforcing agency

Environmental legislations may be made by states, countries or international agencies and organizations. Internationally, breaking any environmental law is considered as an environmental crime. Under international law, countries that have ratified (signed or agreed to) treaties are responsible for carrying them out. But governments cannot do this all by themselves. Governments must take the support of business houses, media and ordinary citizens to ensure that the treaties are followed properly.

In 1920, there were only about ten international treaties. By 2000, the number had gone up to 240. This shows how the world is getting more serious about the environment. *Some important international treaties* are:

- 1971 — Ramsar Convention on Wetlands of International Imoportance — 123 parties (signed by 123 countries)
- 1972 — London Convention on the Prevention of Marine Pollution — 78 parties
- 1973 — Washington Convention on International Trade in Endangered Species of Wild Fauna (animals) and Flora (plants) - CITES — 152 parties
- 1987 — Montreal Protocol on Substances That Deplete the Ozone Layer — 175 parties
- 1992 — Convention on Biological Diversity (CBD) — 178 parties
- 1997 — Kyoto Protocol on Climate Change — 30 parties

These agreements cover many issues like:

- Conservation of wetlands
- Biodiversity
- International fisheries
- Ocean pollution
- Atmospheric pollution, etc.

Enforcement of international environmental treaties is done by many institutions and officials. Secretariats are staffed by international civil servants and are often managed by UN Agencies like United Nations Environment Programme (UNEP). Governments of the countries that signed the treaties are in general very co-operative and provide the necessary data.

For example, a network of national organizations called TRAFFIC was created in 1976. It monitors international wildlife trade. It sees how different nations are enforcing the legislation on wildlife trade, finds out the routes of smuggling of wild animals and plants and passes on the information to CITES. International trade in more than 900 endangered species, like great apes, sea turtles, crocodiles, some orchids, etc. is monitored by CITES.

In India, there are many NGOs like Bombay Natural History Society (BNHS) that help the affected people to take matters to court. Responsible citizens often file Public Interest Litigations (PILs) in court. In many towns and cities, there are citizen forums which pursue matters of pollution, conservation, etc. and take the offenders to court. We also have consumer courts for dealing with cases in adulteration. In villages also, where at least some people are educated and are aware of human rights regarding environment, they form cooperatives to file cases. Small environmental crimes can be dealt with at local levels also by the village elders, Panchayats, Lok Ayukths, etc. This shows that *environmental awareness of the public is very important in preventing environmental crimes*.

Environment Impact assessment: Wherever there is development, some damage to the environment is done. When population is increasing in our country every minute, demands on the space available is becoming more and more. The land area remains the same, but the number of people who have to share this area is increasing extremely rapidly. Just to build more homes, imagine how many trees will have to be cut. To put up industries, to make dams, to construct roads and railway lines, environment has to be disturbed.

However, *development cannot be stopped to protect the environment. The solution to this problem lies in a procedure called Environment Impact Assessment (EIA).* By this procedure, all possible effects on all aspects of the environment will become clear to the persons or industrialists or agencies that are planning a particular activity in a particular locality. After seeing the report, the proposed activity can be started as it is or in a modified form or the locality can be changed. If the activity is very harmful to the environment, it will have to be dropped altogether and some other proposal will have to be thought of.

The Ministry of Environment and Forests has made EIA compulsory before permission is given to start any project — private or public. After the enactment of EPA in 1986, EIA was made compulsory for any project.

Many organizations and agencies have been given the power to do EIA studies. The person or persons who propose the project have to give an application to the SPCB first. After going through the project carefully, the SPCB will instruct the proposers to get an EIA done. The next step is for the proposers to approach a certified organization or agency to prepare an EIA report.

Preparation of an EIA report will take many months. This is because all the effects on all the aspects of the environment will have to be studied thoroughly. Mostly, studies are done on the effect of the industry on the air, water and soil of the locality. But in fact, this is not al all enough. The locality may have great biodiversity. By setting up of the industry or building complex there, this may be lost. This is especially true of large projects like dams, roads or railway lines through forests, etc. If the organization does not have competent staff in all areas of environmental studies, the EIA report will be sketchy and not complete.

As it is the proposers who pay for the reports, the reports may be incomplete too. The agency may not list all the expected problems so that the proposer's project will be cleared by the Ministry.

In spite of these drawbacks, the rule that EIA report is a must for new as well as old industries planning for expansion is a very good procedure to help in environmental protection. *The report finally goes to the Ministry of Environment and Forests which is the impact assessment authority.*

In 1997, the Ministry announced that *a public hearing is also essential.* The SPCB will advertise in one or more local newspapers the date, time and venue of the public hearing. Local people are supposed to attend the meeting and express their opinions. They will also be shown the EIA report.

This system works very well in developed countries, but in our country, most of the villagers are not aware of the fine details mentioned in the report. Therefore, they cannot identify the seriousness of the impact of the proposed project on their locality. This drawback can be rectified only by increasing the level of education and environmental awareness in the common people.

The damage to the environment may be small or big, temporary or permanent, restricted to some areas within the proposed area or spread everywhere. In some cases, the damage can be partly undone over a few years. All these must be clearly mentioned in a good EIA report. The effect of the project on the local people must be given importance too. Especially, when big dams and other such structures are built, thousands of people are displaced, and this creates very big rehabilitation problems. In case of industries dealing with toxic chemicals or radioactive substances, special study is necessary to judge the long term effects of the emissions and effluents on human health.

As water is a major natural resource and is in great demand, exact information on how much water will be needed for the industry and how much of it will be from the under ground aquifers must be looked into. *The local people must not face water shortage due to the setting up of the industry*. It has been made mandatory in many states that unless an industry has made provision for rainwater harvesting, permission will not be given to it for starting the enterprise. Similarly, any new industry must have a foolproof system of waste water management so that effluents will not be discharged into water bodies nearby. The maximum noise that will be produced by the proposed project has to be seen too as noise pollution has serious consequences on human health.

Only some aspects have been discussed here. Plans made by the proposers for minimizing all types of pollution and all types of conservation methods to be adopted will be studied in detail by the organization or agency conducting the EIA.

Many organizations have been certified as competent by the Government for conducting EIA studies. One such institute is the National Environmental Engineering Research Institute (NEERI), Nagpur. As per the Annual Report of NEERI, twenty three EIA reports were prepared by the Institute in 2004-05.

To explain how the procedure works, one example is given here. The study was Impact Assessment of diverting irrigation water from Pakadiguddam Tank Project (PTP) to an industry. The State of Maharashtra wanted to divert water from PTP in Rajura Taluka, Chandrapur District, Maharashtra, to cement and power plants of M/S Maratha Cement Works. Based on writ petitions filed, the Nagpur Bench of the High Court, Bombay directed NEERI to submit a report on the impacts on agricultural operations and suggest mitigation measures. NEERI scientists studied the water conducting system, PTP site and nearby areas, discussed the issues with the farmers, State Government officials and the industry.

The study showed that management by the Irrigation Department was not satisfactory and that water supply to the industry had caused depletion of irrigation. Suggestions were given as to how maximum benefit could be obtained for both the industry and the farmers.

This shows that EIA study is done not to obstruct industrial or any other type of development. It is meant to give suggestions to all parties as how to utilize natural resources fully without causing harm to the environment.

PUBLIC AWARENESS

Pollution of air, water and soil, and mindless utilization of natural resources have raised environmental degradation to dangerous heights across the globe. One of the very best methods to reverse this trend, undoubtedly, is creating public awareness in environmental issues.

Methods of creating public awareness have been discussed at national and international levels. The goal is to develop a world population that is aware of and concerned about environment and its associated problems, and committed to work individually and collectively towards solving the current problems and preventing new ones. To achieve this target, an action plan is needed at all levels of implementation — global, national and local.

I] At the International Level: There are many environmental organisations and agencies at the international level, active in different fields of work. Some examples are:

- *United Nations Educational, Scientific and Cultural Organisation (UNESCO)*: This UN agency was founded in 1945. Among other activities, it also supports activities related to environmental quality, human settlements, training to environmental engineers and other socio-cultural programmes related to environment.
- *International Union for Conservation of Nature and Natural Resources (IUCN):* It is an autonomous body founded in 1948. IUCN always involves the local people in any project, thus spreading awareness.
- *United Nations Environment Programme (UNEP):* This is a UN agency established in 1972. It acts in collaboration with different governments for environmental monitoring and protection.
- *Man and Biosphere Programme(MAB):* This programme, launched by UNESCO in 1971, has 14 projects. India is also a member country. Indian National Committee on MAB is under the Department of Science and Technology, New Delhi. Latest activities of MAB include sustainable tourism, eco-job training for young people, etc. Such activities attract the youth, which is a sure way of spreading awareness.
- *Earthscan:* This is an agency founded by UNEP in 1976. It collects original articles on environmental matters and sells them to newspapers and magazines, especially in developing countries, creating public awareness.
- *Earthwalks:* These are designed to focus international attention on environmental issues. Children are taken out by seniors for programmes like "Meet a Tree", "Leaf Slides", etc. Such nature studies create environmental awareness in young children.

II] At the National Level: All countries have different governmental and non-governmental organisations to promote public awareness in environmental issues.

A. *Governmental Organisations in India*: The Government naturally yields the maximum power and does most of the environment related work.

- *Ministry of Environment and Forests (MoEF) of the Central Govt.*: It was set up in 1980. From 1986 onwards, MoEF has been organising National Environment Awareness Campaign (NEAC) every year. The objective is to create public awareness at the national level. Proposals are invited from NGOs, educational institutions, etc. for conducting various activities like seminars, workshops, camps, padayatras, competitions, festivals, training courses, rallies, public meetings, exhibitions, demonstrations, preparation of educational resource materials, folk songs and dances, street theatres, science fairs, preparation or use of CD-ROMS and other multimedia tools, etc.
  Many Institutes work under MoEF. One of them is the Centre for Environmental Education (CEE), Ahmedabad. CEE is a resource agency for National Green Corps (NGC) scheme of the Govt. of India. In this scheme, NGC promotes formation of eco-clubs in schools. Rs. 1,000/- is given to each school that starts an eco-club. Thousands of schools have already made use of this facility. To create awareness in the youth of the country, CEE introduces courses in colleges and taps youth energy through volunteer activities. CEE has published books in 15 languages on environment related issues. It uses media to raise awareness of the general public. It works with the industrial sector also to convince them of the need for waste minimization. CEE involves students and communities in water quality monitoring with people's participation in local environmental issues.
- *Autonomous institutions like Indian and National Institute of Technology (IIT and NIT):* These institutions conduct different types of research works on environmental issues and popularise their findings. It is as a result of such activities that many villages in our country now have gobar gas plants, smokeless chulhas, solar cookers, etc.
- *The University Grants Commission, Universities and Colleges*: UGC has recently introduced a syllabus on environment for undergraduates in all sections — Arts, Science and Commerce.

(This book is meant for this particular course). Earlier, Environmental Science was taught only to Science students. Research work done in many universities is published for public awareness. In many colleges, eco-clubs have been established which conduct various types of public awareness programmes.

- *State Governments*: Governments of all states have Ministries of Environment and different Developmental Authorities. There are the Corporations, Municipalities and Panchayats also. All these implement various programmes related to environment in India. These help to create awareness from the top rung to the grassroots level. Celebration of Van Mahotsav twice every year is one such event. Ever since Rabindranath Tagore presented to the nation concepts and practice of "Vriksharopana" in 1927, tree planting has become a common practice in India. Slogans highlighting importance of trees can be seen everywhere. Area under plantation forests has increased greatly in recent years.

  Spreading awareness about the need for family planning and the need to increase literacy can never be over-emphasised. Non-formal methods of education like Adult education, Total Literacy Mission, etc. have been taken up by both Central and State Governments and also various NGOs in India. These have been partially successful. The fact that maximum environmental awareness is shown by Keralites proves how important the role of education is.

B. *Governmental or Government supported Agencies in some other countries*: Many such institutions are active in all countries. A few of them are briefly described here.

- *Environmental Protection Agency (EPA):* The best known of all is this independent Federal Agency of the US Govt. established in 1970. EPA believes that for work on environmental awareness, partnership is needed among the states, tribal and local governments, research communities, NGOs and industry.
- *Texas Environmental Awareness Network (TEAN):* This is an association of state agencies and environmental and educational organisations dedicated to creating public awareness in environmental issues.
- *Universities:* Some established Universities, especially of the developed countries, make use of latest technologies like remote sensing, DNA tracing, etc. to identify problems and solve them.

In one recent application, scientists from the University of Auckland in New Zealand confirmed with DNA tracing that whales are being sold in fish markets of Japan. Wide publicity was given to this finding and soon corrective action was taken and the culprits were punished.

- *Governmental Ministries and Departments*: In Denmark, the Govt. has promoted wind energy since the adoption of the national energy plan of 1976 and has established wind cooperatives. There are subsidies for wind and taxes on polluting fossil fuels in Denmark.

III] At the local level: Though governments have great powers, they depend on NGOs, businessmen, media and citizens for help.

- *Non-Governmental Organisations (NGOs):* The role of NGOs in creating public awareness is most significant. Kerala Sastra Sahitya Parishad (KSSP), Bombay Natural History Society (BNHS), World Wildlife Fund (WWF) India, etc. do valuable work in this field. The number of NGOs all over the world has gone up greatly.
- *Business houses*: These have an important role too. Consumers are increasingly getting interested in green products and are willing to pay more for them. As a result, organic industry has blossomed globally. Waste reduction is another common greening business strategy. Such actions by business houses create awareness in society.
- *Media:* Mass media like radio, TV, newspapers, documentary films, hoarding boards, etc. have a great influence on the people. We regularly get information on population growth and disappearing forests from the media. They also help in giving the public information on how to construct rain water harvesting systems, where to get eco-friendly products, etc.
- *Individuals*: Efforts by individuals are no less commendable in creating awareness among larger sections of society. Rajendra Singh, the Magsaysay Award winner and Medha Patkar who leads "Narmada Bachao Andolan" are well known. Much less known are people like Simon Oraon, the 71 year old uneducated tribal in a village near Ranchi, who persuaded his villagers to construct check dams to conserve water, to practise rain water harvesting and to follow multiple cropping for better utilisation of soil. Sarah Jewitt, who did her doctorate from Cambridge University, stayed

with Simon for 9 months to learn his methods. A copy of her thesis is Simon's prized possession though he cannot read it. Setting practical examples like this goes a long way in creating public awareness.

Environmental Calendar: The Environmental calendar can be used for organizing appropriate activities. This is done in all countries to create public awareness in environmental issues. Some days which can be celebrated for motivating the society are also included in the following list.

1. January 12 — National Youth Day
2. February 2 — World Wetland Day
3. February 28 — National Science Day
4. March 8 — International Women's Day
5. March 21 — World Forestry Day
6. March 23 — World Meteorological Day
7. April 7 — World Health Day
8. April 18 — World Heritage Day
9. April 22 — Earth Day
10. May 31 — Anti-tobacco Day
11. June 5 — World Environment Day
12. June 11 — World Population Day
13. August 6 — Hiroshima Day
14. August 9 — Nagasaki Day
15. September 8 — World Literacy day
16. September 16 — World Ozone Day
17. September 27 — World Tourism day
18. October 1 to 7 — Wild Life Week
19. October 4 — World Animal Welfare Day
20. October 16 — World Food Day
21. December 1 — World AIDS Day
22. December 10 — Human Rights Day

Various activities like seminars, debates, quiz, elocution, essay, drawing, painting, slogan, poster, exhibitions, etc. can be organized indoors by students in schools and colleges on these days. Outdoor activities can be street plays, rallies, public speeches, folk songs and dance, etc. All such and similar activities will help create public awareness on various environmental issues.

QUESTIONS

A] Very short answer type questions:

a) Multiple choice type - Tick the right answer:

1. In the world's population, India accounts for:
   (a) 10% (b) 5%
   (c) 25% (d) 16%
2. In 1992, the Earth Summit was held in:
   (a) Geneva (b) Montreal
   (c) Rio (d) New York
3. Coal, petroleum and natural gas are:
   (a) Fossil fuels
   (b) Renewable fuels
   (c) Alternate energy sources
   (d) None of these
4. Silviculture is cultivation of:
   (a) Herbs (b) Medicinal plants
   (c) Vegetables (d) Trees
5. A good soil contains organic matter by:
   (a) 25% (b) 40%
   (c) 10% (d) 5%
6. One example of alternate energy is:
   (a) Coal (b) Petroleum
   (c) Wood (d) Solar energy
7. *Ex situ* conservation is done in:
   (a) Zoos (b) Botanical gardens
   (c) Aquaria (d) All of these
8. Pisciculture is rearing:
   (a) Fish (b) Aquatic plants
   (c) Pearls (d) None of these
9. To neutralize acidity of soils damaged by acid rain, add:
   (a) Potassium nitrate (b) Urea
   (c) Gypsum (d) Superphosphate
10. The lowermost layer of atmosphere is called:
    (a) Mesosphere (b) Ionosphere
    (c) Troposphere (d) Exosphere
11. Bhopal gas tragedy was caused by:
    (a) Methane (b) Carbon monoxide
    (c) Arsenic (d) Methyl isocyanate

12. World Environment Day is celebrated on:
    (a) 5 June (b) 11 June
    (c) 12 January (d) 8 September

b) Define:
1. Sustainable development
2. Conservation
3. Sivliculture
4. Afforestation
5. Unsustainable development
6. Reforestation
7. Forest conservation
8. Climate
9. Biodiversity
10. Germplasm
11. Troposphere
12. Greenhouse effect
13. EIA
14. Reforestation

c) Explain briefly:
1. Environmental problems created by unsustainable developmental processes
2. Sustainable and unsustainable development
3. Agroforestry
4. Social forestry
5. Methods of improving agricultural practices
6. Sand mining from rivers

d) Name the following:
1. Three alternate energy sources
2. Three aspects of forest management
3. Three meaningful methods of water conservation
4. Five means to conserve soil
5. Three emissions that can be checked by Pollution Control Boards
6. Three methods by which deforestation can be checked
7. Reasons for cutting down the use of fossil fuels
8. Three greenhouse gases
9. Important International Treaties
10. The person who leads 'Narmada Bachao Andolan'

e) Briefly explain the following in 2 or 3 sentences:
1. Sustained yield concept
2. *In situ* conservation
3. Methods to prevent over fishing
4. Provisions of Air Act 1981
5. *Ex situ* conservation
6. How burning fossil fuels cause atmospheric pollution
7. Disadvantages of major dams

8. Greenhouse effect
9. Sources of carbon dioxide in the atmosphere
10. Global warming
11. Formation of acid rain
12. Ozone hole
13. The scope of EPA 1986
14. Hazardous substances regulation
15. Environment Impact Assessment (EIA)
16. Coastal zone regulation
17. Ecomark
18. Centre for Environmental Education, Ahmedabad
19. Environmental calendar
20. Reforestation

f) When and where was the Earth Summit held?

g) What can we do as an individual to fill the ozone hole? Give reasons for your answer.

B] Short Answer type questions:

Write short notes on:

1. Earth Summit
2. Fossil fuels
3. Conservation of water bodies
4. Interlinking of rivers
5. Range management
6. Energy conservation
7. Conservation of germplasm
8. Effects of climate change
9. Man made causes of climate change
10. Global warming and its effects
11. Harmful effects of acid rain
12. Solutions to the problem of acid rain
13. Ozone hole and its effects
14. Central and State Pollution Boards
15. Bhopal gas tragedy
16. Powers of the State Pollution Control Boards with regard to Water (Prevention and Control of Pollution) Act 1977

C] Long answer type questions:

a) Give a detailed account of climate change and its effects on environment.

b) What do you understand by Acid Rain? Explain in detail.

c) Write an essay on Ozone layer depletion.

d) Give an account of EPA 1986.

e) Explain in detail methods to create public awareness on environmental issues

f) Write long notes on any two:

1. Water Act 1977
2. Wildlife Protection Act 1972
3. Forest Conservation Act 1980
4. Enforcement of Environmental legislation

Unit 4

Human Population and the Environment

At the World Food Conference held in Rome in the year 1974, Henry Kissinger, then U. S. Secretary of State, declared that by 1984 no man, woman, or child would even go to bed hungry. Thirty two years have passed since; not just ten as Kissinger had wished; more than a billion people in the world go to bed hungry even today. One main reason for this is definitely the uncontrolled population growth in many parts of the world.

POPULATION GROWTH

As per the 'Pro-population school' of thought, most of the ills that the present generation is suffering are because of the extremely fast rate of population growth. They also predict that condition of the future generations will be worse, if this trend is not reversed.

In fact, the rate of population growth was less in the earlier centuries than compared to later centuries, though there has been some downward trend of late. Let us examine some figures (Last three figures estimates by UN, 1978).

Year (AD)	1800	1850	1900	1930	1950	1970	1980	1990	2000
World Population (in millions)	919	1,091	1,571	2,070	2,513	3,678	4,415	5,275	6,199

From these figures, it is clear that population growth was comparatively slow till 1950. It was about 0.8%. From 1950 to 1970, growth rate was about 2%, while after that the decline started slowly.

Between 1970 and 1990, the growth rate reduced to about 1.9% while it fell further from 1990 to 2000 to about 1.6%.

Population growth is not the same in all countries of the world. It is very much uneven. Populations have been growing very fast in the developing countries (Less Developed Countries or LDCs). In the developed countries (More Developed Countries or the MDCs) populations have been growing very slowly. The difference is so much that world figures show high population growth, as is clear from the table.

Year	World	MDCs	LDCs
1950	2,513	832	1,681
1960	3,027	945	2,082
1970	3,678	1,050	2,628
1980	4,415	1,131	3,284
1990	5,275	1,206	4,069
2000	6,199	1,272	4,927

The table also shows that LDCs accounted for about 70% of the world's population in 1950. This increased to 79% by 2000. In the MDCs, on the other hand, rate of population growth has declined since 1970.

Within both developed and developing regions, population has been increasing fast in the urban areas. Increase in urban population has led to formation of slums, pollution and environmental degradation. Increasing urbanization also results in economic imbalance, which leads to social problems also.

	1950 AD	1980 AD	2000 AD
World	724	1807	3208
Developed Regions	449	834	1092
Developing Regions	275	972	2116

Overpopulation results in economic poverty as the available resources are shared by too many people. Moreover, poor countries with large populations are selling their natural resources like forest produce to the rich countries. Because of this, the rate of deforestation has increased greatly in the tropical countries. 20 million hectares of forests are lost per year in these countries.

Brazil is one of the LDCs. Its economic condition is so poor that it is forced to export huge quantities of timber from its rain forests to MDCs. The dense Amazon forests are losing millions of hectares of rain forest every year. The fast depleting natural resources of Africa, Latin America, South and South-East Asia has changed vast tracts of natural vegetation into wastelands.

Exploitation of natural resources to support the ever increasing human population is having many adverse effects on the environment like:

- Reduction in forest cover due to unmindful felling of trees
- Excavation of land for mining purposes
- Reduction in cultivable land
- Water shortage due to excess withdrawal of ground water without recharge
- Pollution of air, water and soil and all the after effects of pollution

Social effects of population growth have also begun to show in different countries and different regions within the same country, like:

- Illiteracy and Unemployment
- Migration from rural to urban areas
- Increase in violence and crime
- Deterioration in hygiene and health, etc.

POPULATION EXPLOSION

Human population in the world is not just growing but we can say that it is exploding. The six billion mark was reached on October 12, 1999. World population has doubled since 1960.

More than 95% of population growth is in the developing countries. Meanwhile, population growth has slowed down in Europe, North America and Japan. The Population Division of the UN Department of Economic and Social Affairs has estimated that world population will grow to 8.9 billion by 2050.

The poorest countries often have the highest population growth rates. In 62 countries of Africa, Asia and Latin America, more than 40% of the population is under 15 years of age. Population is exploding fastest in the poorest countries. Whereas within countries, it is the poorest families that mostly tend to be the largest ones. The share of the global population of the poor countries has been increasing steadily for more than 40 years now.

Population explosion results in:

- Internal instability
- Natural disasters

- Poverty
- Social disruption
- Energy crisis
- Global climatic changes
- Shortage of water
- Water Pollution
- Increase in waterborne diseases
- Reduction in fish resources
- Soils becoming infertile due to intensive farming
- Increased air pollution
- Desertification due to overgrazing
- Destruction of coastal ecosystems, the most delicate ecosystems
- Marine pollution leading to destruction of coral reefs
- Encroachments into forests
- More floods caused by mining of sand from rivers for building construction
- Increase in number of accidents as traffic becomes unmanageable
- Inflation due to shortage of essential articles
- Increase in crime due to unemployment and frustration
- Reduction in discipline and other such values
- Increased waste generation and resulting spread of diseases, etc.

An estimated 13 million refugees have fled their own countries to escape from harassment and discrimination, armed conflict or violence. A considerable number has been forced to leave for social or environmental reasons though they are not refugees. Millions are displaced within their own countries. Most of them have added to the number of the urban poor.

Asia	Africa	Europe	Latin America and Caribbean	N. America	Oceania	Total
3,684	800	728	518	306	31	6067

As is clear from the following table, the population of the world has exploded so much that we have started to refer to the numbers in billions!

	Africa	China	India	Rest of Asia	Latin America and Caribbean	Europe	USA and Canada
Population in 1995 (in billion)	0.7	1.2	0.9	1.3	0.477	0.728	0.297
Population in 2150 (in billion-Projected)	2.8	1.6	1.7	2.8	0.916	0.595	0.424

Let us have a look at the present situation. Many figures have been given by the Population Reference Bureau. Some selected figures from this 2006 World Population Datasheet are given here, just for us to realize how urgent it is to put a stop to the population explosion.

	World	MDCs	LDCs
Population	6,555,336,000	1,216,198,000	5,339,138,000
Births per year	137,035,288	13,496,866	123,538,422
Births per minute	261	26	235
Deaths per minute	107	23	84
Infant deaths per minute	13	0.2	12.8

The statistics are very clear. Differences between the MDCs and the LDCs are too contrasting not just in the total population, but also in the number of births per year. The number of births per minute has been compared here for the readers to understand the gravity of the situation. Population is so high in the LDCs that poverty is more, resulting in poor health care. This is shown by the contrasting figures in deaths per minute. 78.5% of the total deaths per minute in the world take place in the LDCs. In the case of infant deaths per minute, the difference is still more. 98.5% of the infants who die belong to the LDCs.

Even after so many infant deaths, the total population is very high in the LDCs. This shows how extremely high birth rates in these countries are. Let us analyze some statistics.

The Indian subcontinent is adding about 21 million people a year, the equivalent of another Australia. Population of India has nearly tripled over the last 50 years. It has grown from 350 million in 1950 to 1 billion in 2000. According to projections by the UN, India will add 515 million more people by 2050, that is, almost twice the population of the US at present.

Pakistan is expected to double its population in the next 50 years, by adding 189 million people by 2050.

Bangladesh is expected to add 83 million people by 2050.

The subcontinent is thus expected to add another 787 million people by the middle of this century.

From the above account, it is clear that for removing poverty, it is essential to curb the population explosion going on in the less developed countries of the world. For this, family planning is a must. Moreover, it is necessary to improve the overall health of the entire family. In healthy families, infant mortality will be minimized and life expectancy will be more. All these are included in what is known as 'Family Welfare Programmes'.

When people are healhy, they will be stronger with more stamina which will lead to more efficiency. All this will result in overall increase in the prosperity of the nation. There will be only a few people to share the resources. Also, healthy people will be more happy. For all this to happen, it is imperative to practise and popularise family welfare programmes.

FAMILY WELFARE PROGRAMMES

To curb the present population explosion, especially in the developing countries, it has become imperative to adopt family planning measures. Along with family planning, other health related activities are also a must. All these together are called 'Family Welfare'. It is good to see that the developing countries have started working seriously in this direction.

The most essential part of any family welfare programme is family planning. As the words indicate, family planning is planning to keep the family small. Small families are in general happier and healthier.

Family Planning: By practising family planning, couples can plan the number of children and their time of birth. Family planning programmes are conducted mainly by governments and NGOs. The central as well as state government in India have Health and Family Welfare ministries. In addition to birth control, these ministries look into birth spacing, breast feeding, antenatal (before delivery) care of pregnant women, postnatal (after delivery) care of mothers, parental care of infants, immunization programmes for infants, etc.

There are many methods of birth control.

1. *Sterilization:* This is a minor operation done under local anesthesia. It can be done in males and also females. There are no post-operative problems. Both are permanent methods of birth control. However, in case the need arises later to have babies, these can be revoked by a second operation.
   a) *Vasectomy*: *Sterilization operation in males is called vasectomy*. Here, the tubes that carry sperms are tied so that sperms are not able to go out. There is a common misunderstanding in men that this operation is injurious to health and men lose their strength and stamina after vasectomy. The truth is that *vasectomy does not affect men's sexual ability or general physical or mental strength*. Sperms are still produced in the testis, but they are absorbed back into the cytoplasm of the cells. The fluid that comes out during intercourse is just a mixture of some other secretions which do not contain sperms. Making Indian men aware of this truth will lead to more men going for this sure method of birth control.
   b) *Tubectomy*: *Sterilization operation in females is called tubectomy*. In this operation, the fallopian tubes that carry the egg (ovum) to the uterus are tied with the help of surgical threads. Women who have undergone tubectomy produce ova as usual, but they die within the fallopian tubes after 3 or 4 days. The ovum can be fertilized only if it reaches the uterus. Now-a-days, tubectomy has become easier after the invention of the laproscopic method.
2. *Condoms:* These are rubber sheaths used by males during intercourse. Sperms are not able to enter into the vagina, thereby preventing pregnancy. One condom can be used only once. Use of condoms is the most widely used method of birth control in India. Condoms made by different companies are available in the market. The trade name of condoms made by the Govt. of India is 'Nirodh'. It is provided free of cost, and is part of the family planning programme of the Government of India.
3. *Intrauterine devices (IUD):* There are a few types of IUDs, but Copper is the most common device. A Copper T is a small T shaped object which is inserted into the uterus by a doctor. This

prevents the fixing of the ovum on the wall of the uterus, even if it gets fertilized. The fertilized ovum can grow further only if it is attached to the uterine wall. It is essential to change the Copper T every 3 years. Copper T does not cause any discomfort and the woman can do all her works as usual. However, rarely, the woman may have excess bleeding during menstrual periods or may develop back ache. In such cases, she should consult her doctor. If her doctor advises her to discontinue using IUD, she can do so and go for other birth control measures. However, this happens only rarely. Most of the women who use Copper T are very comfortable with it. Other IUDs like loop, spiral ring, etc. are also available, but Copper T is the most popular.

4. *Oral pills*: Many brands of oral contraceptive pills for women are available in the market, like Mala D, Saheli, etc. Such pills prevent the formation of ova in women. While using these, care should be taken not to miss even a single dose. *Before starting pills, a doctor should be consulted.* Oral pills may have side effects like high blood pressure, irregular bleeding, nausea, etc. Yet many women do use them. Pills to be taken daily and those to be taken weekly — both types are available.
5. *Following the safe period*: In the monthly biological cycle of women, there is what is commonly called as 'safe period' and 'danger period'. In one month, a woman produces only one ovum. If this ovum is not fertilized in time, the woman will not get pregnant in that month. The ovum is active and ready for fertilization during the middle of the monthly cycle. During these days, intercourse should not be done. Starting from the first day of menstrual period, the first ten or eleven days are 'safe'; the last ten or eleven days are also 'safe'. The days in the middle is called the 'danger period', as there is the risk of pregnancy if intercourse is done on these days. This is actually the fertility period. *The ovum is usually formed on the 14th day, counting from the first day of menstruation*. It is to be safe that 3 or 4 days are left on both before and after this day, because there may be slight variations in the day of formation of the ovum. This method works well with women who have regular menstrual cycles. For those whose periods are not regular, this method is not advisable.
6. *Use of spermicides*: Spermicides are chemicals that kill sperms. These are available in the form of pastes, jellies, etc. under

different trade names. The paste is to be applied in the vagina before intercourse. These are usually harmless, yet women should see that they are not allergic to these formulations.

7. *Abstinence*: This is the only method which is 100% sure of results. Abstinence means abstaining from, i.e. avoiding intercourse. This is not possible always, yet is done by some couples at times.

Abortion: All the methods described above are meant to prevent pregnancy. In case a woman conceives even after following any of these methods, she can go for abortion. Abortion is medical termination of pregnancy. It should be done in early pregnancy, before the foetus reaches a maximum of 60 days (preferably 45 days). The age of the foetus is calculated from the first day of the latest menstrual cycle. Some couples go for abortion as a birth control measure, but it is always better to use contraceptives or go for sterilisation. *Abortion raises many ethical issues; it also affects women's health, especially in the long run.*

The methods described above can be used to space the children also. The ideal difference of age between two successive children is considered as three or four years. The idea is that it is better to have a second baby when the first baby may start going to school. Spacing of babies is essential for the health of both the mother and the child.

In India, quite often we see that parents' *craving for male babies is a main reason for not adopting family planning measures*. This attitude is deep rooted in many societies, but time is changed now. Like many other social evils, it is the high time to change this attitude too. Whatever the sex of the first two children, no third baby should be given birth to. If every Indian follows this rule, we can at least stabilize our population. It is the duty of every educated person to practise family planning in his or her own life and also to convince others to do the same.

Family planning measures have been very effective in China, Indonesia and Brazil. But in India and Africa, it has had moderate or poor results. This is because family planning services are not readily available in many rural areas. According to UN estimates, one-half of births in LDCs are unplanned and one-fourth is unwanted. If services reach every corner of the country, thousands of abortions can be prevented.

Some countries use the incentive-disincentive strategy to promote family planning. This is a sound method, by which families who follow the national policy are rewarded and those who do not are not rewarded or even punished in some manner. For example, the first two children will get fee concessions, while the third child will have to pay extra fees, or government houses allotted to families with two children can be bigger than houses allotted to families with more children. Such measures are often effective.

To reduce births, it is essential to empower women. In the LDCs especially, most women are fully dependent on men and do not take any decision in any matter, including the number of children they should bear. This situation will change only if women become educated and financially independent. All educated men with great modern views are helping them to achieve this goal.

In general, family planning measures are more concentrated on women. An opinion is gaining ground that more research is needed to find out pills and other methods for men.

As mentioned earlier, *family welfare includes not only family planning, but also other health care practices.* Pregnant women must take nutritious food and the exercises advised by the doctor. They must also have mental peace. Babies born to such women will be healthier. For delivery, it is best for women to be admitted to nearby hospitals. After delivery, women must again have good food and the prescribed dose of iron tablets.

Babies must be given all immunizations as required. These can be given either at hospitals or at family welfare clinics. The first vaccines to be given to a new born baby are BCG and the first doses of Oral Polio vaccine and Hepatitis B vaccine. Further doses of Polio and Hepatitis B are to be given at required intervals up till the age of 5. Other vaccines to be given in time are DPT, Measles and Tetanus. Till the child is 15 to 16 years of age, the vaccinations will continue at fixed intervals. Immunization cards given from the hospitals/family welfare clinics will have all details. Parents must keep the cards safely and get their children immunized as per the instructions given in the cards.

It goes without saying that healthy children grow into healthy adults. In making this possible, contribution of both the parents is essential. Fathers must not think that immunizing children is the duty of mothers alone. As an old saying goes, 'Health is wealth'.

ENVIRONMENT AND HUMAN HEALTH

Human health is closely related to the environment in many ways. Environment has both physical and mental influence on us.

Physically, there are many diseases caused by unhygienic living conditions. These afflict mostly the poor and the uneducated people. One main reason for the spread of such diseases in our country is the overcrowding due to the population explosion. Too many people live in crowded localities without proper sanitation. People throw garbage here and there and there is no proper garbage removal system. There is no waste water management either and dirty water accumulates in patches here and there. Cattle and other animals are also reared by some, and there is no drainage system from their sheds. Ponds in which people take bath are also used to bathe animals. Ponds and other water bodies are polluted by the throwing of garbage. Lack of toilets is another serious problem in such localities. Smoke from 'chulhas' pollutes the air. In short, soil, water and air are polluted in localities of overcrowding.

The main reason for such unhygienic conditions seen in villages and also slums in cities is lack of education. It is essential to include lessons on hygiene in school text books. The family welfare workers and NGOs also helps a lot in making people realize the importance of hygiene.

Many diseases are caused by lack of sanitation and the resulting environmental pollution. The most important of these are:

- *Water- and food-borne diseases*: *Diarrhoea, dysentery, cholera, amoebiasis, jaundice,* etc. are the common gastrointestinal diseases found mostly in the LDCs. These diseases are caused by drinking unclean water or by eating food without washing hands. Thus these diseases are clearly associated with hygiene. Millions of children die due to diarrhoea every year. In India, as per one estimate, one out of every five deaths of children is due to diarrhoea. Lack of education compounds the situation because most of the deaths in all these cases are due to dehydration. An uneducated person does not understand the importance of supplementing water intake to compensate water loss from the body through loose motion and vomiting. *Poliomyelitis* and *Viral hepatitis* are also waterborne diseases. *Typhoid* spreads by drinking water or eating food containing the disease causing bacteria.
- *Insect borne diseases*: *Malaria, filaria, encephalitis (brain malaria), dengue, chikungunya,* etc. are spread through mosquito bite. Mosquitoes breed in accumulated water. This is a direct result of poor water management. Moreover, mosquito eradication schemes are not very successful in the LDCs due to overpopulation and lack of awareness and education of the poor people. Moreover, when insecticides like DDT are banned in many places due to its

non-biodegradable toxic nature, other effective alternatives have not been made available.

- *Diseases caused by parasitic worms*: Children of poor families often play in the mud and then eat food without washing hands properly with water and soap. As a result, they ingest eggs of disease causing worms like *round worm*, *thread worm,* etc. along with food. *Hook worms* enter into the bodies of people who walk barefooted on infected soils. *Tape worms* infect those people who eat pork that is not well cooked. All such cases are much more common in the LDCs as they are all concerned with poverty and lack of awareness.
- *Plague*: This is also a disease associated with sanitation. Plague is a primary disease of rats. Fleas that bite infected rats get the disease. Human beings get plague if they are bitten by the fleas.
- *Kalazar*: This disease is spread by sandfly. It is seen in rural areas. It occurs when cattle sheds and poultry farms are built near houses.
- *Respiratory diseases*: The commonest respiratory disease is *asthma*. Other examples are *bronchitis*, *lung tuberculosis,* etc. These diseases are either caused by or are aggravated by air pollution. Inhaling smoke from chulhas, dust from the air, exhaust fumes from automobiles, gases released from industries etc. all lead to various respiratory diseases. It is essential for people to know the importance of ventilation in houses. As far as possible, fresh air must be inhaled. *Lung cancer* is caused mainly by tobacco smoke. The smokers themselves (active smokers), and also others closely in contact with them (passive smokers) may get cancer.
- *Diseases caused by noise pollution:* Very loud noise is harmful to human health. Loud noises are common in factories, during festivals, in busy roads, near aerodromes, etc. In addition, some people have the habit of keeping the TV or radio on high volume. All this loud noise damages the ears, and cause *deafness*. Other problems are *sleeplessness*, *nausea*, *ulcer*, *lack of concentration*, *irritability*, *high blood pressure,* etc.
- *Diseases caused by pesticides:* Farmers in general sprinkle too much pesticides on their crops to protect them from various diseases. But they do not realize that in the process, they are exposing themselves to various disorders. Inhalation of pesticide fumes affect *hormonal balance* and even *cancer*. In addition to this, the fruits and vegetables that we buy from shops almost

always have pesticides on their surfaces. Therefore, it is most important to wash them well before use.

- *Diseases caused by herbicides:* Herbicides are chemicals used to kill small plants (herbs) which are weeds growing with crop plants. These cause *skin diseases* in human beings.
- *Cancer due to exposure to radioactive substances:* People working in factories or projects dealing with radioactive materials are at a risk of getting radiation induced diseases. There is a maximum level of radioactive substances which can be tolerated by humans. If these levels are crossed unknowingly or by neglect on the part of the authorities, workers get diseased. Cancer is one of them. Mutations may be produced in the sperms and eggs of workers. This will affect future generations.
- *Diseases due to exposure to hazardous chemicals:* Toxic chemicals are manufactured in many factories. Most of these factories are in the developing countries. This is because the developed countries know that such factories pollute the environment. In the developing countries, environmental laws are not strictly followed; offenders often go unpunished, thus the corruption is also more. Therefore, the MDCs try to set up their polluting units in the LDCs. They try to send their old ships containing tons of hazardous waste to countries like India for breaking down. Labour is cheap in India. Population and poverty are so high that people are prepared to work in unhealthy situations, that too for lesser wages. Constant exposure to toxic chemicals causes different diseases, including cancer, tuberculosis, skin diseases, etc.

Detection of new diseases and reappearance of controlled diseases: As and when man detects diseases, research works start and in most cases, drugs are found out to cure those diseases and in many cases, to prevent them too. These drugs act on the disease causing organisms (parasites). However, as time passes, some of these germs become immune to the drugs by changing their own structure by mutation. Many of these diseases spread fast in the slums and overcrowded localities and sometimes become epidemics.

The great flood in Mumbai in the year 2005, it was found out soon, was caused mainly by drains clogged by plastics. People had to wade in water to go to work for even 3 or 4 days. Many got infected with a strange disease which was diagnosed as 'Spirillosis'. This disease was caused by a species of Spirillum (a bacterium).

Diseases due to climate changes: There is no doubt that global climate is changing due to human activities. Temperatures are rising, rains have become unpredictable, incidence of floods and droughts, heat waves and cold waves have increased. Human health is decreasing due to all these. As more regions become warmer, malaria, which is a tropical disease, will certainly spread more to north and south of the equator. The *El Nino,* which causes periodic warming, may affect rats and this may increase diseases in humans.

Death and destruction due to climate changes brought about by human interference in the world environment will be discussed under Disaster management.

Efforts to prevent environment related diseases: Efforts in this direction should be taken, first by the government and secondly by NGOs and other social welfare organizations. Citizens also have a duty to live clean lives and try to prevent diseases from spreading.

- Provide clean drinking water to all
- Provide separate localities for rearing domestic animals
- Set up many more public lavatories in all localities
- Run adult literacy programmes sincerely
- Prepare text books for adult education including chapters on health and hygiene in detail
- Make education for children compulsory
- Change patterns of agriculture in such a way that dependence on chemical pesticides is highly reduced
- Popularise biopesticides like neem, etc.
- Do not allow industries to come up near residential areas
- Control traffic
- Be strict with automobile servicing and checking rules
- Arrange for tree plantation drive so that air will be purified
- Make rainwater harvesting compulsory for upcoming structures to solve water problem
- Punish people polluting water bodies
- Encourage use of alternative energy sources like solar energy
- Make more strict family planning programmes
- Set up health clinics in all villages and towns based on the number of residents in each area
- Make mobile clinics available to the people

- Construct roads in villages; if roads come up, all other facilities are sure to follow. This will remove the disinterest shown by qualified doctors at present to practise in villages.
- Give importance to educating women and empowering them; health of the whole family depends a lot on the women of the house.
- Set up family welfare clinics with qualified doctors, nurses and midwives in all villages. Many women in India die due to complications related to child birth. Clinics are also helpful in the immunization programmes of children.
- Punish polluting industries and projects severely
- Arrange frequent cleanliness drives in villages and also slums in cities
- Make industries using wood as raw material plant trees
- Conduct various programmes to create environmental awareness among the common people
- Award a certain percentage of marks to school and college students who spread environmental awareness, literacy, etc.
- Enforce all environmental legislation with full force
- Poverty is closely related to health; implement poverty eradication schemes.

This is not a comprehensive list at all. Central and State Governments, NGOs, self help groups, social service and family welfare agencies, etc. must all pitch in with full vigour and enthusiasm to prevent environment related diseases in society.

HUMAN RIGHTS

It is an established fact that Vedic India had a strong tradition of respect for human rights. The principle was 'constant desire of giving to every man what is due to him'. The king was duty bound to enforce the law of the land and also the rights and claims of his subjects. Later, Kautilya asserted in Arthasasthra, "In the happiness of the subjects, lays the happiness of the king". *Matsyanyaaya* (the law of the fish, i.e. big fish eating small fish) was never allowed in the state. In other words, the strong and the powerful were not allowed to trouble the weak. Traveler historian Megasthanese wrote, "Kings employed spies to detect violators of human rights".

In the modern times, it was *in 1948* that *the Universal Declaration of Human Rights was adopted by the UN General Assembly.* This was the result of an emerging view that the manner in which countries treat their own citizens must be subject to international standards. There was a long list; some are reproduced below:

- Equality of rights without discrimination
- Equal protection of the law
- Social security
- Work under favourable conditions
- Food, clothing and housing
- Health care and social services
- Special protection for children
- Education, etc.

Few governments deny the fact that these are legitimate rights of any citizen. They agree that good life means suitable means of sustenance, good health, clean environment, etc. As all human beings deserve dignity, their claims to these and other basic necessities of life are legitimate.

All over the world, the belief is growing that man is part of nature (this was the basis of all environmental studies in ancient India; we were much ahead of others in this respect, as in many others). Therefore, nature must be preserved for man's wellbeing. This led to the opinion that there must be a global effort to protect the environment. Many international agreements were drafted for this purpose. Most of the countries of the world ratified these agreements. Human right to equity in different countries, in one's own country and society was widely accepted. So were the rights of future generations to a clean environment, rights to biodiversity, rights to protection of indigenous cultures, etc.

There is however a doubt that globalizing environmental issues will not be beneficial. Though the causes for environmental degradation are indeed global, corrective measures are to be taken locally. Hence the slogan, "Think globally, act locally".

Some workers use the term, "Ecological rights" to denote human and environmental rights, which include the following relationships:

- Relationships of human beings of the same country with one another
- Relationships of human beings of the rich and poor countries with one another, and
- Relationship of all mankind with natural environment

Many environmental issues are closely associated with human rights. One of them is equal distribution among the people of:

- Wealth

- Resources, and
- Energy

In all the three, we see that disparities are very high between:

- The rich and the poor
- Men and women, and
- Present and future generations

Disparities between the rich and the poor countries: As is clear, *in each category, there is an advantaged group and a disadvantaged group.* The burden of polluted air, water and land, and of dwindling natural resources always falls on the poor. Often, it is the poor who are compelled to encroach into a forest just to survive; or keep their surroundings dirty as they cannot afford to keep them clean.

The challenge of bringing in an era of equity is very large indeed. To make this happen, there must be sincere efforts at the national level and also at the global level. Responsibility for the present position of the planet has to be shared between the rich and the poor countries. Therefore, for finding out solutions and implementing them, the rich and the poor countries must work together. For this, trade talks and flow of money is not enough social and political ties between governments and civil societies must be strengthened.

Disparities between the rich and the poor exist within countries also. In fact, the disparities are more in India than in America. Some waste resources and live lavishly while some have almost nothing to eat. Resources belong to all, irrespective of class. But when rich contractors are given permission to fell trees, it is the poor *vanvasi* who suffers. In addition, the poor forest dwelling tribal people are accused of destroying the forests. Same is the case with poor fishermen. The ocean does not belong to anybody. But when mechanized trawlers catch fish, the poor fishermen suffer losses and even go hungry. The poor fishermen have as much right to the fish in the sea as the rich and powerful trawler owners have.

Rehabilitation of displaced rural masses has always been neglected in many countries while big projects like major dams are taken up. The land where the original settlers lived belongs to them. Without paying them adequate compensation, they should not be displaced. In fact, the best system would be finding them alternate arrangements and then begin the project. But this is never done. The human rights of the poor people are always sacrificed for the benefit of the powerful.

Disparities between men and women: *Unequal status of women is seen in many cultures*. As per the Universal Declaration on Human Rights, human rights apply to all people equally, "without distinction of any kind such as race, colour, sex, language....or any other status". But the truth is that women are sufferers, doubly. They, like men, are victims of human right violations; in addition, they are discriminated against as women. Women of all countries, races, classes, cultures, religions and nationalities are treated unequally, though in some cases, the treatment is worse.

The Convention on the Elimination of all forms of Discrimination Against Women (CEDAW) or the International Women's Human Rights Treaty was adopted by UN in 1979. CEDAW was the first document to address women's rights.

Most of the world's displaced people and refugees are women, and most of the world's poor are women. Still, as they have been given a subordinate status in society, their human rights violations are continuing, as always.

Nutritious food and health are human rights. Yet it is unimaginable how many women are discriminated against in these fundamental rights. In many Indian households, it is the women who take food last, after the men have finished food. In low income households, mostly, the quantity left is not enough for the women. Regarding health issues also, in many houses, the women do not get what is their due. Studies have shown that in India, it is the women and the girl children who are most anaemic.

Disparities between the present and the future generations: We are depleting the natural resources belonging to the future generations. Of all the creatures of nature, only human beings are capable of planning change. However, we have been using this extra mental capability to change the course of our own development! This is because human beings have a natural instinct to overuse resources and very little incentive to conserve them. Human activities are thus destroying forests, draining rivers dry and raising the level of the seven seas!

It is important that we think of the future generations and not degrade the Earth's life support systems. It will give them options for dealing with the problems they will face. If our ancestors had left us much less of what we have today, what would have been our position? We would not have been able to enjoy life as we are doing now. So also, it is our duty to leave the earth in good shape to our descendents. Our descendents have a human right to a clean and rich environment. On the other hand, we have no right to destroy what is their property. The land, the water, the air, the variety of plants and animals — they all equally belong to them as much as they belong to us at present. We must hand over them with these natural resources as they are. We cannot be greedy and use resources.

VALUE EDUCATION

Values are principles based on which we decide what is right and what is wrong. Just being literate does not make one educated. Even one who has a degree, for that matter, need not be educated in the right sense of the word, unless he has been taught the values of life on earth. If an educated person is intolerant, selfish, greedy and dishonest, what is his worth in society? *An educated person must be able to differentiate between right and wrong; only then can we say that he or she has had value education.*

The search for harmony, knowledge and truth had inspired the ancient Indian minds more than their counterparts in other parts of the world. More than 5,000 years ago, ancient Indian philosophers and thinkers had put forward a theory of morals and values. They were humanists and moralists. They evolved a system of legal theory based on higher values and ideals. Every aspect of life was regulated by *Dharma*, i.e. righteousness. Kings and subjects alike were bound by the law of *Dharma.* It was a concept of values helping mankind to live in harmony with each other and with nature, free from conflicts, exploitations and misery. This system was in vogue for thousands of years.

It is true that values declined gradually in Indian society. Therefore, it has become all the more important to spread value education in today's India.

There are different types of values. They help us live fulfilling lives. By thinking of the welfare of others, we feel happy ourselves. By expanding this theory a little, the coming generations can be included in the 'others', and then, we will stop exploiting nature. This will automatically result in sustainable development.

Environmental values are important as environment is all that surrounds us. How we should treat nature and pass it on to our descendents is what environmental values teach us.

Environmental values were etched strongly in the minds of ancient Indians. People knew and felt that environment, as the provider of all, must be conserved. They felt one with nature. They never thought of themselves separate from nature or as superior to other creatures of nature. All were environmentalists. There were great thinkers and astronomers among them. The ancient Indians therefore knew not only about our earth, but also about the universe in general. In many Vedic prayers, we see that peace is sought for the whole universe. Plants and animals were given a lot of importance in the Vedic period. The concept of 'Pancha bhoothaas' was prevalent. The 'Pancha bhoothaas' or the five elements are Prithvi (earth), jal (water), agni (fire or energy), vaayu (air) and aakaasham (space). These were known to make the earth and therefore were respected. All 'chara' (movable) and 'achara' (immovable) objects were given due respect. In short, the Vedic period was the golden period of man-environment understanding.

On the other hand, western cultures always considered man as the master of the universe. The belief was that God created everything in nature for the sake of man. This approach gave no importance to other living or non-living things. Man could exploit nature to the maximum as per this concept and this he did, for years, in the West. Now also, the belief remains the same, but people in the West have begun to realize that natural resources are not unlimited; and hence they must conserve these resources. So the West is trying to teach the new doctrine to itself, though the wasteful style of living continues there, especially in the affluent industrialized West.

In general, all over the world, seeing the results of man induced destruction of the earth, people have become worried and hence cautious. This has resulted in the branch of studies called 'Environmental Education'. The aim of this study is to become aware of all environment related issues oneself and also to create awareness among the general public.

- *Man as part of nature*: To realize the environmental values, one has to just go out of the house and look around. The sky, the soil, the plants and animals including man all are part of nature. There is oneness in everything. When this feeling comes, it becomes easy to conserve nature and difficult to even think of destroying it. This is the first environmental value.
- *Realizing the harmful effects of our actions*: It is not only that we must not cut trees. We will have to cut some, for our various needs. But we should replace the cut ones. We must not allow soil to be washed away — soil takes years to form. We must not throw garbage here and there — it stinks, it causes diseases; it also makes the beautiful nature look ugly. We will not be able to appreciate the beauty of nature if we ourselves dirty our surroundings. In short, while we need not and can not stop development, we must take care that the environment is not degraded.
- *Appreciating the positive effects of our actions*: Suppose we

stop wasting energy, then we can be happy that we are saving precious energy, brought to our house by the efforts of many. While wasting water, let us remember the village woman walking for 10 km a day to fetch a pot of water. While caging a bird, let us ask ourselves, "How would I have felt if I was locked in a tiny room day and night?" We won't feel like wasting energy or water or cage a free bird. These are small things indeed, but when the message spreads and more and more people begin following such environmental values in life, the overall change will be big.

- *Enjoying the aesthetic values of each part of nature we see*: We go about our work the whole day. We get up in the morning and begin brushing. Do we listen to early morning birds chirping? Can we identify birds with their chirps? (Except the cawing of the crows, of course!). Do we notice the dew drops on the grasses? How many times do we look at the changing shapes of the clouds or the leaves of some trees closing in the evenings? How many times do we stop to smell the scent of opening flowers? Let us enjoy the sights, sounds and scents of nature. We feel one with them; we also experience tremendous peace of mind.
- *Valuing our roots*: A person who values his place of birth or the place where he spent his childhood, will feel like protecting that piece of earth to perfection. This is because he is emotionally attached to the place and cannot see it destroyed.
- *Valuing the holistic character of nature*: Nature is one whole entity. It is not 'a total of the sum of its parts'. We should realize this even while learning of its parts individually. Every part of nature depends on other parts and other parts depend on it. To know and appreciate nature, this realization is essential that all its parts are connected to each other and interact with each other.
- *Valuing all cultures*: All cultures have the right to exist and they have a right to live life their own way. We should not interfere in their way of living. For example, some try to change the life style of tribal people in an attempt to bring them into the main stream of life. This should not be done. What they need is education and health care. Without disturbing their culture, these advantages of modern society must be made available to them.
- *Understanding how the earth works*: If we understand how the earth works and sustains itself, we can use such earth wisdom to

guide our lives. With the same wisdom, we can live peacefully on earth.

- *Value of Social Justice*: The underprivileged should not be neglected. It is the duty of the privileged to see that everyone is taken along in the path of the forward march of humankind. If this does not happen, there may be revolts in the long run, which will be harmful to all.
- *Learning how to evaluate and resist advertisements*: In the modern world, millions are spent on advertisements. These are meant to tempt us to buy things that we may or may not need. If we learn to analyze the advertisements and resist buying what we do not need, a lot of resources ca be saved.
- *Valuing heritages*: We have a rich history. It is the duty of each and every one of us to preserve it for our own sake and also for the sake of the generations to come. They are priceless symbols of the past and cannot be reconstructed if damaged. This is why it is essential to preserve the Taj Mahal, our ancient temples and other such heritages.
- *Valuing ecosystems*: Rivers, streams, forests, grasslands, deserts, oceans, croplands, etc. are all ecosystems. It is our duty to protect them. Ecosystems are interdependent and are connected in space and time. Destroying one of them will have a negative effect on all the others. Therefore we should not do anything that will degrade any ecosystem by our activities.
- *Valuing fresh air*: Air is everywhere. We do not need a penny to breathe air. However, if it gets polluted, how we suffer health-wise and monetarily. If we do not want others to pollute the air we breathe, we have no right to pollute it ourselves too. Let us keep the air fresh.
- *Wishing sincerely for a better world*: *Value education should not stop at learning the values of nature and all it contains. Its purpose will be served only when each and every student decides to do something about conserving nature.*

HIV / AIDS

The word disease can be split into two — *dis* and *ease*, i.e. not at ease. Human beings are susceptible to different types of diseases. Most of them are caused by microbes like bacteria and viruses. Fungal infections also lead to some diseases. Some diseases are due to deficiencies of minerals—

these are called deficiency diseases or physiological diseases.

Examples of viral diseases are common cold, influenza, measles, mumps, polio, rabies, chickenpox, smallpox, etc. and the most serious of them all, with no documented medication found yet — the dreaded AIDS. Viruses are neither fully living nor fully non-living but share the characteristies of both. They thus form a connecting link between the living and the non-living world. Different types of viruses cause diseases in plants, animals and human beings. All viral diseases are infectious.

AIDS is the short form for Acquired Immune Deficiency Syndrome. It is the progressive destruction of the immune system of the body by the Human Immunodeficiency Virus (HIV). It was first diagnosed in 1981 in a small population of homosexual men. Soon it spread to heterosexual males and females with multiple sex partners, habitual intravenous drug users, infants and small children.

The virus HIV has been recovered in the saliva of infected persons, but its concentration in the saliva is very low and so does not cause infection. Insect vectors have not been found yet, which can spread the disease.

The virus binds to T lymphocytes, also called T_4 cells and destroy them. This results in gradual non-functioning of the immune system. The result is that the host becomes susceptible to all disease causing microbes like viruses, bacteria, fungi and other parasites. The host thus becomes more and more vulnerable to infection.

HIV can also be seen in certain cells of the intestine and in the bone marrow. It can also cross the blood-brain barrier, infecting the cells of the central nervous system, such as the brain and the spinal cord.

AIDS runs a slow course in many stages. This course is very unpredictable. There are no symptoms immediately after infection. This period of the 1st stage lasts for 6 weeks to 1 year. Only after this period, the virus can be seen in the blood. At this 2nd stage, the patient may not show any symptom or may feel weak, or may have fever, swollen glands or rashes. These symptoms disappear within a few weeks. The symptom of swollen lymph glands however lasts for 3 to 5 years.

In the 3rd stage also, serious symptoms are not seen, but tests show that the immune system is gradually breaking down.

About one and a half years after this stage, many unrelated symptoms develop. Together, these are called ARC, i.e. AIDS Related Complex. It includes unexpected fevers, night sweats, chronic diarrhoea, etc. Disorders of the central nervous system may develop. Cancer of skin, tongue, etc.

and pneumonia have been reported. Throughout this period, the patient is susceptible to many infections that are normally controlled by the immune system of the body. But now as the immune system has broken down, the patient becomes susceptible to all types of infections. This results in death.

As HIV controls all the functions of the host cells, drugs that destroy the viruses kill the host cells also. This is why it has become so difficult to find out any cure. No effective vaccine is also available.

As of now, education is the only best way to check the growing number of AIDS cases. Early diagnosis and medical care may delay complications. It also helps to change the patient's behaviour and prevent transmission.

Some cheap drugs have arrived in the LDCs. But a terrible side effect has been noticed in many cases. The antiviral drugs cause the immune system to recover; and then old hidden infections of leprosy make an appearance. Patients develop painful ulcers on their faces, and their fingers become numb. The first such case was reported in 2003 in London. This has proved to be a major drawback of the antiviral drugs discovered.

Some Important Facts About AIDS

A] It is not exactly known whether everyone infected by HIV develops AIDS. 10 to 30% of infected persons develop AIDS within 5 years. The percentage increases with time. HIV infection is believed to be life long.

B] HIV-infected people who do not have symptoms of AIDS can spread the virus. They are the healthy carriers. They can be infected for years without knowing it and can transmit the disease for many years.

C] *HIV is transmitted mainly by 4 methods.*

1. Sexual intercourse
2. Transfusion of HIV contaminated blood
3. Through injection needles, skin piercing instruments like ear piercing or tattoing needles, etc.
4. A mother can infect her child in the following ways:
   - During pregnancy
   - During birth
   - Through breast feeding

D] *HIV is not spread by*:

- Casual contact in schools, colleges, at offices, markets, public transports, telephone booths, etc.
- Taking bath in swimming pools

- Sharing toilets
- Hand shakes
- Hugs
- Kissing on the face
- Eating from the same dish
- Drinking from the same glass
- Massaging each other's bodies
- Mosquitoes or other insects
- Sitting next to infected persons
- Coughing or sneezing

E] *One can protect oneself in different ways*:

1. Do not have multiple sex partners
2. Use condom — it does not guarantee protection, but reduces the risk greatly
3. Do not share items that could have blood on it, like
   - Razors
   - Tooth brushes
   - Any skin piercing instrument

F] *Persons who are already infected or think that they may be infected should seek medical help immediately*. They should inform their partner that they have contracted AIDS. They should not donate blood or sperm or body organs to anyone.

In the year 2000, there were 24 million people all over the world, who had AIDS. Of all the places in the world, Africa has the highest incidence of AIDS. Even by 2000, HIV infection rates had become 20% in South Africa, 25% in Zimbabwe and 36% in Botswana. These are countries with limited government health services.

The role of government programmes is very important to curb AIDS. For example, in 1990, both Thailand and South Africa had less than 1% of pregnant women infected with HIV. By 2000, the percentage increased to 20% in South Africa while it was just 2% in Thailand. This was because of extensive public education programmes, and a 100% condom campaign in Thailand. The results were quick to see that condom use increased to 90%, there was a sharp decline in premarital and extramarital sex and visits to prostitutes.

In India, the first case of AIDS was reported in 1986. Since then there has been a gradual increase in HIV infection. Of the total AIDS cases in India, 89% are in the age group of 15 to 44 years. 79% of HIV positive persons are males while 21% are females. Heterosexual contact is the commonest method of spread, accounting for 74% of the transmissions.

Central and State Governments and various NGOs in India are trying to spread awareness about AIDS among the public. Inclusion of this topic in the UGC curriculum for college students is itself a proof of this.

WOMEN AND CHILD WELFARE

Women and children are at the receiving end in many cases, especially as they are not physically as strong as men. Many injustices are meted out to them in different parts of the world.

Plight of women in many countries of the world: Women account for half of the population in the world. They do almost all domestic work and child care. Rural women also help in agriculture, and bringing firewood and drinking water from distant places. They neither get holidays nor the payment for the work they do at home. After all this, it is the women who are treated badly in most cases.

In villages and lower income families of the towns and cities, women eat last, after serving food to all other members of the household. Many a times, there is not enough food left by that time. They get up earlier than the other members of the household every morning and sleep last. While the men folk get holidays to relax and rest, the women work 24 hours a day, 7 days a week. In poor urban families, women remain inside poorly ventilated tiny houses or shanties the whole day and night and inhale the smoke coming from chulhas. They get no chances to rest or relax. In families where family planning is not done, women give birth frequently.

All this have a major effect on their health. *Women suffer more malnutrition than men and get less medical care than men.* They become anemic and get infections faster. About two million women, 95% of them living in LDCs, die every year from pregnancy related diseases, which could have been easily prevented. In many poor households, men do not work or work less. So the burden of looking after the entire family falls on the women. On top of this, men come home drunk in the evenings and beat and ill treat their wives. In short, poor women have to bear the brunt of every possible inequality.

The inequality is seen in all spheres. Women account for two thirds of the illiterate persons in the world. In many LDCs, percentage of literate women is only half of that of men. *Adult female illiteracy rate is as high as 57% in India.* Women do not have legal rights to own land or borrow money in some cultures. The social evils of child marriage, dowry system, the non-acceptance of widow remarriage, etc. are all practices that go against the welfare of women.

Plight of children: Population levels and poverty are so high in many LDCs that children are forced to work. There are laws banning child labour in many countries. In India also, the law is of course there, but implementation has always been poor. Recently, the government has decided to enforce the law strictly. All children below the age of fourteen are being taken off from their work places and sent to their homes.

But how far this entire exercise will have the desired result is debatable. There is no guarantee that the parents will take them back. They again have to go out in search of work because of poverty. Without poverty being removed, how the parents can suddenly support their children and give them education is exactly not known. It has been proposed that children whom their parents refuse to accept will be handed over to NGOs. The desirability of this scheme is also controversial. The problem is that there are too many problems in the LDCs and they are all interrelated. To solve one, the others have to be solved first. Much thought therefore needs to be put while trying to come to a conclusion on the priority list of things to be done.

Inclusion of the *Directive Principles of State Policy in the Constitution of India was aimed at protecting the basic rights of individuals*. Those who framed our constitution also wanted it to become an effective instrument for social revolution. Article 37 declares that the Directive Principles are "fundamental in the governance of the country" and that "it shall be the duty of the State to apply these principles in making laws". It is thus clear that the Directive Principles must be treated as the human rights provisions of the Indian Constitution.

However, the Directive Principles did not give any legal rights to individuals in case of their violation. This was because the availability of resources and the readiness of the society to accept the changes had to be taken into consideration.

Anyway, even after 59 years of independence, some of the provisions of the Directive Principles remain unfulfilled. For example, Article 39 says that the state shall direct its policy in such a manner that all men and women have the right to an adequate means to livelihood, that the health and strength of workers — men and women, and the tender age of children are not abused, and the childhood and youth are protected against exploitation. Article 45 relates to the free and compulsory education of all children up to the age of 14.

Education of children is a luxury for many parents even now. Due to poverty, boys are compelled to help their fathers in grazing cattle or working in the field. Sometimes it is due to the absence of schools nearby.

In many cases, boys are sent to schools while girls are not. Even if they are sent, girls are withdrawn from schools soon for helping mothers in household jobs or looking after younger siblings. The daughters also get less food and rest than sons. Percentage of anemic girls in the LDCs is alarming.

Tens of thousands of children die annually in India due to various diseases and malnutrition. Diarrhoea, measles, malaria, pneumonia, etc. are the main causes of death. Many children become victims of breathing trouble because of breathing polluted air. Contaminated water and food causes stomach and intestinal diseases more in children.

Blindness in children is mostly due to malnutrition. In addition to not getting enough quantity of food, intake of food items low in nutrient value also results in malnutrition. In the case of infants, early cessation of breast feeding causes malnutrition. The period between stopping breast feeding and starting normal food is more risky for infants. They do not get proper nutrition at this time if mothers do not take extra care. But, for this, mothers need to be educated and families must be able to afford nutritious food.

In fact, *unhygienic environment, poverty and lack of awareness are the main causes of many diseases, especially in women and children.* This clearly shows how urgently we need reforms in India to correct these imbalances. Steps have been taken in this regard. Ministries of Women and Child Welfare have been set up in almost all the states. Various NGOs are taking active part in organizing awareness programmes like the Breastfeeding Week and the like.

It has been estimated that *a minor increase in breastfeeding can prevent up to 10% of all deaths in children under five years of age.* This is due to two reasons: 1) Mother's milk is the purest form of milk and it contains all the nutrients needed for infants till about 6 months of age. 2) When breast feeding is stopped early, infants are given outside food and water which are easily contaminated. This leads to diarrhoea or dysentery and the infants die due to dehydration. Therefore infants should be breastfed as long as possible. In later stages, additional food must of course be given, but not before 6 months. For nursing mothers to produce sufficient milk, they should also know what to eat. Methi seeds are known to increase milk production in nursing women.

Educating the girl child is one major step towards woman empowerment. Another is formation of Self Help Groups (SHGs) for women. Over 5.4 million women have organized themselves into 380,000 SHGs in the state of Andhra Pradesh. The model of Andhra Pradesh is being highly appreciated all over India. Most of the other Indian states also have SHGs.

Self Help Groups of women in India have been recognized as an effective strategy for women empowerment. These are operative in rural as well as urban areas. They bring together women from all walks of life. The aim is to fight for their rights or for a cause. Many issues like health, nutrition, agriculture, forestry, etc. are all dealt with through Self Help Groups. The system is based on micro credit, i.e. women deposit small amounts daily or monthly (the amount may be as small as one rupee) to office bearers selected from themselves and take loans from the collected amount when in need. The money is used in income generating activities. The women thus make profit; more than that, their self-esteem and self-confidence increase to a great extent.

The Nobel Prize for Peace 2006 was won by Muhammed Yunus, Bangladeshi Economist and micro-credit pioneer, who started the Grameen Bank more than thirty years ago. It provides credit to the poorest of the poor in Bangladesh. At present, the bank has 1092 branches in 36,000 villages. The bank has 5 million borrowers, of which 96% are women. This silent revolution in Bangladesh is sure to inspire many other countries.

ROLE OF INFORMATION TECHNOLOGY IN ENVIRONMENT AND HUMAN HEALTH

Information Technology (IT) is concerned with the use of technology in managing and processing information.

IT especially deals with the use of electronic computers and computer software to Convert, Store, Protect, Process, Transmit, and Retrieve information.

Information technology is also called Information Services (IS) or Management Information Services (MIS). Its new name is Information and Communication Technology (ICT).

Information Technology has been a revolution, just like electricity. It brings about changes in economies and societies all over the world.

IT also helps in environmental issues in different ways, like:

- Changing the way how goods are produced and consumed, and
- Finding out new means to address new changes.

Computers have become common now even in villages. In many developed states in India, a large quantity of data regarding ownership of land, agricultural practices, details of institutes and agencies which sell seeds, fertilizers, etc. and various government communiqués regarding subsidies, etc. are stored in computers. The villagers get all informations regarding their crops from the internet. Weather forecasts are very important for farmers. These are also available on the net.

Information brings about:

- Equality
- New work environment
- New work ethics
- New work standards
- Objectivity, and
- Analytical capabilities to decision making and development

Information technology constantly focuses on performance. In the present system of information revolution, the Indian system cannot and should not remain isolated. Indian institutions should be connected to international networks and have access to international data bases and information on new technologies. Such an Information system has a few components like:

- Management Information System
- Project Information system
- Personnel Information system
- Financial information system, and
- Library Information system.

A lot of research is going on in different laboratories and educational institutions on environment and health. The best way to pass on information regarding the results of these experiments is through the Information Communication Technology system.

For example, farmers must know *latest developments in tissue culture, genetic engineering, biotechnology, etc.* as early as possible for procuring new seeds, etc. and for adopting new methods of cultivation as early as possible. IT helps the farmers in this. All informations are put up on the websites of the laboratories and institutions as soon as any breakthrough is made. Computer literate farmers surf the net daily. Thus they get all the information they need immediately.

Similarly, environmentalists around the world are in touch with each other through the IT system and exchange all types of new knowledge.

Early warning system of disasters can work best along with IT. Predictions of floods, droughts, storms, cyclones, etc. and meteorological information on the forecast of weather will be very helpful to farmers and the common man alike.

There are many *websites on health and medical information* these days. Anyone can go to these sites with the help of search engines like Google, Yahoo, etc. People in far flung places can get up to date information on any health problem from these medical sites. For example, medexplorer.com is one of the several Websites for medical information on the Internet. Who.org is the website of WHO, i.e. World Health Organisation, one of the important wings of the UN (United Nations).

The World Wide Web (WWW) is the most powerful tool found out in many years now. Any information on any subject is available at the click of a mouse. (See Appendix - 3 for more details). Therefore, being computer literate helps a lot. As far as India is concerned, many villagers have learnt how to use Information Technology and gather information from the websites. The craze for computers is catching up fast.

Satellite technology has advanced greatly. Of late, meteorological satellites are being used more and more to forecast weather in India also. Use of computers has played a major role in improving the accuracy of predictions. This is because complex calculations can be done very fast using computers. Weather supercomputers can provide an enormous quantity of weather related data to the forecasters.

Mobile phones have become very common. They have penetrated to even villages with very little development. Information on current market prices, weather and other information is passed on through mobile phones (cell phones). In Kerala, owners of fishing boats are constantly using this facility. Sitting at home, they find out every day how the catch was and decide the price of fish for that day.

Internet kiosks have come up in states like Tamilnadu. These include internet services, telephones and other related equipments. Each kiosk connects 300 to 400 villages.

Let us hope awareness about IT (ICT) and its benefits spread far and wide in the shortest time possible.

Case Studies

Cases of Information Technology spreading far and wide in India are so many. In the West, it had spread a few decades earlier, no doubt, but India is moving forward very fast. Some case studies in this regard are given here. Some are from other countries.

1. In Hyderabad, two NGOs—SKS and SHARE are helping women borrowers use electronic 'Smart Cards' to increase savings and find out details about micro-credit services that were available to only the rich clients earlier.

2. In Bangladesh, the Grameen Phone Project gave cellular connectivity to 190,000 subscribers, covering 55% of the country by 2002. Poor women entrepreneurs started earning $700/- per annum, which is more than double the per capita income of the country.
3. A website *www.ewg.org* *<http://www.ewg.org>* is keeping track of all subsidies given by the US Government to their farmers. The amount was $170billion in 2002. The site is getting millions of hits. The system gives information to the citizens while at the same time gives transparency and accountability to the Government.
4. M.S. Swaminathan Research Foundation has created a multimedia database on ecological farm families in India. This database contains many details of farms, like management practices, new technologies, farm economics, etc.

Case studies in other Topics

- India's population has grown to 110 crores. 65% of India's population depends on agriculture. The farm population is increasing by 1.84% per year. The average farm size is becoming smaller each year. Marketing infrastructure is poor, especially in perishable produce. As a result of improved agricultural practices, yield of all crop produce has increased tremendously in India; however, there is no good storage facility. For example, while Indian farmers are suffering huge losses by the rotting of tomatoes, some sauce and ketchup companies in India are importing fresh tomatoes from China. This is most unfortunate. A recent NSSO survey has revealed that 40% of the farmers would like to quit farming, if they have the option to do so.
- A study in Italy has shown that diets high in refined cereals, especially white bread, is associated with kidney cancer. The study was done by Francesca Bravi and his associates. 767 cases were studied for the purpose.
- For every 4 malnourished adults in the world, 5 more are overweight, according to a study by the World Health Organisation (WHO). In other words, out of the 6 billion people in the world, 1 billion are overweight while 80 million do not have enough to eat.
- Government investments in education and health care have

increased greatly in some countries. The results are beginning to show. For example, the share of children in the secondary schools of Brazil went up from 46% in 1980 to 66% in 2001. During the same period, the proportion of illiterate women in Brazil has fallen from 27% to 16%.

- There is a very large health gap between the developed and the developing countries of the world. Polluted water is one of the main reasons for this. Although infant mortality rates have dropped from 25 to 50% in many countries from 1990 to 2000, 43 infants die per thousand in China and 70 per thousand in India as per the 2000 estimates. Most of these deaths are due to waterborne diseases.

INTELLECTUAL PROPERTY RIGHTS (IPR)

Intellectual property is a product of the brain. The right to protect this property, or the right to prevent others from copying this property is called intellectual property rights (IPR).

Intellectual property is property in any field. IPR includes rights in many spheres like:

- Trademarks
- Copyright
- Patent
- Design, etc.

These may be related to literary, artistic or scientific works, inventions in all fields, industrial, scientific or technological designs, commercial names, etc.

A] Trademarks

A trademark helps to identify the source of the goods of one owner from those of others. It may be a:

- Brand
- Heading
- Label
- Name
- Signature
- Word
- Letter
- Numeral
- Shape of goods

- Packaging
- Combination of colours, etc.

For example, PEPSI is a trademark. This is different from a service mark. *A service mark distinguishes the source of one service of one owner from those of others.* DHL is a service mark.

Registration of trademarks: The owner or his lawyer has to go to the Registrar of Trade Marks and get the application form by paying the fees. Then he has to fill in the form and submit it to the Registrar to get the trade mark.

Benefits of registration: Registration is proof that the owner has the IPR for his product. If any case is filed against the owner, or if anyone violates his IPR, he can show the registration in the court. If there is no registration, the owner will have to prove that he has acquired the right by use and / or reputation. Violation of somebody else's trademark is a criminal offence.

Period of validity of registration: Validity period is 10 years. After that, renewal can be done every 10 years.

B] Copyright

Copyright is the right of an author to control the reproduction of his intellectual creation. Copyright protects original works in the following fields:

- Literary works like
  a) Books
  b) Computer programmes
- Dramatic
- Artistic works, like
  a) Pictures
  b) Paintings
  c) Architectural works
  d) Engineering drawings, etc.
- Musical works like
  a) Compositions
  b) Sound recordings
- Films or cinemas
- Performer's rights
- Broadcast rights, etc.

A copyright need not be registered. On creating a work that is within the scope of the Copyright Act, copyright arises automatically. It can be registered if the author so wishes. Or if it is a book, the author may insert a copyright note on the work in the form given below:

©, *Name of the author, Year of creation*

Process of registration: The owner or his lawyer has to buy the application form and apply in the copyright office in the format obtained. Registration is for the lifetime of the owner plus sixty years.

The owner of the copyright has both civil and criminal rights under the Indian law.

C] Patent

A patent is a legal right granted by the government to the patentee under Indian Patent Act (1970) for an invention for a fixed term. The patentee can use the invention for himself and/or by an authorized person/s. He can also prevent others from making, using or selling his invention.

If an inventor doe not get his invention patented, anyone can use it without restriction. So to get monopoly over the invention, the inventor must get it patented.

The purpose of granting patents is to encourage inventions and to promote new technologies in industries and other fields. It inspires research and development.

Process of getting a patent: The Indian Patents Office has branches at Chennai, Mumbai, Delhi and Kolkata. The inventor can file his application in any of these offices. The application will be published 18 months after filing. After that, the applicant must file a request for examination. The Patent Examiner will then examine the application and see whether the application is justified. For justification, the patent must have utility and novelty. Moreover, it should not come under any non-patentable categories described under Sections 3 and 4 of the Patent Act. If the examiner is satisfied, he will recommend the case to the Controller of Patents. The patent will be granted by him. In the meantime, if anyone has any objection, he can file opposition to the patent that it is not new or not useful, etc. If the patentee is able to prove that he did not have any knowledge about the existence of the invention earlier, no fine need to be paid. But the patent cannot be used.

Benefits of patent registration:

1. Patents give exclusive rights to the patentee. The invention can be used and exploited for 20years from the date of filing the application.
2. The inventor must have spent time and money on the invention. Now he/she can sell the invention and get the money back.

3. If the inventor has got his/her invention patented, he/she can bargain while selling it and get maximum profit.
4. As nobody else can use or sell the invention, the patentee will become the main seller in the market.
5. The patentee can sell his invention if he so wishes, instead of exploiting it himself. This can be a regular income.
6. If the inventor has not patented the invention, somebody else who has made the same invention after him/her may patent it. If that happens, then the other person will have all exclusive rights over the invention.
7. Patenting boosts up the image of the company in the eyes of partners, investors and share holders.

Biopatent: *Patenting of life forms or biological material is called biopatenting*. Before 1980, life forms could not be patented. Later, Dr. Chakraborty, a Non-Resident Indian (NRI), genetically engineered a bacterium of the genus *Pseudomonas* which could consume oil. He patented this bacterium. It is named 'super bug', and is used to clean oil spills. Another scientist in the US got a patent for a genetically engineered mouse carrying human cancer gene.

Naturally occurring plants or animals or their parts cannot be patented. But transgenic plants and animals (where genes have been transferred) can be patented. Similarly, natural DNA sequences cannot be patented. But some gene sequences have no biological function; they are useful only in industries. Such genes can be patented.

Terminator technology: Some seeds have been found out and patented by the US government, which will grow into high yielding plants. But the seeds borne by these plants cannot germinate. This has been named as terminator technology. India has banned the entry of such seeds.

Biopiracy: *Biopiracy may be defined as the theft of natural products and selling them after getting them patented, without giving any financial or technological benefit to the original inventors.*

It is most essential that we get our herbal medicines patented fast. Otherwise, foreign countries may patent them. The importance of this issue came to be known to us when patent claims were filed in the US for the use of Neem seed oil for biopesticide in 1992. These have also been used in India for thousands of years. We won the Neem case in 2000.

A patent was issued to the University of Missisipi Medical Centre for the 'use of turmeric on wound healing'in 1995. We Indians have been using turmeric (haldi) for healing wounds for thousands of years. But we did not have a patent for the same. So others took advantage of this. We had to fight a legal battle to revoke their patent. Based on objections by Council for Scientific and Industrial Research (CSIR), New Delhi, with documented proof, US patent for turmeric was revoked.

These are examples of misuse of IPR. But biopiracy cannot be allowed to continue. *We Indians must take up the matter urgently and get all our ancient medical preparations patented.* Work in this regard is progressing.

D] Designs

Designs of finished articles, whether manual or by machines, can be registered. The Designs Act was passed in 2000.

Requirements for registration:

1. The design must be new or original, not used earlier in any country before the date of application.
2. The design should relate to the shape, pattern or ornamentation of the article. So industrial plans cannot be registered.
3. The design should apply to articles of industrial processes. Thus paintings cannot be registered as they are not produced in mass. (However, they can get copyrights)
4. The features of the design must appeal to the eye, as judged solely by the eye.
5. The design should not have any trade mark.

The process of registration: The inventor has to file an application in the Indian Design Office along with the fees. It will be examined by the Examiner first. On being satisfied, he will forward the application to the Controller of Designs who will grant the registration.

Duration: Registration for design is for 10 years from the date of filing the registration form. It can be extended for 5 more years.

Benefits: By registering, the creator can be sure that others will not copy the original design. He/she will get a copyright in the design. Only the creator can apply that particular design to the product for which it is registered. If anybody copies a registered design, that person can be taken to court and damages claimed.

Thus, *the 4 main areas covered by IPR are trademarks, copyrights, patents and designs.* IPR laws are not very well known in India. There is very little awareness about these laws. However, our patent laws are being brought in line with international standards. Research and development is therefore sure to expand. This will create a demand for trained persons who can draft patents.

QUESTIONS

A] Very short answer type questions:

a) Multiple choice type-Tick the right answer:

1. Present world population is:
   (a) Less than 6 billion (b) More than 6 billion
   (c) More than 7 billion (d) 6 billion
2. One of the diseases caused by lack of hygiene is:
   (a) Dysentery (b) Chickenpox
   (c) Pneumonia (d) Cancer
3. A disease which is spread by mosquitoes is:
   (a) Influenza (b) Typhiod
   (c) Jaundice (d) Malaria
4. Kalazar is spread by:
   (a) Tapeworms (b) Rats
   (c) Mosquitoes (d) Sandflies
5. Universal Declaration of Human Rights was adopted by UN General Assembly in:
   (a) 1984 (b) 1992
   (c) 1948 (d) 1990
6. The virus that causes AIDS is:
   (a) HIV (b) TMV
   (c) HMV (d) None of these
7. IT helps in providing information on:
   (a) Tissue Culture (b) Health and Medicine
   (c) Disaster Warning (d) All of the above
8. IPR includes rights in:
   (a) Trademarks (b) Copyright
   (c) Design (d) All of the above

b) Define:

1. Values
2. AIDS
3. Information Technology (IT)
4. Intellectual Property Rights (IPR)
5. Microcredit
6. Trademark
7. Service Mark
8. Copyright

c) Explain briefly:

1. Why Amazon forests are losing millions of hectares of forest every year
2. Population explosion in India
3. Family welfare
4. Functions of Family Welfare ministries
5. Insect borne diseases
6. Diseases caused by worms
7. Plague
8. Muhammad Yunus and Grameen Bank
9. Diseases caused by pesticides and herbicides
10. Components of Information System
11. How IT can help farmers
12. Biopatent
13. Terminator technology
14. Biopiracy

d) Name the following:

1. Three harmful social effects of population growth
2. Five basic human rights
3. Three benefits of patenting

e) What does AIDS stand for?

f) In which year did world population cross the 6 billion mark?

g) What is the full form of HIV?

B] Short answer type questions:

Write short notes on:

1. Results of population explosion
2. Sterilisation
3. Types of pollution in overcrowded localities
4. Water and foodborne diseases
5. Respiratory diseases
6. Cancer
7. Methods to prevent environment related diseases
8. Disparities between men and women in society
9. Environmental rights of future generations
10. Symptoms of AIDS
11. Self Help Groups
12. Trademarks
13. Copyright

14. Patent
15. Design

C] Long answer type questions:

a) Describe in detail Family Welfare Programmes.
b) What are the common methods of birth control?
c) Which are the diseases caused by environmental pollution? Elaborate.
d) Give an account of Human Rights and the various types of their violations.
e) Write an essay on the different types of 'Values'.
f) Describe in detail about AIDS.
g) Give an account of the plight of poor children in the developing countries.
h) What are the various ways in which women are discriminated against?
i) Discuss in detail the role of IT in environment and human health.
j) What do you understand by Intellectual Property Rights? Elaborate.

Unit 5

Environmental Pollution

Life on earth depends on the environment. There is a balance between the living organisms and the environment on the earth. If any component, living or non-living, is disturbed, it causes a harmful effect on the entire system.

DEFINITION, CAUSES AND EFFECTS OF ENVIRONMENTAL POLLUTION

Any physical, chemical or biological change in the environment that leads to its deterioration is called environmental pollution.

The agent which causes environmental pollution is called a pollutant. Kitchen refuse, garbage, smoke, pesticides, noise, emissions from industries and automobiles, effluents from industries, etc. are all pollutants. Pollutants are mostly residues of things we make, use and throw away. They are by-products of man's actions. They rise with a rise in human population. Pollutants disturb ecological balance and create serious problems on earth.

Pollution is a man made problem. Developed countries are exploiting available natural resources to produce luxurious goods, using them and also exporting them to the developing countries. During this process, many things are dumped in the environment. In the developing countries, industrialization is catching up, and because of this, pollution is increasing. Moreover, due to lack of proper management measures, waste materials accumulate in the environment. Garbage dumps become breeding grounds for mosquitoes and flies which cause many diseases.

In short, *pollution is an undesirable change in the air, water or land of our environment that harmfully affects human, animal and plant life in different ways.*

Classification of Pollution

Pollution can be classified in many ways.

Based on the part of the environment being polluted, pollution is of three types.

1. Air pollution
2. Water pollution, and
3. Soil pollution

Based on the nature of the pollutant, pollution can be of several types like:

- Lead pollution
- Carbon dioxide pollution
- Smoke pollution
- Noise pollution
- Pesticide pollution
- Radioactive pollution
- Mercury pollution
- Plastic pollution, etc.

Pollution can also be classified into:

a) *Natural pollution*: There are some pollutants in nature itself. These are oil and natural gas; mercury, lead, sulphur, etc. in river waters; lava from volcanoes, etc.

b) *Man made pollution*: Wastes from factories, exhausts from automobiles, pesticides from farms, smoke from chimneys, etc. cause man made pollution.

Based on the area affected, pollution is of different types:

- Local
- National
- Continental, or
- Planetary

Air and water pollution have no boundaries. Sulphur dioxide from chimneys of factories spread far and wide. Rivers, carrying pollutants, may flow through different countries before emptying into the sea.

Pollutants can be classified into two:

a) **Biodegradable pollutants:** *Pollutants that can be easily decomposed by natural processes are called biodegradable pollutants*. Examples are various types of kitchen wastes like skin and seeds of vegetables and fruits. Sometimes, however, they too accumulate and create problems. This happens when more is dumped in the environment than can be degraded by nature at the rate of dumping.

b) **Non-biodegradable pollutants:** All *materials that do not degrade or degrade only very slowly in nature are called non-biodegradable or non-degradable pollutants*. Aluminum cans, mercuric salts, long chain phenolic compounds, plastics, DDT, etc. are non-biodegradable. Similarly, the detergents that we use for more lather cannot be degraded by natural processes. The fumes produced during burning of plastics remain in the atmosphere, as nature cannot degrade them. The wastes produced during the manufacture of many luxury goods are blown by wind or carried by water to far off places. These non-degradable pollutants are not recycled naturally and get accumulated.

Environmental pollution results in different types of diseases in humans, animals and plants. It makes the environment dirty and damages various articles of our use. Smoke and dust in the air reduces visibility leading to traffic accidents. Pollutants cause genetic changes in microbes also, which become resistant to various drugs. Many plants and animals fall prey to pollution and become extinct, reducing biodiversity and disturbing the balance of nature. Pollutants damage all types of buildings including national heritages, etc. It leads to decrease in tourism.

Pollution is a very costly affair. Its cost can be measured in different ways:

1. Loss of resources due to wasteful exploitation
2. Cost of pollution control — money, manpower, etc. to dispose off the pollutants
3. Cost of medical care on different diseases
4. Loss in working hours due to illness
5. Cost of repairing buildings or monuments damaged by pollutants.

How to Minimize Pollution

Various strategies can be adopted to reduce the level of pollution in the world at present.

1. *Educating people at all levels:* To a great extent, pollution occurs because man makes too much use of natural resources and throws away too many unused things into the environment. This upsets natural balance. To maintain balance, intake and output should be equalized. Equilibrium should be maintained.

   People should be made to realize that they can do quite a lot to prevent pollution. Some methods to educate the people about pollution:

- Reading materials and audiovisuals like newspapers, radio and television have a major role to play in creating awareness among the general public.
- Conservation clubs can be established in rural and urban areas.
- Anti-pollution weeks can be observed.
- Demonstrations can be organized at sources of the pollution.
- Youth can impart information at school and college levels.

2. *Finding ways to reduce the already existing pollution hazards*: Concentrated efforts must be made to minimize the existing pollution levels in both rural and urban areas. For this, governments, NGOs and the people must work in union. Here is a list of what can be done. The list is in no way complete. It is just an indicator.
   - Sewage treatment plants have been set up in some cities like Jamshedpur; many more must be set up in all towns and cities.
   - Industries should not be allowed to be set up near residential areas.
   - A chain of monitoring stations and equipments should be set up around the sources of pollution.
   - There must be a massive drive to pressurize industries to adopt pollution preventive measures at all levels.
   - New projects should not be cleared if they have not included provisions for pollution prevention.
   - Offenders must be punished without any discrimination.
   - Proper traffic control must be done.
   - Automobile engines that pollute less must be designed.
   - Farmers must be encouraged to use organic manures and biopesticides.
   - Solid waste management must be done in rural as well as urban areas, according to the nature of waste generated.
   - Organizers of festivals must be told firmly not to cross permissible levels of sound.
   - Tree planting must be encouraged in all suitable locations.

AIR POLLUTION

Air is a mixture of many gases, and is composed of 3 components — major, minor and trace. Major components are Nitrogen, occupying 78.09% and Oxygen, occupying 20.94% of air by volume. Minor component — Carbon dioxide occupies 0.03% of air by volume. Argon is also a minor gas. The percentage of water vapour may vary greatly from 0.1 to 5%. In addition to these major and minor gases, there are many gases present in traces in the air, in different proportions such as Hydrogen, Helium, Methane, Ozone, Nitrogen dioxide, Sulphur dioxide, Carbon monoxide, Ammonia, Neon, etc.

Air is never found clean in nature as it contains various natural and man made pollutants. Gases like Carbon monoxide, Sulphur dioxide, Hydrogen sulphide, etc. are continuously released into the atmosphere due to natural activities like volcanic activity, decay of dead plants, forest fires, etc.

However, the gases released into the air by the activities of man are present in much higher concentrations. They upset the natural equilibrium and cause great harm to plants, animals and human beings. These are the results of chemical and biological processes used by man.

There are 5 primary pollutants in the air. Together, they account for 90% of air pollution on the earth. These are:

1. *Carbon monoxide* (CO): This is a colourless, odourless poisonous gas. It is released on incomplete combustion of coal, wood, natural gas, etc. and also from automobiles.
2. *Nitrogen oxides* (NO×): These are found in the exhaust of vehicles.
3. *Sulphur oxides* (SO×): On the burning of fuels that contain sulphur, sulphur oxides are produced. Bursting of fire crackers during festivals release large amounts of SO_2 into the atmosphere.
4. *Hydrocarbons* (HC): These are compounds made of carbon and hydrogen. They are produced during incomplete combustion of various fuels.
5. *Particulates* (tiny particles of solids in the air): Particles suspended in the air are known as aerosols. These include particles of smoke, dust, asbestos, ash, soot, etc.

Industrialization is a major source of air pollution. Industry is ever expanding in all countries. In the developing countries, especially, many industries have been set up in or near cities. Large quantities of polluting gases are released into the air by industries. They are also carried to far off places by air currents. For example, a large power station produces 500 tons of Sulphur compounds and dust each day. 10% of these pollutants may be carried over a distance of 5 kms. Atmospheric pollutants produced in Texas have been detected in Ohio (both in the US), 1,600 km away.

Automobile exhausts are another major source of air pollution. There has been an enormous increase in the number of automobiles in the past few years. These are essential for

- Personal conveyance
- Passenger traffic, and
- Goods transport

The exhausts from the internal combustion engines of the motor vehicles foul the air. Motor vehicles contribute about 60% of air pollution. Many harmful gases are emitted by vehicles. Most important of these are Carbon dioxide, Carbon monoxide, several unburnt hydrocarbons and various compounds of Sulphur, Chlorine and Nitrogen. Atmospheric levels of tetraethyl lead, a compound added to petrol before selling, are increasing rapidly. Each vehicle emits about 1 kg of lead per year.

Oxygen content of the atmosphere is about 20% by volume. A single aeroplane uses 35 tons of precious oxygen in crossing the Atlantic Ocean. The plant cover of the earth is also being depleted fast. The major consumers of carbon dioxide in the air are the green plants. When they are destroyed, there is an increase of carbon dioxide in the air. When the carbon dioxide emitted by the vehicles and factories is added to this, the total amount becomes highly disproportionate to the levels of oxygen, now depleted by aeroplanes in the air and automobiles on the land.

Carbon dioxide content of the air is slowly increasing. Over the past century, it has increased by about 15% (15% of 0.03%). Some amount of CO_2 in the atmosphere is absorbed by the oceans. But as industrialization and the number of automobiles increase, rate of CO_2 release has become more than the rate of absorption by the oceans (see biogeochemical cycles in Unit 7).

Normally, carbon dioxide is 0.03% by volume in the atmosphere. In such normal condition, the temperature at the surface of the earth is maintained properly. The sun rays that strike the earth and the heat that is radiated back into the atmosphere are balanced. However, when there is an increase in CO_2 concentration, it prevents heat from being re-radiated out. The thick layer of CO_2 functions like the gas panels of a greenhouse. It allows sunlight to pass through, but prevents the heat from re-radiating out. This is called the *greenhouse effect.*

The result of this greenhouse effect is that the air becomes warmer near the earth's surface. The earth also becomes warmer by re-radiation from the air. The whole process is called *global warming*. Global warming will increase the temperature of the biosphere as a whole. This can melt the polar ice caps, leading to flooding of the coastal countries (See Unit 3).

All gases which contribute to the greenhouse effect are known as *Greenhouse Gases or GHGs.* These are:

- CO_2 (Carbon dioxide)
- N_2O (Nitrous oxide)
- CH_4 (methane)
- HFCs (hydrofluorocarbons)
- PFCs (Perflourocarbons)
- SF_6 (Sulphur hexafluoride)
- SO_2 (Sulphur dioxide)
- CFCs (Chlorofluorocarbons), etc.

These are discharged into the atmosphere mainly by industries. Some greenhouse gases like carbon dioxide, nitrogen oxides, sulphur dioxide, etc. are also emitted by automobiles. Both nitrogen oxides and sulphur dioxide are harmful to human health.

Nitrogen oxides cause:

- Irritation of the lungs
- Asthma
- Chronic bronchitis, etc.

It causes irritation of eyes and nose and also makes the affected persons more susceptible to influenza, common cold and all other types of infections of the respiratory system.

Sulphur oxides, especially sulphur dioxide, have the following harmful effects:

- They irritate the respiratory tissues.
- They damage lungs and cause problems like bronchitis.

SO_2, NO_x and other acid forming chemicals are emitted by automobiles and industries. In the air, these gases combine with water vapour to form acids. These fall back to the earth along with rain. This is called *acid rain.* Acid rain increases the acidity of soils, damages crops and forests and causes harm to plants and animals living in water. This problem is seen more in the industrialized West (See Unit 3).

Diesel and petrol produce similar pollutants on burning, but petrol produces more gases. The gases given out during idling and acceleration are more harmful than those produced during moderate speed.

One of the main pollutants emitted by automobiles is carbon monoxide (CO). It has many negative effects on human health:

- Headache
- Irritation of mucus membranes
- It combines with haemoglobin of blood, reducing its oxygen carrying capacity. This results in:

  a) Difficulty in breathing

b) Bronchial asthma
c) Fainting
d) Coma, etc.

- Decrease in thinking power
- Slowing down of reflexes
- Drowsiness
- Dizziness
- Higher concentrations may be fatal

Incomplete combustion in the internal combustion engines of automobiles releases 3, 4-benzpyrene, which causes *lung cancer*.

Control of air pollution by automobiles: Air pollution by automobiles can be controlled by various methods:

1. By traffic control
2. By designing automobile engines with reduced amounts of unburnt hydrocarbons
3. By fitting vehicles with exhaust emission controls
4. By using additives like Barium compounds in fuels
5. By control of idling
6. By developing non-polluting engines for automobiles. (Recently, Hydrogen fuelled engines have been introduced on an experimental basis in Germany and Japan. They will emit only water vapour into the air)
7. Using Compressed Natural gas (CNG) in vehicles. This has reduced pollution to a great extent in New Delhi.

Control of air pollution by industries: Following are some methods to control air pollution by industries.

1. Factories should be far away from residential areas.
2. Old machines should be modified and new more efficient ones should be set up.
3. Pollution control equipments like scrubbers, dry and wet collectors, filters, etc. must be installed at the design stage itself.
4. Chimneys should be high so that the fumes are emitted as far away from the ground as possible.
5. Raw materials that cause more pollution should be replaced by less polluting ones.
6. Round the clock measurement of pollution must be done.
7. When tolerance limit is reached, the factory must be shut down.

Smoke Pollution: Smoke is a major pollutant of air in rural as well as urban areas. This is called smoke pollution. *About 10 to 15% of air* *pollution is caused by smoke.* Smoke is produced from various sources. Smoke from chimneys of industries, automobiles and from homes cause air pollution. It is mostly a man made chemical pollution of air. Smoke contains Lead (Pb), sulphur dioxide (SO_2) and carbon monoxide (CO). All the three are poisonous to animals and man. SO_2 gets oxidized in the air to sulphuric acid (H_2SO_4). This is seen in the air in the form of small droplets. SO_2 is formed mostly during combustion of coal and fuel oils. Incomplete oxidation of domestic fuels gives out CO.

Smog formation: One very bad effect of smoke is *smog formation.* Dust and soot (powdery carbon) particles from factories provide the nuclei (central cores) for smog formation. Particles of coal dust, SO_2 and CO form the nuclei around which other chemicals collect to form smog. *Smog is a fog made darker and heavier by smoke and chemical fumes* (Smoke + Fog = Smog).

Photochemical smogs have also become common now. When hydrocarbons and nitrogen oxides react in the atmosphere in the presence of the ultraviolet (UV) rays in the sunlight, ozone (O_3) and Peroxyacetyl nitrate (PAN) are formed. These together produce a photochemical smog. Both ozone and PAN are bad for health; their harmful effects are:

- They cause watering in the eyes.
- They lead to respiratory diseases.
- They are toxic to plants.
- PAN blocks the Hill reactions of photosynthesis (making of food by green plants).

Smoke and smog both are very bad for health.

- They irritate the eyes and nose.
- They constrict throats.
- Lung diseases like bronchitis, asthma, etc. are widespread in areas where there is smoke pollution.
- Smoke, smog, dust, soot, etc. scatter and absorb light, which reduces visibility and is a major cause of traffic accidents. In December 1952, more than 4,000 people died in London in a 5-day-long smog.
- Smog also damages agricultural crops.
- It damages livestock.
- Corrodes:
  a) Metals
  b) Building materials
  c) Painted surfaces

d) Textile
e) Paper
f) Leather, etc.

The dust and soot particles in the smoke

- Make our surroundings dirty.
- Spoil our material goods.
- Rob us of our pleasure, etc.
- In many localities, rooms and all the articles in them get covered by a layer of black dust every morning. This is the soot deposit. Those who walk barefoot get their soles blackened. Dusting household articles becomes a problem.
- Soot also settles on uncovered water surfaces, thereby polluting water.

Effect of air pollution on plants: Some air pollutants break down the cuticle (waxy covering layer) of leaves. This has many harmful effects on the plants.

a) It makes the leaves susceptible to various types of infections, drought, frost, etc.
b) More water is lost from the plants.
c) Photosynthesis is reduced and leaves turn yellow and fall.
d) Growth becomes stunted.

Air quality monitoring: Pollution by various pollutants in the air must be monitored properly; only then can the right pollution control measures be taken. In India, all air pollutants are not being monitored at present. Only common ones like carbon monoxide, sulphur dioxide, nitrogen oxides, etc. are monitored effectively. Benzene, small particulates, etc. are not monitored.

National Environment Engineering Research Institute (NEERI) located at Nagpur has prepared a database of ambient air quality in different cities of India. But this does not give a complete picture. The State Pollution Control Boards (SPCBs) of various states also collect data and send it to The Central Pollution Control Board (CPCB). Even then much more work — more extensive and wider spread — is to be done in our country in air quality monitoring.

WATER POLLUTION

The quality of water is very important to mankind. It is directly connected to human health. Pollution of drinking water is the single major cause of waterborne diseases like diarrhea, dysentery, cholera, jaundice and many others. These diseases kill many children in the developing countries. Adults also fall prey to these diseases.

Water which has lost its pure nature is called polluted water. Polluted water can be distinguished from pure water by observing one or more of the following signs:

- Mild or stronger colouration (yellowish, brownish, etc.)
- Bad taste
- Offensive smell
- Too many water plants in the water body
- Reduction in the number of fish and other aquatic animals
- Oil and grease floating on water surface

Polluted water is harmful not only to human beings but also for animals, plants and industries. Pollution affects freshwater bodies like ponds, lakes, streams, rivers, etc. and also marine water bodies like seas, oceans, estuaries, etc.

The major sources of water pollution are

- Domestic wastes, mainly sewage
- Industrial wastes, and
- Agricultural wastes

A] Pollution of fresh waters: Ground water, streams, rivers, lakes, ponds, etc. are polluted in different ways.

1. *Pollution by sewage*: *The pollutants of sewages may be solid wastes or liquid wastes*. Household sewage and laundry detergents add sulphates (SO_4), nitrates (NO_3) and chlorides (Cl) to water. This dirty water flowing down the sewages and drains contaminates ground water also before it reaches rivers. Sewage water contains large numbers of coliform bacteria.

In many localities, the supply of drinking water is obtained from wells — either dug wells or tube wells. Such water is unfit for drinking unless purified or boiled. Many water borne diseases like cholera, dysentery, typhoid, jaundice, diarrhoea, amoebiasis, etc. are caused by drinking such polluted water.

Filtration is not enough as only the suspended impurities can be removed by filtration. Purification is a better process, if done effectively. By boiling, germs are killed, but some useful minerals are lost. Moreover, water develops a certain taste by boiling, whereas pure water is always tasteless.

In sewages, some chemicals promote the growth of small non-flowering aquatic plants — blue-green algae. When they grow very fast, the whole water surface is covered by them. This is called *water bloom*. The algae consume a large quantity of oxygen in the water for their respiration. Some blue-green algae also release certain toxic substances into water.

The algae release oxygen during photosynthesis. This oxygen is used up by bacteria to oxidize organic matter in the sewage. This also reduces the oxygen content in the water. Gradually, aerobic bacteria (bacteria which need oxygen for respiration) are totally excluded from the sewage. Only blue-green algae and anaerobic bacteria (bacteria which do not need oxygen for respiration; they use other chemicals for energy) can live in such dirty sewages. These emit a foul smell due to rotting organic matter.

In villages, towns and cities where sewage treatment plants have not been set up, this stinking sewage is released into rivers, streams or other water bodies. The water in these water bodies thus gets polluted. If poisonous substances are also present in the sewage, which does happen frequently, many aquatic plants and animals die.

To fight sewage pollution, septic tanks and specially designed sewage treatment plants must be set up to remove the excess organic matter. If septic tanks are not properly maintained, water contaminated with faeces leak out and gets mixed with ground water. Drinking such water causes many waterborne diseases.

Sewage Treatment Plants: In a typical sewage treatment plant, purification of water is done in 2 stages. First of all, the sewage water coming from various sources collects in a stilling chamber. Here, the velocity of the flow is reduced. A flow meter checks the velocity. From here, the sewage moves on to the screening chamber.

Stilling Chamber →Screening Chamber →Aeration Tank →
Secondary Clarifier →Clear water discharge in to River
↓
Sludge Drying Bed
↓
Manure to Parks and Gardens

The screening chambers are fitted with mechanical bar screens. Here large sized suspended matter such as stones, plastic pieces, sticks, pieces of cloth, etc. are filtered out. This whole process is called *primary treatment*. After filtration, the sewage is taken to large aeration tanks, where huge fans are rotating continuously. This churns and aerates the sewage, making aerobic bacteria grow fast and oxidize organic matter. From the aeration tanks, water moves to a secondary clarifier or settling tank, where the remaining impurities settle down. The clear water from the top part can now be released into nearby rivers, after chlorination if necessary. This series of processes together is called *secondary treatment.*

The sediment waste matter in the secondary clarifier is taken to sludge drying beds, where it is dried and made into manure. Thus a sewage treatment plant serves a dual purpose.

This is the basic principle of Sewage Treatment Plants (STPs). In some plants, there is provision for tertiary treatment also.

2. *Pollution by industries*: While setting up mills or factories, owners see to it that there is a water body like a river or stream nearby. Into almost all rivers in the LDCs, industrial wastes are dumped continuously. For example, this is done at short intervals in the Yamuna River, which is one of the primary sources of water supply in Delhi.

When industrial wastes are emptied into a river, its self-purification mechanism is lost. Common industries polluting river water are petrochemical complexes, fertilizer factories, oil refineries, plants for making drugs, rubber, fibres, plastics, etc. A synthetic drug factory's wastes may contain up to 70 different chemicals and many concentrated acids. These wastes pollute the river water, and also filter through the soil to pollute the ground water.

River water pollution by different industries is very injurious to the aquatic animals and plants as many chemicals are poisonous to them. This indirectly affects human beings also, through a food chain from one organism to another. For example, the effluents from paper mills contain mercury (Hg). When dumped into rivers, Hg enters the body of the fish that man eats. Hg causes nervous disorders in man.

Coke works and plastic units produce many phenolic compounds as wastes. Metal finishing plants release heavy metals and even cyanides. Caustic soda and chlorine factories produce wastes rich in Hg. Many aquatic organisms die due to these toxic industrial pollutants. This upsets the natural balance of the rivers.

Several mollusks (snails) in streams take in copper (Cu), cadmium (Cd), etc. and accumulate them in their bodies. This happens because of *bioaccumulation*. This affects their metabolism (life activities in the cells). Rivers take these toxic substances to the seas or oceans. Cadmium (Cd) and chromium (Cr) are toxic to many marine animals.

Wastes produced during the manufacture of paper, sulphur, cement, sugar, etc. when released into streams or rivers get oxidized into various compounds, utilizing a large quantity of oxygen in the water. Due to this oxygen deficiency, many fish die. This problem is very severe in Kali

Nadi in Meerut. Several wastes are released by Paper mills into River Gomti in Lucknow.

Industrialisation leads to *thermal pollution* of water also. Many industries use water to cool the plants. Water becomes hot during this process. This hot water is usually released into rivers. The *rise in temperature of water leads to thermal pollution.* Power plants which use turbines to generate electricity take water from water bodies. The steam that leaves the turbines is condensed into water. This water, which is at least 15 degrees hotter than normal water, is released back into the water body. The solubility of oxygen in water is reduced when water becomes warmer. Such thermal pollution also disturbs the life of aquatic organisms.

3. *Pollution by nitrates from fertilizers:* Chemical fertilizers are used widely to increase farm yield. These seep into ground water below the farms. It has been estimated that since the 1950s, use of nitrogen fertilizers has been increased 20 times all over the world. It was much later that a study in China showed that plants absorb, on an average, only 40% of this increased nitrogenous fertilizers applied. In the US, between 33% and 50% of the applied nitrates are lost. All the unused nitrates dissolve in rain and irrigation water and filter through the soil to finally reach the underground water.

In Punjab and Haryana, where nitrates are added in large quantities to crop plants, nutrient levels of well water has been found to be 5 to 15 times higher than the safe limit. Extra nitrate in water leads to many problems.

- Certain type of anemia in babies which causes suffocation and death
- Miscarriages (natural abortion) in women
- Cancer of digestive tracts
- Too much of nitrogen is also not good for plants. The plants become susceptible to diseases fast. Their roots are damaged, and their stems and branches become thin.

4. *Pollution by pesticides*: *Chemicals which kill pests are called pesticides*. These chemicals are sprayed in croplands. In India, many farmers use pesticides indiscriminately. Moreover, much higher quantities than needed are sprayed in the fields. Both these practices are harmful to the environment.

Pesticides were first introduced in the 1940s. But it was only after many years that it came to be known that they were highly injurious to non-target organisms including human beings. Ground water pollution by pesticides has become a serious problem now.

Pesticides are used not only in farmlands to kill insect pests attacking plants, but they are also used to kill harmful insects like mosquitoes. So the effect of pesticides is seen in rural and urban areas. They accumulate on the soil surface or surface waters where they are sprayed to control mosquitoes. From here, they seep down and pollute underground water. Sometimes, they remain in the underground water long after usage of the pesticide is stopped. For example, even after 30 years, of ban DDT (earlier used to kill mosquitoes) is still found in the underground water in some parts of US.

DDT is an organochlorine pesticide and has been banned all over the world as it is a non-biodegradable compound. In Bihar and West Bengal, CPCB (Central Pollution Control Board) has found that DDT is present in ground water many thousands of times the permitted level. This has happened after stopping the spraying for a few decades.

Even low concentrations of organochlorine compounds inhibit aquatic photosynthesis in phytoplanktons (microscopic algae on the surface of water). As phytoplnktons are the food of zooplanktons (microscopic animals on the water surface), they decrease in population. The population of small fish which depends on the planktons falls as a result. This affects the population of big fish. In this way, the whole food chain is affected and finally, we suffer due to shortage of fish.

One consumer consumes many producers. So, the amount of pesticides in one consumer is always more than that in the producer. This goes on increasing at all levels of the food chain. This is called *biomagnification.* It is very harmful to the organisms at the top level of food chain. In many food chains, we, human beings, are at the top.

In the recent decades, many new pesticides have been found out which are highly toxic. Drinking water contaminated with organochlorines affect hormonal balance of human beings. It may also cause cancer.

Natural recovery process in rivers: Usually, if the pollutants are not much in quantity, the rivers have a natural recovery process. This is by a combination of dilution and decay by bacteria. Thus rivers which are not polluted much recover fast from degradable, oxygen demanding wastes. This is what is happening at present in the rivers of US and many MDCs.

Control of fresh water pollution: After the water pollution control laws were passed in the 1970s, the number and quality of sewage and industrial waste water treatment plants have increased greatly in the more developed countries. Still, some incidents do occur, like accidental spillage, or rarely, deliberate releases or malfunctioning of treatment plants.

In the 1950s, the famous Thames River in Great Britain was highly polluted. But because of consistent efforts by the government and the industries, the river has recovered now. Commercial fishing is going on and many water birds have reappeared. There is an urgent necessity for the same strategy being worked out in India.

- We have the Ganga Action Plan, Yamuna Action plan, etc. But these plans should be implemented with more force and dedication.
- Action plans should be formulated and implemented for all rivers in India.
- Waste water treatment plants must be set up in all towns and cities.
- New industries should not be given permission to start production unless they have set up waste water treatment plants.
- The general public must be educated on the need to keep water clean by not polluting water bodies, etc.

Estimation of water pollution: The result of water pollution can be estimated by determining the Biochemical Oxygen Demand (BOD) of sample water. (*BOD is the milligrams of oxygen needed to decompose the organic material in 1 litre of waste water.*) Dissolved oxygen (DO) in water is an another indicator of pollution. The optimum value of DO for water of good quality is 4 to 6 mg/l. Lower DO values indicate water pollution.

The distribution of aquatic organisms is influenced by several factors, but dissolved oxygen and organic matter content of the water are the most important. *Habitats rich in organic matter are called saprobic habitats. Organisms which can live in saprobic habitats are called saprophiles.* From the frequency of saprophiles in a particular zone of a river, the pollution level of that zone can be determined. Based on this, different saprobic zones have been recognized. This gives an indication of the river pollution.

1. *Oligosaprobic zone*: Clear unpolluted water near the springs from where the river originates and the upstream parts, i.e. the starting point of the river, is called the oligosaprobic zone.
2. *Polysaprobic zone*: The part of the stream or river which is very much polluted and is often reddish in colour, with an unpleasant smell, is the polysaprobic zone. Due to lack of oxygen, the sulphur

present in the water in this zone is converted to Hydrogen sulphide (H_2S) instead of sulpahtes (SO_4).

3. *Mesosaprobic β-zone*: The zone where water is mildly polluted, and is greenish in colour, is called the mesosaprobic β-zone. Here, all ammonium (NH_4) compounds are converted to nitrates (NO_3).
4. *Mesosaprobic α-zone*: The zone which is downstream, at some distance away from the source of pollution, and shows some recovery is the mesosaprobic α-zone. These waters are poorly oxygenated, yellowish and still contain some ammonium salts and nitrates.

Each of these zones has its own characteristic aquatic organisms — both plants and animals. By identifying these organisms, the level of pollution of different parts of the river can easily be found out. This is because such biological indicators are the most reliable indicators of pollution.

B] Pollution of Marine Water: Many waste materials are deposited in the seas and oceans. These are mainly municipal wastes, oil, minerals, radioactive substances, etc.

- Municipal waste and sewage from small towns are often discharged directly into the sea.
- Lubricating oils used for the machines in various industries are discharged into the sea.
- Accidental spilling of oil from the oil tankers is a matter of grave concern. Oceans and beaches are becoming more and more polluted by oil residues. In 1968 alone, 5 million tons of oil was deposited in the sea. Once, in an island near France, over 100 tons of oil waste was seen on a single day.
- Fertilizers and pesticides from agricultural lands are washed away finally into the sea.
- Ships carrying toxic substances may meet with accidents in the sea.
- When ships are washed and oil tankers emptied, the sea water is polluted.

Many birds are affected by marine oil pollution. The future of birds like divers and sea gulls is threatened by such oil waste discharges in the sea. If this continues, oysters, mussels and some edible fish will ultimately become unfit for human consumption.

Fish stock may also decline through destruction of eggs and fry (baby fish) and interference with food chains, especially a reduction in planktons.

Fishing nets are also damaged by oil.

Control of marine pollution: There are some methods to reduce marine pollution:

1. Setting up of sewage treatment plants at the required sites
2. Spraying chemical dispersants on the oil on the surface
3. Use of slick lickers - materials that absorb the oil slick, i.e. the layer of oil on the water surface etc.

SOIL POLLUTION

Soil is polluted by wastes produced in industries, automobiles and agricultural and domestic practices.

Soil pollution by industries: Combustion of sulphur containing fuels in factories emits sulphur dioxide. This is converted to *sulphates* which are deposited on the soil. *Nitrates* from industries are also deposited on soil. In highly industrialized Poland, people living in areas where the soil was highly polluted by mercury had to be evacuated.

Sulphur dioxide and nitrogen oxides emitted from industries dissolve in water vapour, and are converted to sulphuric and nitric acid, respectively. These fall on the soil along with rain water. This is called *acid rain*. Acid rain makes soils acidic. This is a serious problem in the West. In some parts of India also, there have been reports of acid rain. On acidic soils, many crops cannot grow. Such soils become degraded and need extensive reclamation programmes to make them cultivable again.

Soil pollution by automobiles: *Lead* added to petrol (to prevent knocking) is emitted in vehicle exhaust. It is toxic and settles along soil on both sides of the roads with heavy automobile traffic. It causes *paralysis* and many other disorders. Tissues of people living near roads have a higher lead content than those living in other places. In the Netherlands, Great Britain and some other countries, it has been reported that *cancer* in towns is twice as much as in villages.

Soil pollution due to wrong agricultural practices: Pesticides, herbicides, etc. are retained in the soil for many years. They seep into the soil and become part of the environmental cycle.

a) *Pesticide pollution*: Till a few years ago, DDT (Dichloro Diphenyl Trichloroethane) was used as a pesticide on crops and also a powerful chemical that kills mosquitoes. DDT is a non-biodegradable compound and gets fixed in the soil. It also collects on the surface of plants. It is dispersed in the whole area through water and air currents. DDT and also other synthetic organochlorine compounds like polychlorinated biphenyl have been in use since 1940.

The initial success of organochlorine compounds encouraged the belief that pesticides can solve any pest related problem. The possible ecological consequences were ignored. The wide use of pesticides is very harmful to man and other animals and plants. Some of their adverse effects are as follows.

1. Many harmless and often beneficial species are harmed. This reduces the diversity of species in the habitat. Honey bees and other insects which pollinate different plants have been killed by pesticides.
2. Resistant strains develop which become increasingly difficult to control.
3. Residues accumulate in the soil which may persist for many years.
4. Fish are very susceptible to DDT.
5. Bird population, especially birds of prey has been reduced by pesticide poisoning.
6. In birds, ovulation (formation of ovum or egg) is delayed and gonad (reproductive organs) development is inhibited.
7. Mosquito control programmes using the wrong type of pesticide may be one of the reasons for the reduction of fauna (animals) in the treated areas.
8. Pesticides in the soil are consumed by earthworms. Earthworms, called 'friends of farmers' increase soil fertility by their worm castings. They also make the soil porous. Porous soils contain more air and are better for seed germination and also root growth. Mass killing of earthworms reduces soil fertility.
9. The dead earthworms are eaten by birds; these birds also get poisoned and die.
10. Pesticides destroy many aquatic organisms by killing their larvae.
11. Use of pesticides kills parasites of some other pests.
12. Pesticides from the soil seep into underground water polluting it and making it unfit for drinking for humans and animals.

To demonstrate the harmful effects of pesticides, some incidents are described here.

- In the highlands of Scotland, the proportion of breeding pairs of Golden Eagles was 72% of the total pollution in the 1940s. The percentage fell to 29% by the 1970s.

- The Bald Eagle has become threatened in the US. Fatal doses were found in 25 of the 26 species
- Once, the Clear Lake in California was sprayed with TDE (a compound similar to DDT) at 0.014 ppm (parts per million). Soon, zooplanktons had 5 ppm, small fish feeding on them had 22 to 221 ppm and large fish feeding on small fish had 40 to 2,400 ppm of TDE in their bodies.
- In Spain, pesticides were used to control some pests of oak trees. This programme was successful, but parasites of other pests were also killed. In the absence of their parasites, these pests increased greatly in numbers, and turned out to be a big problem.
- In 1961, a mean value of 925 mg/person of organochlorine residues were recorded in the adipose (fat) tissue. In France, this value was 370. Such concentrations do not kill or harm immediately, but result in chronic (slow, continuous) poisoning, just like poisoning by fluorides and some carcinogenic (cancer causing) hydrocarbons.

b) *Herbicide pollution*: Like pesticides, herbicides have also become major pollutants of soil. They are frequently used in clearing the sideways of roads and rails, to clear new tracks for new roads and rails, for weed control in agriculture and for forest management. After the discovery of natural phytohormones (plant hormones), many organic compounds have been synthesized artificially resembling the natural hormones. These are widely used in agriculture and horticulture (cultivation of fruits and vegetables). Weed control is one of them.

Herbicides have selective action. For example, the synthetic herbicide 2,4-D kills weeds with broad leaves, but do not kill plants with narrow leaves. So it is used in fields of rice, wheat, corn, etc. where the crop plants have narrow leaves. Other herbicides having the opposite effect are also available which kill plants only with narrow leaves and do not kill those with broad leaves. Such herbicides are used to kill grasses (which are narrow-leaved) in fields of vegetables like potato, brinjal, cabbage, cauliflower, tomato, etc. In both these cases, death is due to increased respiration.

Herbicides cause great harm to plants, animals and humans.

- Herbicides like Monuron, Simazin, etc. interfere with photosynthesis.
- 2,4-D and 2,4,5-T block transport of food in plants.
- Herbicides kill plants selectively, thus modifying natural communities.

- When plant communities are changed, herbivores (animals eating plants) also change; and then carnivores (animals eating other animals) also change in a community.
- Herbicides modify the functions of important systems of all organisms.

In short, pesticides and herbicides can be considered to be harmful drugs introduced into nature and natural systems. Both these are sprayed on plants for specific benefits; but they pollute air, water and soil.

(See Units 3 and 6 for conservation, land degradation and soil erosion)

Control of soil pollution: Some methods of controlling soil pollution are:

- Prevention of erosion
- Soil conservation
- Prevention of land degradation
- Right agricultural practices, minimizing use of artificial fertilizers and chemical pesticides

NOISE POLLUTION

Unwanted or undesirable sound is noise. It is harmful to human ears. Noise is increasingly becoming a major pollutant. As noise pollutes the air, noise pollution can in fact be included in air pollution. Just like smoke, dust, toxic fumes, etc. pollute air, noise also pollutes air. This is why noise is included as an air pollutant in Schedules 3 and 7 of EPA (Environmental Protection Act). On February 14, 2000, Noise Pollution (Regulation and Control) Rules were enforced, supplementing EPR. These rules stipulate ambient air quality standards with respect to noise. Industrial, commercial and residential areas have been covered under these rules. Some areas have been fixed as Silent zones also. As per the Rules, there are restrictions on the use of loud speakers and public address systems. It is the duty of District Magistrates and Police Commissioners to enforce the Noise Pollution Rules. They must see that noise levels do not exceed the stipulated norms.

The sound measured at the source is meaningless. There must be a distance at which the hearing process is taking place. This is because the sound becomes lesser and lesser as the source is farther and farther away. Therefore, all measurements must also indicate the distance. Thus there is no single number of measurements — it is always decibels along with distance.

Bel is a unit of sound invented by the engineers of the Bell Telephone Laboratory.

It is the decrease in the level of sound over 1 mile (1.6 kms) length of standard telephone cable. It was initially called transmission unit (TU). But later, to honour Alexander Graham Bell, who had invented the telephone, this unit was renamed as Bel. As the Bel was too large for every day use, decibel (dB) which is equal to one-tenth of a Bel (0.1Bel) became popular.

dB	Source with Distance
194	Theoretical limit for a sound wave at 1 atmosphere environmental pressure; more intense waves function as 'shock waves'
180	Krakatoa (an island near Java) volcanic explosion at 1 mile in air
160	MI Garand (a semi-automatic US rifle) fired at 1 metre
150	Jet engine at 30 metres
140	Low caliber rifle at 1 metre
130	Civil defence siren at 30 metre — threshold of pain
120	Train horn at 1 metre — pierces ear drum
110	Football stadium at kick off at 50 yard line
100	Jack hammer at 2 meters; inside Disco
90	Loud factory, heavy truck at 1 metre
80	Vacuum cleaner at 1 metre
70	Busy traffic at 5 metres
60	Office or restaurant inside
50	Quite restaurant inside
40	Residential area at night
30	Theatre, no talking
20	Whispering
10	Human breathing at 3 metres
0	Threshold of human hearing with healthy ears; Sound of a mosquito flying 3 metres away

Measurements of noise that damage the ear are taken near the ear itself. Measurements of ambient noises do not need a distance, as it is spread uniformly in a small area. In general, *healthy human ears cannot perceive sounds of less than 3 dB.*

Noise pollution is partly a result of technological progress. Aggressive noise is produced by road and rail transport, in factories, business centres, industries, space rockets, aeroplanes, sirens, different machines, etc.

Televisions, radios, telephones, alarm clocks, calling bells and even kitchen appliances sometimes produce sound too loud to our liking. If produced continuously, these irritate us.

The commonest direct effect of noise pollution is temporary deafness. This is medically known as Temporary Threshold Shift (TTS). People affected with this problem cannot hear sound of low volume. It is seen in people living very close to 'pooja pandals'. After a month or so, the symptom disappears. *Noise Induced Permanent Threshold Shift (NIPTS) is much more serious. People affected with this problem lose their hearing capacity for ever.* In addition to these, noise pollution has many negative effects on human health. These are as follows.

1. Loud noise strains the senses and the nervous system.
2. The most evident effect is on the central and sympathetic nervous system.
3. Refreshing, undisturbed sleep becomes impossible.
4. Causes neurosis and head aches.
5. Blood pressure rises.
6. Heartbeat becomes faster.
7. Leads to palpitation.
8. Constricts stomach muscles.
9. Flushes skin.
10. Results in depression.
11. Produces ulcers.
12. Leads to general fatigue.
13. Affects emotional aspects.
14. Reduces intelligence.
15. Sleeplessness leads to tiredness.
16. People become short tempered.
17. Memory is affected.
18. Concentration power is lowered, etc.

In general, intermittent (discontinuous) noise is more harmful than continuous noise.

Control of Noise Pollution and its effects: There are many methods of controlling noise pollution.

1. *Reduce noise at source*: This is the best method. To reduce the noise of *machineries*, they can be covered by sound absorbing material, placed in sound proof enclosures, kept high on absorbent mounts, etc. Machineries must be maintained properly. Noise from *vehicles* can be controlled by maintaining them properly, by proper traffic control, etc.

2. *Block the path of noise*: Noise from *construction equipments* can be blocked by temporary barriers. To block noise from *vehicles*, along the sides of highways, vertical barriers can be built.
3. *Increase the length of the path of noise*: Machineries, construction equipments, etc. can be kept at far distance from residences. Alternative routes are usually given to trucks. Airports are always constructed away from residential areas.
4. *Protect the recipients*: Workers working in high sound situations should wear ear plugs and ear muffs. Good ear muffs can reduce sound up to 40 dB.
5. *Plant trees*: Trees can prevent noise pollution. Therefore, plant trees around houses.
6. *Use absorptive finishing materials*: Use finishing materials that are highly absorptive on ceilings, walls and floors inside houses, offices, etc.

NUCLEAR HAZARDS

Nuclear hazards are a result of radioactive pollution, which affects air, water and soil. During their decay, some elements emit high energy radiations called ionizing radiations. This is because electrons are removed from stable atoms in these elements. *There are 3 types of ionizing radiations of ecological importance*:

1. Alpha (α) particles
2. Beta (β) particles, and
3. Gamma (γ) rays.

α-Particles travel only a few centimeters and when stopped, produce a lot of local ionization. These are formed from the nucleus of the Helium atom and are +vely charged.

β-Particles are high speed electrons and are –vely charged. They penetrate further, traveling a few feet in the air, giving energy over a long path. In living tissues, they travel up to 8 cm.

γ-Rays are non-particulate and are electromagnetic radiations like the X- rays. As they have no electrical charge, they penetrate deep. They have shorter wavelengths and ionize biological materials through which they pass.

The effect of radiation depends on its energy and intensity. The intensity is determined by the number of particles emitted in unit time. Radiation intensity is measured in Curies (Ci). Smaller units used in biology are millicurie (mCi) and microcuries (μCi).

The radiation dose received by an organism is expressed as the amount of energy absorbed by a given mass of an organism. Units used are roentgen (R), rad (radiation absorbed dose) and rem (roentgen equivalent, man).

The maximum dose to which man can be exposed without injury is 0.3 rad/week for repeated exposures and 25 rad for a single dose.

Radiosensitivity varies between one organism and another. The lethal dose is 10^6 rads for bacteria, 10^5 rads for insects and 10^3 rads for mammals. It varies with age. Children under 4 are more sensitive. They accumulate Strontium-90 more rapidly than adults.

Isotopes of elements that emit radiations are called radioactive isotopes or radionuclides. These escape frequently into the atmosphere.

Wastes from nuclear power reactors add such isotopes to water reservoirs, rivers, etc. Nuclear tests emit large quantities of isotopes in to the air.

The most important pollutants are Strontium-90, Iodine-131 and Caesium-137. Non-radioactive forms of all these elements occur in the human body. Iodine (I_2) is found in the thyroid, while Strontium accumulates in the skeleton as its chemical properties are similar to those of Calcium (Ca). Accumulation of radionuclides in human tissues leads to development of cancer. Mutations (changes in the structure of genes) in the reproductive cells are hereditary. This results in the birth of abnormal babies with malformed organs.

This danger is made more serious by the concentration of radionuclides along food chains. This is very clear in the tundras of Lapland, where the food chain is very short. The food chain starts with lichens which grow slowly getting their nutrients from air and water. These lichens are eaten by reindeer. The reindeer are killed and eaten by the Lapps themselves. The Lapps eat up to 2 kg of reindeer meat per day. Thus the Lapps are the most affected by radioactive substances left in the environment when nuclear tests are conducted in the Arctics. Extensive nuclear tests were conducted in the Arctics in 1961 and 1962. The fall out from these tests contained Caesium-137, Polonium-210 and Iron-59. Caesium was the moist abundant. Even if tests are conducted far away, people are affected, at least slightly.

The quantity of radioactive substances within the bodies of different organisms goes on increasing along the food chain by biomagnification. When radioactive Phosphorus (P) was added to a pond, insects concentrated it by 500 times, and the ducks which consumed the insects concentrated it further 75,000 times. The concentration in duck eggs was up to 200,000 times the original concentration.

Radioactive pollution results from industrial wastes also. Disposal of wastes in the sea in water tight barrels is also a dangerous practice, as corrosion by sea water over the years can liberate the wastes before radioactivity has fallen to a safe level. Burying it in the soil is also not a solution. Once, isotope of Ruthenium-106, placed in a waste pit, moved through the soil into oak trees 80 metres away. The trees concentrated the radionuclides in leaves and branches. Levels were higher in the leaf litter (fallen and decomposing leaves).

Thus *waste disposal is a serious problem.* It is because of this problem that full utilization of nuclear (atomic) energy for peaceful purposes has not been possible yet. Moreover, a large quantity of heat is also generated in the nuclear power reactors. *Thermal pollution is thus a result of radioactive pollution.* Thermal pollution increases water temperatures. At higher temperatures, aquatic organisms become more susceptible to toxic materials. Also, less desirable blue-green algae replace the desirable green algae.

Biological effects of ionizing radiations on humans are mainly of three types.

1. 'Acute radiation sickness' results from brief exposures to high doses.
2. Chronic exposure to low doses has many effects like cancer.
3. Production of mutations in sperms and eggs affects future generations.

Bombing of Hiroshima and Nagasaki: A powerful atom bomb containing Uranium-235 was dropped on Hiroshima, Japan in 1945. This was the greatest man made disaster in the history of the world. The Second World War had begun in 1941 and was continuing without any sign of ending. It was then that the American forces dropped the first ever atom bomb used against civilians on the city of Hiroshima on August 6, 1945. 70,000 people died on the spot. Radius of total destruction was 1.6 km. Resulting fires engulfed 11.4 sq km. 90% of buildings in Hiroshima were destroyed. By December 1945, thousands had died from injuries and radiation poisoning.

Just three days after the bombing of Hiroshima, the American forces dropped a second atom bomb in Japan — this time in another city, Nagasaki, on August 9, 1945. The radioactive isotope used was Plutonium-239. 40,000 civilians died on the spot whereas 60,000 were injured. On account of radiation poisoning, thousands died in the following months.

The number of deformed children in Japan after these two bombings also ran into thousands. Even in 2005, 266,000 exposed people were living in Japan.

The world war ended after these two bombings. But what happened in the war was a crime against humanity. Atom bombs or fission bombs (Nuclear Fission is the splitting of the nucleus of the atom) generate enormous amounts of energy in terms of air pressure and heat. In addition to this, is the significant amount of γ-radiations, which kill and maim and mutate genes.

Chernobyl disaster: Forty one years later, on April 25, 1986, a series of explosions took place in a nuclear power plant in Chernobyl, Ukraine, then Soviet Union. As a result of the explosion, the heavy roof of the reactor building was blown off. Radioactive waste materials and radioactive dust containing Caesium-137 flew high into the atmosphere. These spread gradually over the entire earth. The reasons for this horrible disaster were poor design of the reactor and also human error. The after effects were terrible.

- Thirty one people died on the spot from massive radiation exposure. 259 people were hospitalized due to acute radiation sickness.
- After the accident, 65,000 people were employed to build a concrete shell around the destroyed reactor. They were also exposed to dangerous levels of radiation. Thousands who worked on this project died.
- Thousands were affected with cancer. Four mullion people are estimated to be suffering from the effect of constant low levels of radiation. Due to weakening of immune systems, exposed people get various infections.
- All plants in the neighbouring areas were poisoned. Thousands of cows and goats had to be shot because they had eaten radioactive grass. Poisoned cats and dogs were abandoned when people were evacuated. 135,000 people were evacuated a few days after the disaster. However, in 1991, when the authorities realized that this was not enough, 125,000 more were evacuated. The severity of the disaster had not been fully understood.
- Twenty months after Chernobyl, radioactive snow fell in Antarctica.

- Lichens, the main food of reindeer were contaminated. Thousands of reindeer of Lapland had to be killed.
- Even 8 years after the disaster, some lambs could not be sold in 1994 as they were too radioactive.

A nuclear accident may take place anywhere in the world; the harmful effects are seen in very vast areas around the site. This is the main lesson from nuclear hazards. This is also the major obstacle in attempts to tap nuclear energy as an alternate energy source.

SOLID WASTE MANAGEMENT

Waste materials may be solids, liquids or gases. These waste materials pollute our environment. Therefore all types of wastes must be disposed off harmlessly, or better still, utilized for various purposes. This is called waste management. If not managed properly, solid wastes accumulate in the environment and cause several problems. Solid wastes are of different types. They may be domestic, industrial, agricultural or plain natural. The most common examples are:

- Garbage/rubbish such as wastes from kitchens, slaughter houses, broken glass, plastic, crockery, etc.
- Large wastes such as bricks, concrete, sand, furniture, etc. from demolition/construction works
- Sludge/settled components of sewage
- Industrial wastes like chemicals, paint, etc.
- Mining wastes from coal mines
- Agricultural wastes like crop residues, pesticides, farm animal wastes, etc.
- Ash
- Biomedical wastes
- Dead animals and animal excreta
- Fallen leaves, branches, etc.

Generation of solid wastes is an important issue and it has to be given due importance, because:

- The volume of solid wastes generated is very large
- They are major sources of pollutants, and
- They are unhygienic and are a health hazard.

Solid waste management must be practised in all rural and urban areas, because:

- There are many known eco-friendly methods of management.
- Enough land is available for the purpose.

- The cost of disposal is not high, compared to the direct and indirect costs that will have to be borne if management is not done.
- In industries, management of wastes by recycling, recovery, reuse, etc. generates revenue, thereby increasing profit.

Solid wastes are produced in rural as well as urban areas. Some important solid wastes and their place of generation are listed below:

- *Municipal (domestic) solid wastes* are produced in all places, rural and urban.
- *Industrial solid wastes* are produced in various factories, located mostly in urban areas.
- *Biomedical wastes* are generated in hospitals and clinics, seen mostly in urban areas.
- *Agro-wastes* are produced in rural areas where agriculture is done widely.

Methods of management of these solid wastes are highly varied.

Urban and Industrial Wastes

Towns and cities are included in urban areas. Cities are more highly populated than towns. People live in towns and cities crowded together. In many, due to shortage of space, buildings are high rises instead of being single storey houses. Garbage generated from all the residents in any urban locality amounts to hundreds or even thousands of tons per day. To keep the city clean and to maintain the health of the residents, management of these solid wastes is most essential. If there is no proper garbage management systems in urban areas, the garbage heaps become breeding grounds for mosquitoes, flies, rats and other agents which cause diseases.

Waste materials found in urban areas in India are:

- Empty bottles
- Broken crockery
- Plastic containers
- Polythene and other packing materials
- Automobile spares
- Damaged machines
- Cycle parts
- Wastes from building construction and demolition
- Dead animal skeletons
- Rotten fruits and vegetables and their skins, etc.

Solid wastes were causing problems in effluent countries of the West also. Along highways were seen:

- Discarded paper cups and glasses
- Empty cigarette packets
- Paper cartons
- Magazines
- Maps
- Coffee cans
- Shirts
- Tyres
- Bags
- Shoes
- Comic books, etc.

But of late, regular voluntary companies have been set up in the US and European countries for collecting these waste materials and taking them to recycling plants.

Used vehicles were also creating a serious problem in the Western World. In New York City, as early as 1966, 40,000 abandoned cars were counted by the Police!

In India, several million tons of solid wastes are dumped along roadsides, rail tracks and other places in towns and cities. In Class 1 cities, an average of more than 2 million tons of solid wastes is thrown in the environment every year. In Class 2 cities, it is about 0.3 million tons.

Major steps in municipal solid waste management in urban areas:

1. *Collection*: Collection is done mostly by trucks. Much time and labour are spent in loading, transporting and unloading, which increase the cost of collection. To avoid this, some methods have been suggested. These are being implemented in the developed countries. We are also using some of them.

   To reduce the volume of the wastes, garbage grinders are used. In US, Japan, Sweden, etc. the garbage from homes are sucked through underground pneumatic pipelines. Compactors have been set up in many urban commercial establishments. In larger communities, transfer stations are set up. In one city, there are many transfer stations. Trucks from different parts of the city or town bring the refuse to the transfer stations. From here, bulldozers cram the refuse into large vans, which take it to the final disposal site, which are mostly landfills.

$$\text{Town} \xrightarrow{\text{Trucks}} \text{Transfer Station} \xrightarrow{\text{Large Vans}} \text{Landfills}$$

2. *Disposal*: The compressed garbage is used to fill huge gaping pits formed during mining away from the city, or in sanitary landfills. Sanitary landfills are seen near many cities in the world.

A modern sanitary landfill is not just a low land, but a depression in an impermeable soil layer with an impermeable membrane. This is important because in the absence of such layers, pollutants from the lower layers of the wastes leach into the ground below and pollute the ground water. There must also be a proper ventilating system, so that the gases formed escape safely into the atmosphere. However, in some advanced countries, methane, which is the major gas formed in landfills is collected, purified and used for various purposes. In such countries in the west, vast areas are demarcated for this purpose.

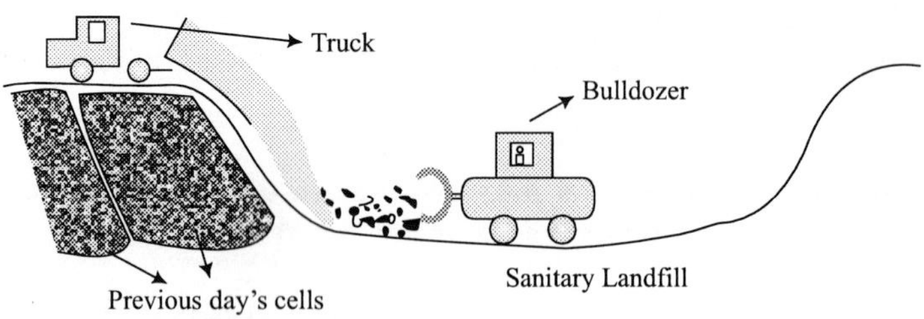

Sanitary Landfill

To reduce the volume of the wastes, before depositing in the landfills, different methods are in use:

a) *Incineration*: *Incineration means burning at a specific temperature in specific conditions*. An incinerator is a suitably designed furnace. The collecting trucks unload the wastes in a pit. From the pit, cranes lift the waste and put it in the combustion chamber. Most of the refuse burns. The gases formed are collected, separated and used for different purposes. The unburnt wastes are compressed and used as manure. Incineration can reduce the volume of solid wastes by 90%. Disadvantages are that some gases formed may be harmful, and that the ash might contain toxic heavy metals. So proper arrangements should be made to dispose off these end products.

b) *Shredding*: Shredding is also called pulverizing. Here, solid wastes are torn to small pieces. Then they are spread in fields.

c) *Pyrolysis*: Combustion (burning) in the absence of oxygen is called pyrolysis. The end products of this process are combustible gas, tar, charcoal, etc. These are used as raw materials in different industries. Pyrolysis reduces volume and also produces stable end products.

3. *Resource Recovery*: Resource recovery is commonly called recycling. Paper, metals, glass and organics are usually recycled. The first step is collection of garbage by hawkers. The hawkers sell garbage to the industries. The industries clean it, sort it out, recycle it and resell it to customers.

   Recycling is beneficial in many ways. It helps in:
   - Conservation of resources
   - Reduction in volume of waste to be disposed
   - Enabling selling of remanufactured items in the market

Resource recovery of certain solid wastes

a) *Paper*: Recycling of paper can partly save forests. Recycled paper must be made popular. Use paper should be torn and not crumpled. Recycled paper made out of crumpled paper is not of good quality as fibres tear during crumpling. Collection, sorting and transportation of waste paper cost a lot; that is why recycled paper is costly as of now. Better technology is sure to be developed in future.

b) *Metals*: Metals can be easily recycled from industrial scrap. This is a main source of metals, next to extraction by mining. Metals are recovered from old and useless automobiles and Aluminum cans. Ship breaking yards recover tons of metal from scrap.

c) *Glass*: Recycling of glass is also done. But cost of making new glass bottles and making them from recycled glass is almost the same. So there is no profit.

d) *Organics*: The commonest recovery process from organic wastes is composting. This is an aerobic process, i.e. it takes place in presence of oxygen. In this process, first of all, metals, glass, plastics, etc. are sorted out. The remaining organic wastes are ground in a grinder. Then they are sent to the 'wind rows' where they get enough ventilation, are given sufficient moisture and are kept at about 140° F. Aerobic bacteria grow in such condition by themselves and convert the wastes into compost manure.

Sludge from Sewage Treatment Plants (STPs): Sewage is the wastes carried by water. It is generated from homes and food processing factories. Domestic sewage contains human excreta, cloth, soap, detergents, etc. Sewage should never be directed into rivers or any other water body. It should be treated in sewage treatment plants. The clear water on the top can be released into water bodies. The sludge which settles at the bottom is dried and made into manure.

Biogas Plants: Biogas plants are very useful in managing solid wastes by converting them into energy and manure. During anaerobic decomposition (decomposition in the absence of oxygen) of organic wastes by bacteria, methane (CH_4) is produced. Some carbon dioxide (CO_2), Hydrogen (H_2) and Nitrogen (N_2) are also released along with methane. All this together is called biogas.

Biogas contains 60 to 70% methane, 25 to 30% carbon dioxide and the rest is Hydrogen and Nitrogen. The composition and the amount of biogas depend on the composition of the wastes and the temperature.

In villages, usually, dung is the main organic matter decomposed in biogas plants. But wastes from kitchens can also be fed into biogas plants. Recently, ARTI (Appropriate Rural Technology Institute), Pune has designed a modified biogas plant in which sugarcane waste produced during the making of sugar cane juice, and such other waste materials are used in biogas plants as raw materials. The biogas produced is carried by pipelines to households. In Pune, this modified version of the original biogas plant has become very popular.

Biogas is an ideal alternate energy source for rural and urban people alike. Use of biogas has many advantages. Some of them are listed below:

- Prevents environmental pollution
- Improves family health by avoiding smoke, dust, soot and coal in houses
- Prevents deforestation, as fire wood is not needed in homes any more
- Improves economic condition

  (See Alternate Energy Sources in Unit 6 for details of Biogas Plants)

Some cities have found out ways to manage municipal solid wastes successfully. Two such success stories are given below in brief.

1st Success Story: Within the city limits of Vijayawada in Andhra Pradesh, 500 tons of solid wastes are generated every day. Of this, the Vijayawada Municipal Corporation (VMC) is disposing off 350 tons of waste every day with the help of two innovative projects. The Corporation has involved the private sector in the process to put into practice the concept of 'Wealth from Waste'.

One project is run by the Excel Industries Ltd. to generate manure from waste. The other is run by Sriram Energy System Pvt. Ltd. to generate power from waste. VMC has given the 2 companies land on lease and the necessary subsidies. It supplies waste to them for a royalty.

2nd success story: This success story is from our neighbouring country, Bangladesh. A survey showed that of the 10 million plastic bags used daily, 9 million were dumped every day in the city of Dhaka alone. Only 10-15% were put in dust bins; the rest finally landed in drains, rivers, ponds and lakes. 80% of the drains in Dhaka got clogged because of plastic bags. The bed of River Buriganga flowing by Dhaka got littered with plastic bags to a height of 5-6 feet. All the water bodies started getting choked with plastic bags.

The government understood the seriousness of the problem and issued a ban on the use of plastics within the city of Dhaka, with effect from January 1, 2002. Later, the ban was made effective throughout the country. The Govt., various NGOs and environmental activists undertook a massive campaign to create awareness. Those breaking the law were punished.

Media also played an important role. They circulated survey reports from the West that workers of plastic factories are more susceptible to cancer, skin diseases and other health problems and that fish and meat wrapped in polythene become poisonous. As people gradually understood the seriousness of the problem, they started heeding the Govt. ban. Due to all these determined moves of the Govt., Bangladesh is now largely free from poly bags, if not fully. Shopping bags made of paper, cloth and jute are catching up fast.

These two success stories show beyond doubt that determination is the key to success.

Solid Waste Management in Industry

In industries, solid waste management is based on the following approaches:

- Source reduction (Waste minimization)
- Segregation (Separation of different types of wastes)
- Recycle
- Recovery
- Reuse
- Environment friendly disposal

Source Reduction: Waste reduction at source has become a 'greening business strategy'. (The term 'Green' has come to mean 'environment friendly'). Since the past few decades, *companies have realized that pollution and waste are signs of inefficiency*. Many companies give incentives to employees who find out methods to prevent pollution and minimize waste in the production process. This saves money.

Waste minimization in industry is based on the following principles:

- Find out and make a list of sources, quantities and types of waste material being generated in the industry
- Collect information on the various operations, the raw materials needed for these operations, the products and the wastes generated in these processes
- Find out what the inefficiencies in the various processes are
- Find out the areas of poor management
- Fix certain targets for minimization of wastes
- Do research and development to find out the best methods of waste management
- Make the workers in the company aware of the benefits of waste minimization

Segregation: Many industries have realized that segregation of waste at source saves time, energy and money. Accordingly, wastes are classified; colour codes fixed for bins/collecting areas and specific departments/ agencies given the full responsibility for the removal of these wastes within a stipulated time period.

For example, the wastes may be:

- Scrap
- Plastics
- Wood
- Used oil
- Rubber material
- Biodegradable wastes like cut grass
- Wastes from canteens
- Electrical wires, etc.

The colour codes may be:

- Any one colour
- A combination of colours
- Any one colour with stripes of any other colour, etc.

The departments/agencies responsible may be:

- Any Engineering department
- Canteen services, etc.

Recycle, Recovery and Reuse: After segregation, wastes are sent to different sites for appropriate recycling, recovery and reuse. These are known as the 3 R's. Industries that practice the 3 R's are more efficient, more cost effective and less polluting. For example:

- Used plastics, paper and some other materials can be recycled
- Metals can be recovered from scrap
- Oil soaked jute and all types of oily wastes can be mixed with coal to be burnt for power generation.
- Used acid from old batteries can be used to neutralize alkaline waste water
- Biodegradable wastes can be made into compost manures for the gardens, etc.

It has been proved beyond doubt that as solid waste utilization goes up in any company, the purchase of raw materials goes down. The companies that do not waste anything are called zero waste companies.

Environment friendly disposal: Different industries use different environment friendly methods of waste disposal/management. In Japan, land fill space is very little. Therefore, Japanese companies have thought of many other waste management plans. Plastics are recycled into carpets; cardboard cartons are reprocessed into paper, etc.

Companies can also use environment friendly waste disposal mechanisms to improve their public image and to reduce the popularity of their competitors who do not use such methods.

Hazardous substances as wastes: These are wastes generated by certain industries like chemical manufacturing companies, paper mills, petroleum refineries, etc. They cause death or very serious health problems to human beings or severely damage the environment permanently. Hazardous wastes can be included in one or more of the following types:

- *Toxic*: Toxic wastes may kill immediately or cause slow poisoning or cause severe illnesses. In Japan, mercury waste from a plastic factory dumped into the nearby Minamata Bay entered into the body of fish. Many people who ate these poisoned fish were killed, many others were seriously affected. From then onward, mercury poisoning is known as Minamata disease. Arsenic is another toxic heavy metal, which is polluting ground water in many parts of the world. In India, Midnapore in West Bengal is a highly affected district. Lead is also a toxic heavy metal. It is deposited on roadside soils as automobile exhausts contain lead.
- *Highly reactive*: Some wastes react strongly with air or water when exposed to these media. They may produce toxic gases on heating. They may explode while handling. For example, Nitroglycerine, gunpowder, etc.

- *Igniting rapidly*: Some waste materials burn by themselves at low temperatures. They may catch fire in storage, during transport or disposal. Examples are gasoline, alcohol, etc.
- *Corrosive*: Acids and bases corrode materials on coming into contact with them. These are examples of corrosive wastes.
- *Infectious*: Many components of biomedical wastes are infectious. Examples are used dressing materials, bandages soaked in blood, injection syringes and needles, human body tissues discarded after operations, wastes from pathological laboratories, etc.
- *Radioactive*: Wastes generated in nuclear power plants are radioactive. These remain in soil, water and air for many years. For example, Uranium ore mined out from the earth is of a low grade; containing only 0.2% of Uranium by weight. It is processed (enriched) so that the Uranium-235 content is increased to 0.7 to 3%. The enriched Uranium is used for nuclear fission. Part of it remains unused. This is the radioactive nuclear waste.

There are many toxic wastes in the articles of daily use, like various solvents and cleaners. Hazardous wastes pollute all parts of the environment, but the most affected is ground water. This is because wastes from the soil also finally reach ground water.

Control of hazardous waste pollution: Hazardous chemicals like Poly Vinyl Chloride (PVC) used in the manufacture of plastics cannot be eliminated. Similarly, many such wastes are bound to be formed during the manufacture of many utility articles. Therefore, this type of pollution can only be minimized and not fully stopped. Methods of minimization are:

- Reduce the quantity of wastes
- Encourage recycling
- Use substitutes
- Educate the public, etc.

E-wastes (Electronic wastes): Electronic wastes are increasing very fast in today's progressive world. Old and useless computers, televisions, mobile phones, etc. are the main e-wastes. In the US, more than 3 million tons of e-wastes are deposited in landfills every year. Some data are given here to create awareness regarding the extent of this problem:

- Between 1997 and 2007, there will be 500 million obsolete computers in the US.
- Total e-waste generated in India at present is 1,46,000 tons per

year. This is expected to go up to 16,00,000 tons by 2012.

- 8,000 tons of computer waste is generated in Bangalore alone every year.

The main constituents of e-wastes and their health effects are given below:

Lead — Damages nervous and circulatory systems and affects kidneys
Cadmium — Accumulates in kidney and liver, causes nerve disorders
Mercury — Damages brain and skin
Barium — Causes muscle weakness; damages heart and liver
Beryllium — Causes skin diseases and cancer
Plastics — Burning releases poisonous Dioxin, which damages reproductive system and hormones

When computer wastes are deposited in landfills, the toxins leach into ground water; if they are burnt, toxic pollutants are emitted into the atmosphere. Therefore, some other methods must be thought of:

- Upgrade old computers as far as possible
- Buy only those brands which can be upgraded
- Give old ones for recycling
- Metals, plastics and other materials can be recovered from e-wastes. Send the e-wastes for land filling only after recovery of reusable substances.

Composting and Vermicomposting

Composting and Vermicomposting are two very good methods of managing domestic biodegradable solid wastes. Both compost and

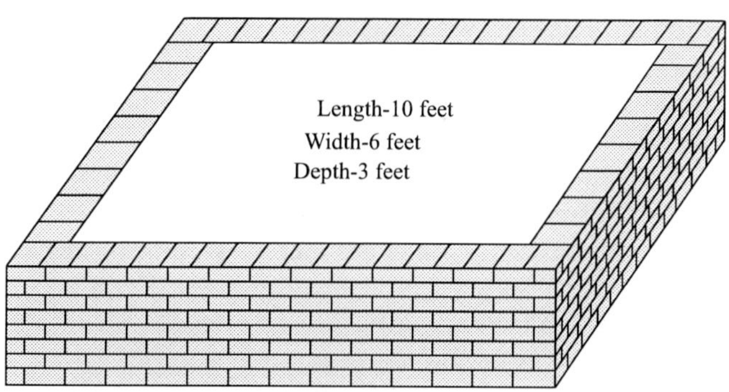

Compost Pit-Maharashtrian model

vermicompost are organic manures. They can be absorbed by plants easily. Moreover, the heat generated during their formation kills many harmful insects, worms and eggs. Seeds of many weeds are also destroyed in this way.

Composting

The most popular method of composting in India is by the Marathi model. This is known as the NADEP method. It was named after the designer, Narayana Rao Pandari Pande, who was addressed by the villagers as Nadep kaka. Steps to make a compost pit of this model are given below (dimensions can be changed as per requirements; but it is advisable to keep the depth as suggested).

1. Select a high level ground, free of water logging.
2. Clear the ground of weeds and stones.
3. Dig a pit of 10′ × 6′ × 3′
4. Spread a layer of bricks at the bottom.

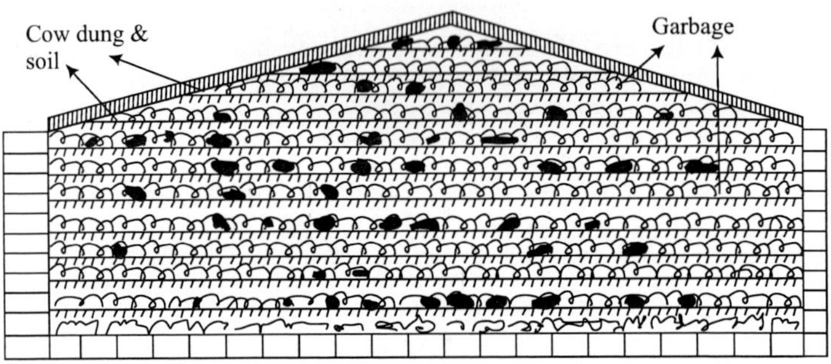

Compost Pit-Vertical Section

5. Bind the sidewalls with bricks.
6. For the side walls, set the bricks without gaps in the lowermost two rows and uppermost two rows. In all the intermediate rows, leave gaps between bricks.
7. Make a low wall of two rows of bricks above ground also, and cement it smoothly.
8. Apply a paste of cow dung in water all over the inside walls and floor.
9. Let the paste dry. This will take 1 or 2 days.
10. The compost pit is now ready to use.

Start putting all types of household garbage in the pit. When it reaches a height of about 6 inches, pour cow dung solution over it and put some good soil on top. Sprinkle water and repeat the process.

In this way, when the waste materials reach a height of about 1.5′ above the ground level, put soil to about 3′ ′ on top and seal with cow dung solution. In 15 to 20 days, the garbage will sink into the pit. Then, repeat the process. This can be done 5 to 6 times.

Compost of excellent quality will be ready in 90 to 120 days.

Vermicomposting

Making compost manure using earthworms is known as vermicomposting.

There are many methods of vermicomposting. It can be done underground or above ground. Sizes of the pits can be varied as per requirement.

A] ***The Underground method*:** Here, the vermicompost pit is underground. The steps in making such a pit are given below:

1. Dig a pit of 2m × 1m × 1m.
2. Put good soil to a height of about 15 to 20 cm in the pit.
3. Water well.
4. Put 100 earthworms in the pit, along with some soil from their habitat.
5. Put some cow dung on the earthworms and then spread some dry leaves or hay to a thickness of about 6 cm.

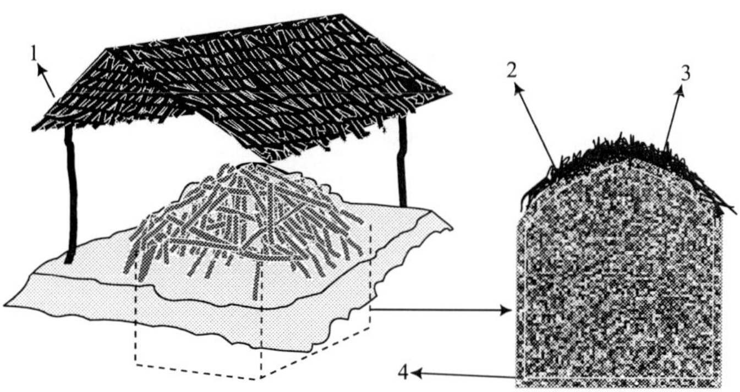

Vermicompost Pit-outside and inside

Vermicompost Pit-outside and inside: 1. Thatched roof 2. Covering of coconut leaves 3. Garbage and cow dung 4. Coconut husks

6. Leave undisturbed for 30 days. Within this time period, the earthworms will multiply. This worm rich bed is called vermibed.
7. On the vermibed, garbage and cowdung in a ratio of 8:1 can be put twice every week.
8. After 2 weeks, mix well, sprinkle water, put cow dung and cover with hay or dry leaves.
9. Continue the process.
10. In 60 to 75 days, the vermicompost will be ready. Take it out without disturbing the worms
11. Dry the vermicompost in the sun and pack in polythene bags.

From the next time onwards, time taken will be only 45 days as the vermibed is already made. Similarly, if more number of earthworms is used in the beginning, at the rate of 1000worms (about 1kg) in a 2m × 1m × 30cm pit, vermicompost will be ready in 45 days.

B] *Over ground method:* Vermicompost pits can also be over ground.

1. Clean the floor of a firm ground, plough it, remove the stones, etc. and press the floor firm.
2. Make low walls all around with bricks and cement.
3. Spread coconut fibres, sprinkle water and put garbage.
4. After 2 days, put 30 earthworms for every sq ft area.
5. Then put wastes and cow dung mixture in a ratio of 8:1.
6. When the wastes reach about 1.5 ft above ground, cover with wet sacs.
7. Moisten the sacs daily.
8. In 2 months, the vermi compost will be ready. Take it out and heap it in the sun in a few places. The worms move down.
9. Take out the manure from top.
10. Redeposit the worms in the pit.

Whatever be the method, some points should be kept in mind:

1. Make a tent over the pit to protect it from sun and rain.
2. Use good species like *Eudrelus euginae, Lampito mauriti, Dravida wills,* etc.
3. Don't pour too much water as it will reduce reproduction.

Making vermicompost in houses: To make vermicompost in small amounts in individual houses, use barrels with both ends open. Tie a plastic sheet with holes in it (for ventilation) at the bottom. Excess water will flow out through these holes but the worms cannot move out.

Benefits of compost and vermicompost:

1. Yield of vegetables, fruits, etc. increases.

2. Quality and taste improve.
3. Soil texture improves.
4. Water holding capacity of soil increases.
5. Soil erosion is prevented.
6. Pollution is minimised.
7. Purchase of costly artificial manures can be minimised.
8. Both composting and vermicomposting can be developed into cottage industries for the economic upliftment of the rural poor.

In today's world of rapid urbanisation and unplanned population growth, accumulation of solid wastes has become a very grave problem in urban areas. This pollutes air, water and soil. It also causes social, aesthetic, ecological and economic problems in society. It causes serious health problems, especially in the slums of big cities. Open garbage dumps are converting more and more land into waste lands. Per capita solid waste generation in India is about 500 gm/day. The composition of this garbage is as follows:

Moisture content — 50 to 55%

Fruit pulp — 35 to 45%

Other organic and inorganic waste — 8 to 15%

Earthworms consume organic substances very rapidly. They have an organ called 'grinding gizzard', which fragment the consumed particles into much finer particles. It is to increase the rate of decomposition that cow dung is added to the solid wastes in vermicompost pits. The most suitable pH to be maintained in the pits differs depending on the nature of the wastes and also the species of the earthworm used. Some species like *Lampito mauritii* increase the pH of the wastes during conversion to vermicompost. The percentage of increase in pH differs with different types of wastes.

Water passed through vermicompost is called vermiwash. This solution also has high nutritive value. *To prepare vermiwash easily at home, the following steps can be taken.*

1. Take a flower pot about 12″ in height.
2. Place pieces of stone or bricks to about 4″ in height.
3. Place a piece of muslin cloth or filter paper on the stones.
4. Above this, deposit a mixture of degradable wastes and cow dung in an 8:1 ratio along with earthworms.
5. Keeping only one hole at the bottom open, close the others with cement.
6. Fix a rubber tubing in the open hole. When not in use, this tube can be folded back.

7. Every day, pour three table spoons of a solution of cow dung in the pot, on top of the garbage. Also, pour 100 ml of water in the pot daily.
8. Collect the vermiwash through the rubber tubing.
9. Pour the vermiwash directly at the base of plants.
10. Vermiwash can also be sprayed on leaves, after diluting it four times.
11. Vermiwash will make leaves greener and plants healthier.
12. By pouring vermi wash in vermicompost pits, formation of vermicompost can be speeded up.

In short, composting and vermicomposting are two excellent methods of solid waste management that can be followed in rural as well as urban areas. *Technology for large scale use of earthworms for converting all types of organic wastes into vermicompost and the final application of the same in croplands* is being developed. This technology *has been named as vermitechnology.*

ROLE OF INDIVIDUALS IN PREVENTION OF POLLUTION

Pollution is a global problem. It has reached dangerously high levels. Therefore, for controlling pollution, we just cannot and should not depend fully on government. Since a few decades, in addition to governmental departments and agencies, NGOs are also doing a lot of work in this direction. Yet the role of individuals in preventing pollution can never be underestimated.

Individuals should take more interest in pollution control activities due to many reasons.

1. First of all, it is ultimately the individuals who have to endure the harmful effects of pollution.
2. By actively participating in pollution control activities, individuals are setting an ideal precedent.
3. The benefits of clean and unpolluted environment will be enjoyed by individuals first.
4. If one or more individuals start any work on controlling pollution, more and more will emulate them and ultimately this will give rise to a Movement.
5. 'Think Globally, Act Locally' has become a famous slogan in all environment related issues. To act locally, individuals and collections of individuals are best suited.

6. Controlling pollution is a huge task. It cannot succeed unless everybody puts in his/her effort.
7. Even if the Government and the NGOs are working whole heartedly towards pollution control, a lot of their efforts will go waste if the individuals in the society are not aware about the seriousness of the work.
8. No individual should think that he/she has only rights and no duties. The society is giving the individuals a lot; it is the duty of the individuals to reciprocate. They must help in keeping the neighbourhood clean and pollution free.

A society, a state, a country, the whole world is a collection of individuals, in the final analysis. Therefore, if each and every individual does his/her duty to perfection, the whole world can theoretically become pollution free. It is the duty of all of us to spread awareness among the people in this regard.

Following is a list of what individuals can do to control different types of pollution and to conserve natural resources as depletion of these resources is one of the most harmful effect of pollution.

1. Plant as many trees as possible in as many places as possible. Saplings can be planted not only in own gardens, but also on many public places like roadsides, near railway lines, canal banks, farm boundaries, around villages, school and college yards, public buildings, places of worship, cemeteries, market squares, etc.
   In many of these cases, permission from authorities may be needed. But this should not be considered as an obstacle. Individuals can always approach the authorities. It will not be difficult to convince the authorities as all know that trees prevent air pollution. What is needed is initiative and determination.
   There are many institutions and agencies which supply saplings at nominal rates or even free of cost. They can be approached for saplings. Saplings can be planted during the Van Mahotsav or on any day in the rainy season. Care must be taken to look after them properly after planting.
2. Paper is made from trees. So wasting paper is a crime. Write on both sides of paper.
3. Use recycled paper if possible. This is made by recycling used paper. Popularise the use of recycled paper among your friends. It can be used not only for writing, but also for making greeting cards, flags, curios, etc.

4. Do not crumple paper before throwing it; tear it into pieces. This is because crumpling breaks fibres and recycled paper made out of this will be of inferior quality.
5. Try to use envelopes more than once by carefully tearing one side and pasting a new label on the address written.
6. Do not use paper napkins. Instead, use cloth napkins.
7. Do not use plastic glasses, cups, plates, etc.
8. It is meaningless to say 'Don't cut trees' or 'Don't use teak for furniture'. Trees will have to be cut for different purposes; teak will be used for doors, windows and furniture as it is hard, durable and resistant to white ants. Therefore the aim is to make use of natural resources, but at the same time to regenerate them. Only in the case of non-renewable resources, we need to be extra cautious.
9. For example, petrol is a non-renewable resource. Burning of petrol also pollutes the air. It releases many harmful gases into the atmosphere. Therefore, use petrol sparingly. If the distance is not much, use a bicycle in place of a scooter, motor cycle, or car. If the distance is walkable, just walk. (In any case, walking is good for health!)
10. Car pools are ideal when people have to go to the same place or nearby places at the same time. Taking out individual cars on such occasions is waste of petrol.
11. Get your vehicles checked for pollution control regularly. If there is any defect, get it repaired immediately.
12. Service your vehicles periodically. This will help to keep them in good condition.
13. Do not keep cars switched on while getting out for even a short time. This state, called idling, releases more harmful fumes into the air.
14. During acceleration also, more harmful fumes are emitted. Therefore, do not increase and reduce the speed of your vehicle frequently while driving. Try to go at the same speed if possible, accelerating only when necessary.
15. Do not honk the horn unnecessarily. This is a very bad habit with Indians in general. There is no meaning honking the horn while stuck in a traffic jam; anyway, everybody wants to move! Just imagine, if every vehicle stuck in traffic jam blows horn, how much noise pollution it will cause! So, be patient; the jam will

clear anyway. It is not going to clear any sooner if you honk your horn!

16. Know the sound levels permitted in different places so that you can act accordingly. Hospitals are silent zones; so are libraries. Do not disturb the peace of these places by shouting or talking loudly.
17. If you know the rules regarding noise pollution, may be you can politely tell organisers of festivals about them.
18. Use pesticides in your homes only when necessary. Try biopesticides instead of chemical pesticides as much as possible. Keep neem leaves between the folds of your woollen garments this winter, in place of naphthalene balls. Even after winter, while storing your woollens away for next year, just use neem leaves. No insects will damage your woollens; they will remain as fresh as ever and will also be free of dust. In your garden also, you can use neem seed oil or oil cake to protect your plants against pests. Popularise these concepts among your friends.
19. Enquire whether organic farming is being done for the vegetables and fruits you buy. If many people make such enquiries, word will spread and finally reach the growers.
20. Do not throw garbage outside your house.
21. Practise segregation at source, i.e. keep different dust bins for biodegradable and non-biodegradable waste materials. In the former, put kitchen wastes like skins of fruits and vegetables, tea leaves, waste paper; etc. In the latter, put plastics, Aluminium foils, broken glass, crockery, etc.
22. If you live in single houses with some gardening space, have compost pits dug for preparing compost manure from kitchen wastes, used paper, cloth, torn jute bags, etc. This way, you will be managing the biodegradable solid wastes also.
23. Vermicomposting can also be done, in suitably sized barrels, in houses.
24. If you are living in flat complexes or in houses with no garden space, convince the landlords/neighbours to have community compost pits in a common area.
25. By managing the degradable solid wastes, you are doing a great service to the whole locality — you are safeguarding yourself, family members and neighbours against sanitation related diseases. You are also keeping the neighbourhood clean.

26. Throw only the non-biodegradable wastes in the municipality dumpers.
27. Do not use plastic bags as far as possible. Take bags of cloth or jute for shopping.
28. Do not throw toffee wrappers, etc. on the roads or footpaths or college grounds if dustbins are not seen nearby. Put them in your bags or pockets and bring them home to be put in dustbins.
29. Do not waste water. Do not keep the tap on while brushing, shaving, etc.
30. If any tap leaks in your house, get it repaired immediately.
31. If you see a leaking tap in your neighbourhood, report it to the authority.
32. Do not throw garbage into the sewer. It will block the flow of the sewage and create problems.
33. Do not throw garbage into any water body. It will harm the aquatic plants and animals.
34. Learn how to do rainwater harvesting and try to get one system in your house. If you are living in a flat complex, persuade the landlord to get one installed. He will most probably agree, collecting a certain amount from all the residents.
35. Do not waste electricity. Switch the lights and fans off when you are not there.
36. If the electric metre in your house is not working, report the same to the authorities.
37. Use solar energy as much as possible. Solar water heaters, solar calculators, etc. are available in the market.
38. While buying new computers, choose equipments that can meet your long term needs or be easily upgraded.
39. During new purchase, trade in your old computer to be recycled.
40. If you are buying a better model, donate your old computer to a school or any other charity organisation, if possible.
41. Try to contact any recycling company nearby to handle your recyclable equipments.
42. For items like glass, aluminium, etc. also try to contact dealers, if possible.
43. Do not buy new batteries every time; use batteries that can be recharged.
44. Do not let land go waste. Even if you have a small plot behind your house, have a flower garden or kitchen garden there. In

addition to enhancing beauty and utility, plants prevent soil erosion.

45. If you see soil being washed off and underlying pebbles being shown above in public places, inform authorities.
46. As far as possible, maintain lawns instead of concreting courtyards. This helps water to seep into the ground.
47. Spread environmental awareness by organising various activities in your locality.
48. Write articles on pollution control and publish in local newspapers.
49. Appreciate the beauty of plants; birds and butterflies — that will make you a nature lover and you will never feel like disturbing nature or any organism in nature any longer.
50. Understand that though we are superior to all other animals in intelligence, still, we are part of nature. By polluting nature, we are harming ourselves along with the other parts of nature.

ENVIRONMENTAL DISASTERS AND THEIR MANAGEMENT

Environmental hazards or disasters are extreme events, natural or man-made, which exceed tolerable magnitudes. The results include loss of lives and property. They become big news at local, national and even international level. Disasters occur suddenly and indiscriminately.

In the world, many environmentally extreme events take place regularly. But we consider them as disasters only if they affect human beings in some way. For example, a very severe cyclone may develop in the middle of an ocean and die off before it reaches the shore and cause damage to human beings. This is not considered a disaster. On the other hand, if a cyclone originating in an ocean reaches a coastal area where people live and cause loss of lives of human beings and animals and destroys plants and affects buildings and other property, it is called a disaster. Similarly, a volcanic eruption in an island where nobody lives is an environmentally extreme event, but not an environmental disaster. Only if a volcano erupts in a well populated area, it is called a disaster.

Thus all environmental disasters are extreme events, but all environmentally extreme events are not disasters.

The intensity of any disaster is proportional to the seriousness of the damage done to

- Lives of human beings, animals and plants, and
- Property

Environmental disasters are broadly of two types — Natural and Man made.

Examples of *natural disasters* are:

- Volcanoes
- Earthquakes
- Landslides
- Cyclones
- Droughts
- Floods, etc.

Examples of *man-made disasters* are:

- Earthquakes
- Landslides
- Nuclear explosion, etc.

Floods and droughts are also sometimes man-made. For example, when dam gates are opened, (many times, this happens without any prior warning to the affected people), flash floods occur in the downstream areas. Floods are also caused by human activities like sand mining from the river beds, siltation of the river banks due to deforestation, etc. Droughts can also be man made, as a result of deforestation or overgrazing. But generally speaking, floods and droughts are natural disasters.

Floods

Flood can be defined as submergence or inundation of vast areas of land under water for many days in continuation.

Floods are generally associated with rivers. In India, the rivers which get flooded most frequently are Ganga and its tributaries like Gomti, Gandak, Ghagra, Kosi, etc. and Brahmaputra, Mahanadi, Krishna, Narmada, Godavari, etc. Rivers Missouri and Missisippi in the USA, Yangtze in China, Irrawaddy in Myanmar, Indus in Pakistan, etc. are also regularly affected by floods.

Floods are the result of accumulation of large amounts of water in rivers which overflow the river banks. These are in fact natural phenomena. But when they cause damage to life and property, they become environmental disasters.

To understand the most suitable disaster management methods to adopt, it is essential to know the causes of any disaster.

Causes of floods: There are many causes of floods. These causes act in various combinations and thus become very complex. It is because of the complex nature of their causes that floods are difficult to manage and control. (See Floods in unit 6 for details)

A] *Natural causes*:

- Heavy Rainfall
- Winding course of rivers
- Occurrence of vast floodplains
- Sudden change in slopes
- Landslides

B] *Anthropogenic causes*:

- Deforestation
- Urbanization
- Changing the course of rivers
- Construction of dams
- Some agricultural practices
- Man-induced landslides

In India, most of the flood prone areas are in the northern parts. The Gangetic plains of UP, Bihar and West Bengal get flooded very frequently. Floods are frequent in Andhra Pradesh also. These states account for more than 60% of the damages caused by floods in our country.

The frequency and severity of floods have increased in India over the years. The losses caused by floods to lives and property and the expenses incurred in trying to control floods have become enormous. Therefore, it is high time for us to take steps in preventing, controlling and managing floods.

Control of floods: There are various methods to control floods. By implementing these, occurrence of floods can be minimised. So these can be considered as flood management measures.

1. *Slowing down runoff to rivers*: After a heavy rainfall, most of the water flows off along the slope. This is called runoff. By reducing the speed of runoff, more water can be directed underground. For this purpose, reforestation and afforestation should be done on a large scale on the hills from where the rivers originate. By this method, soil erosion is reduced. As a result, siltation of rivers is slowed down.
2. *Reducing the number of twists and turns of rivers*: By doing this, water flows straight and fast to the ocean, thereby reducing floods. But this is an expensive process. It has been successfully done in the US.
3. *Constructing diversions in the floodplains*: In this method, the flood water is diverted to low lying areas or artificially constructed

channels.

4. *Constructing flood control reservoirs*: These are large reservoirs to store water. They control floods and also help in irrigation. In India, we have many such reservoirs and dams on many rivers like Damodar, Tapti, etc.
5. *Constructing flood walls*: Artificial walls of earth or stones or concrete are built on the sides of the river, where necessary, according to the volume of water and its force. This helps to prevent water from flooding the plains.
6. *Forecasting and warning*: Central Flood Control Board was set up in India in 1954. Later, State Flood Control Boards were also set up. These boards have implemented many flood control measures in our country. In 1959, flood forecasting and warning systems were started in Delhi. Soon, many such systems were set up in different parts of the country. These Centres continuously monitor the river basins of the Ganga, Brahmaputra and their tributaries, Damodar, Subarnarekha, Tapti, Narmada, Godavari, etc. The Centres collect data on:
    - Rainfall
    - Rate of discharge
    - Flood level, etc.

These data are passed on to the local authorities who warn the local people about the chances of floods well in advance. Thus, people get opportunity to move to safer places with their belongings before the flood arrives. Loss to life is prevented fully and loss of property is greatly reduced by this method.

Earthquakes

An earthquake is a motion of the ground surface caused by wave motion starting from a focal point. It may be a mild tremor or a severe motion, leading to collapse of buildings and formation of deep fissures in the ground. An earthquake is the result of a form of energy transmitted through the surface layer of the earth.

The intensity of the energy released during earthquakes is measured on the Richter scale. This scale was devised by Charles F. Richter in 1935. The scale is made between 0 and 9. A very small tremor, which can be detected only by a seismograph, is taken to be of 0 (zero) magnitude. Depending on the severity of the earthquake, they are described as feeble, moderate, strong, very strong, destructive, disastrous, etc. Disastrous earthquakes measure 7 and above on the Richter scale. Above 8, quakes are classified as catastrophic.

The place of origin of the quake is called the Focus. It is also called Hypocentre as it is always deep, even a few hundreds of kilometres deep, below the surface of the earth. *The place on the earth's surface, right above the hypocentre, is the epicentre*. The seismic waves of energy move in widening circles away from the hypocentre. *The instrument used for recording these waves is called the seismometer.*

Most earthquakes are of small magnitudes. Those of magnitudes 5 or above cause serious damages, while catastrophic quakes, measuring 8 or more on the Richter scale, lead to complete destruction.

The degree of destruction caused by earthquakes depends on many factors:

- Magnitude
- Distance from the epicentre
- Type of ground
- Duration
- Materials used in building construction, etc.

Causes of earthquakes: Any disturbance of the equilibrium in any part of the earth's crust results in an earthquake. Many factors may cause such disequilibrium:

- Volcanic eruptions
- Fractures and faults in the earth's crust
- Pressure of man made dams, reservoirs and lakes
- Movements of plates of earth, etc.

As per the most popular theory, the earth's crust is made of 6 major plates and 20 minor plates. These are constantly in motion. These movements are the result of convective currents within the earth.

For example, the Asiatic plate is gradually moving southwards. At the same time, the Indian plate is slowly moving northward. The resulting collision is the cause of the slow rise of the Himalayas. This rise is taking place at the rate of 50 millimetres per year. When there are sudden variations in this slow movement, earthquakes occur in North India.

Disastrous effects of earthquakes: An earthquake that strikes an area where many people live has many disastrous effects. In the 1934 earthquake in Bihar, the epicentre was located near Darbhanga. A total of 10,700 people belonging to Bihar, Nepal and Tibet lost their lives. Landslides blocked roads for 250 km. Ground was cracked in many places. The magnitude of this earthquake was 8.4. The earthquake at Latur,

Maharashtra in 1993 measured 6.3 on the Richter scale. Losses were great as most of the houses were made of mud.

The disastrous effects of earthquakes are:

1. *Landslides due to slopes being unstable*: When there is an earthquake, from unstable slopes, masses of land fall down. This takes place more in the rainy seasons as soil is soft during the monsoon. Such landslides damage houses and block roads and rails. It may take days to clear the debris.
2. *Damage to man-made structures*: Buildings, roads, railway lines, bridges, industries, bridges, etc. are all damaged to different extents. If the grounds on which buildings are built are not firm, buildings collapse faster. For example, buildings on drained marsh. Mud huts collapse more while buildings of cement and concrete last longer.
3. *Loss of human life and property*: Thousands die in earthquakes. The worst disaster was in China in 1976, measuring 7.8 to 8.1 on the Richter scale. It caused the death of 750,000 people. Whenever there is loss of human life, there is a corresponding loss of property also.
4. *Fire inside houses, mines, etc.*: When earthquakes shake buildings, gas cylinders upturn and burst, leading to fires. Fires are also caused by quakes in factories, mines, etc.
5. *Damages to towns and cities*: When large buildings shake and collapse, many people are buried under the debris. Uprooted trees and broken branches block traffic. This results in the rescue teams reaching the site late. Underground water pipes are bent and spoilt. Sewers are destroyed and blocked at many places. Clean water and dirty water mix and cause epidemics. Electricity supply and telephone lines are cut off. Communication systems are totally disrupted.
6. *Damage to ground surface*: Strong quakes damage the surface of the ground, creating fissures and deep clefts. Masses of ground are displaced.
7. *Flash floods*: Severe earthquakes result in breach of dams, which lead to floods.
8. *Tsunamis*: (In Japanese, *Tsu* means harbour and *nami* means wave) Seismic waves caused by earthquakes under the sea travel through water. This gives rise to very high sea waves, which lash at the coasts. The untold misery and destruction caused by the

tsunamis that lashed the eastern coast of India in 2004 are well known.

The importance of acting on warnings: The quake that occurred on January 26, 2001, in Bhuj, Gujarat, was a killer earthquake, measuring 8.1 on the Richter scale. 90 % of houses in Bhuj were destroyed completely. Almost 100,000 people died. A moderate quake had occurred just a few kilometres away from Bhuj a month ago in December. But no one had paid any heed to this. Damage would have been much less if timely evacuations had been done. This was a failure on the part of the Disaster Management System.

Cyclones

Tropical cyclones are deadly atmospheric storms, blowing in a spiral manner. They are closed low pressure systems with an average diameter of 650 kms. Cyclones are very destructive. They are known as *Hurricanes* in the Atlantic Ocean and as *Typhoons* in the Pacific Ocean. Cyclones are different from local storms which also cause great harm. Tornadoes of USA, thunder storms with heavy rains, dust storms of deserts, etc. are all different types of storms, but are not cyclones.

Characteristics of cyclones:

- A cyclone is a wind blowing spirally inward towards a centre of low barometric pressure
- Wind speeds of tropical cyclones range from 180 to 400 km per hour.
- Continuous very heavy rainfall accompanies cyclones.
- Atmospheric pressure dips very low, leading to a rise in sea levels. Vast areas near the sea are flooded.
- Great damage is done to life and property.

Tropical cyclones trouble the eastern coasts of India and the southern coasts of Bangladesh regularly. The most disastrous such cyclone killed 300,000 people (1000, 000 as per some sources, including missing persons) in Bangladesh and West Bengal put together, in 1970. Most of the deaths were by drowning.

The importance of setting up monitoring, predicting and forewarning systems:

The cyclone which struck the Andhra Coast in 1977 caused sea waves to rize 6 metres high and killed 55,0000 persons. The speed of wind in this cyclone was 175 km/hr. Homes of 2,000,000 people were destroyed. 1,200,000 hectares of croplands became useless due to thick salt deposition after the sea waves receded. It took 3 years to desalinate (remove salt from) this land.

A more powerful cyclone struck coastal Andhra again in 1990. But by then, advance cyclone monitoring system had been set up. From the time it originated in the Bay of Bengal, the movement of the cyclone was closely monitored every minute and predictions and warnings given from time to time. Therefore, though 300,000 people became homeless and 90,000 cattle died, number of human deaths was just 598.

The importance of reaching the warning to all concerned:

In 1999, Orissa had to bear the colossal damages caused by a super cyclone. The cyclone ravaged coastal Orissa for 3 days. It started as deep depression in the Andaman Sea, evolved into a cyclonic storm, became very severe and finally became the super cyclonic storm, moving at a velocity of 260 km/hr.

Indian Meteorological Department (IMD) issued an alarm, but the Orissa Government could not reach the information to the general public in time due to lack of radio network. Finally, the super cyclone took people unawares. The cyclone at 300 km/hr speed remained within 10 coastal districts for 8 hours. Tidal waves 9 metres high travelled up to 20 km towards land.

Damages brought about by the super cyclone of Orissa:

- Though official estimates are 10,000 deaths, unofficial estimates are about 100,000.
- Many hundred thousand cattle lost their lives.
- Croplands spread in 1.75 million hectares were destroyed.
- Innumerable homes were flattened.
- Water and electricity supplies were totally disrupted.
- Communication systems stopped working.
- Road and rail transport stopped completely for many weeks.
- Paradeep port became non-functional for weeks.
- Dead bodies accumulated in several places could not be burnt due to lack of dry wood.
- The decomposing dead bodies of people and animals raised such a stink that relief operations had to be done with difficulty.
- There was no clean water to drink.
- Epidemics broke out.
- Most of the trees were uprooted.
- Total loss of property amounted to Rupees 10,000 crores.
- Suffering from intense mental agony, many people went into

depression.

India has more than one hundred disaster warning systems. There is need to set up more. More important than setting up systems is to maintain them in working condition at all times and to pass on information well in advance to the local people.

Landslides

Tumbling down of huge landmasses is called a landslide. They may be natural or man-made. Landslides may be a natural phenomenon caused by heavy rain and the resulting softening of the soil, earthquakes, etc. They can be anthropogenic (man-made) also. For example, mining on hills makes the ground shaky, leading to landslides. Deforestation on slopes results in severe soil erosion; in the absence of trees and other plants which fix soil in its place by their roots, huge masses of land tumble down.

If landmasses from hill slopes fall into rivers, the flow of water is blocked. This may result in flooding. If the landmasses fall from slopes on to passing vehicles on roads, loss of life and damage to the vehicles result. Roads also get blocked and till the time as the blocks are removed, all transport come to a standstill.

Causes of landslides: There are many causes of landslides.

1. Sometimes, *earthquakes* occur on hilly or mountainous localities. If these hills or mountains are made of comparatively soft and weak rocks, the slopes are unstable and they fall down in a mass, causing landslides. In 1970, 25,000 people were killed in a massive landslide caused by an earthquake in Peru. More landslides occur in the rainy season.
2. Deep underground *mining* also leads to landslides.
3. *Blasting of rocks* for the construction of roads, dams and reservoirs.
4. *Dam building* can also lead to landslides. In major dams, enormous quantities of water are stored. This water causes the building up of great hydrostatic pressure underground. This may result in landslide and flow of mud along the reservoir walls.

Disaster Management

Natural disasters will happen; there is no way to prevent them. But *it is possible to reduce the level of risk by developing efficient early warning systems.* By doing this, the local people will be prepared for the approaching disaster. Plans can be finalised and implemented accordingly. The losses resulting from the disaster can be reduced and relief measures carried out properly. *It will also be a good idea to bring those affected by disasters under insurance cover.* There must be provision to insure not only human life, but also the lives of animals and also houses, household goods, crops, etc.

It has been estimated that about 60% of the area of the Indian subcontinent is vulnerable to seismic activity of different degrees. We have a long coastline of 5,700 km and cyclones occur in large coastal areas. The maximum damage is done by floods in different rivers, which occur with a regular periodicity.

There are mainly four aspects of management of environmental disasters:

1. *To provide immediate relief measures to the people affected by the disaster*: The victims of disasters need immediate help. This must be reached to them as urgently as possible. There may be more deaths and more trauma affected people if help does not reach the victims in time.

If it is an earthquake, there will be enormous quantities of debris all over the place. *Many people may be trapped under the debris.* To rescue them is an urgent need. For this, various equipments will be needed like gas cutters, etc. Skilled people must be present at the sites. Those who are taken out of the debris will be injured and will also be in a state of fear. They need physical as well as psychological treatment. They have to be carried to the nearest hospitals for such medical help.

There will be many others who are injured. They have to be reached to the hospitals too. Many of them will continue to live if help is provided at the right time. It is due to delays on such occasions that the number of deaths shoots up. People who know first aid are of great help on such occasions.

After a disaster, the scenes will be horrible everywhere. *There is need to cremate the dead* too. There is the risk of people who are almost dead being mistaken for dead. So doctors will be required to certify death. The dead must be cremated without much delay. The more the delay, the more are the chances of spread of epidemics. Many animals will also be dead. Disposing of their dead bodies must also be taken on a war footing.

All the survivors will need drinking water, food, clothing and shelter. It may become really difficult to reach potable water to the survivors. Special tankers carrying drinking water will have to be reached to the sites. As for food, the general practice is to drop food packets from helicopters in the affected areas. This is the only practicable method in many cases as roads and rails will be blocked by uprooted trees or debris from fallen buildings, or totally damaged by cracks, deep fissures, etc.

Regarding clothing, it usually takes time. The news of the disaster spreads far and wide in no time in these days of fast communication methods. This happens immediately after the disaster. But as communication links from the site of the disaster are usually completely damaged, news broadcast directly from the site becomes next to impossible in many cases. In such cases, satellite communication may become the most important method of communication.

The role of media in disasters is very important. *It is the duty of responsible media to give correct coverage*. Under coverage is as bad as over coverage or incorrect coverage. If the coverage is not proper, sufficient help will not reach the site. But more often, problems are created by over coverage and incorrect coverage. The worst effect of over coverage is that too many agencies, organisations, other states and even other countries send relief materials to the affected site. This creates a huge problem of storing them safely and distributing them properly. More often than not, large quantities of relief materials are wasted.

There have been cases of *unscrupulous elements taking advantage of the situation* and taking away the relief materials meant for the victims. *The presence of police and army helps to a great extent* to tackle this menace. Unthinking acts of charity cause problems too. Many times, uncooked rice and dal reach the sites in tons, sent by neighbouring states and foreign countries. But there is no way the victims can cook them in the absence of dry fuel, stoves, vessels, etc. Hence *the system of air dropping prepared food packets is the best*, till relief camps with cooking arrangements are set up. Once the camps are set up, raw materials can be reached to the site of disaster. This can continue till the affected people start normal lives, which may take weeks or months.

Just after the super cyclone of Orissa in 1999, clothes had piled up in all the affected districts. For the women everybody had sent saris. Nobody had sent blouses or petticoats. This had become a real practical problem. Such unthinking acts of charity should be avoided.

In some cases, *as soon as there is news of a disaster anywhere, doctors and nurses arrive* from different parts of the country and the whole globe in very large numbers. They also bring with them enormous quantities of medicines. Many of these medicines may not be the ones required. They also cause storing and handling difficulties. The agencies and countries that send the relief act out of their assumptions and not on the real needs. This adds problems to the already existing ones. Once it had so happened that after a disastrous earthquake in Mexico City, Mexico in 1985, an aeroplane full of doctors and nurses landed in the city. Ton loads of medicines also arrived. It all resulted in great confusion as all the doctors and nurses needed places to live; moreover, they did not know the local language and so interpreters had to be arranged for them. Thus well meaning measures had resulted in adding to the woes of the local populace.

To avoid such situations, *it is always best for others to wait for appeals of help from the concerned governments*. However, the government must do its duty efficiently, i.e. gauge the effects of the disaster correctly, and determine the type and quantum of help needed and pass on the information to the organisations, states, and countries willing to help in cash, kind or both.

In this context again, *the media has a most sensible and responsible role to play*. This is because the *social response to disasters depends on the media reports*. If the reporter's assumption of the situation is wrong, people may get the wrong message. The reporters will never deliberately misguide the public; yet their assumptions may not give an accurate picture of the magnitude of the disaster and the types and quantum of help needed. In India, a lot of coverage is given to visuals, showing the scenes of occurrence and the affected people's miseries. This does arouse the emotional part of the human mind, leading to an increase in the flow of relief materials. At the same time, more important is an authentic list of what is needed and where.

Fearing of breakout of epidemics, tons of medicines may arrive, when what are needed urgently may be special tools for rescuing trapped persons, communication equipments, heavy machines to remove debris, water pumps, etc. along with people skilled in such jobs. Epidemics will not break out at all if blocked roads are cleared, people are taken to hospitals and dispensaries according to the seriousness of their injuries, extra water is pumped out, etc. and hygienic conditions are restored fast.

Social responses to disasters are increasing as days go by. Help pours in from all quarters. There are many voluntary organisations, including various NGOs doing commendable work immediately after any disaster. The help provided by the Red Cross can never be over emphasised.

2. *To predict the disaster beforehand and forewarn the people in the area well in advance of the time and intensity of the disaster*: All said and done, preventing disasters is always more desirable than managing them after their occurrence. It is not always possible to predict all types of natural disasters, but there are excellent methods of predicting some like cyclones, floods, etc. available to us. The simplest method is to study past records carefully. From such studies, the periodicity and magnitude of disasters

can be predicted to fairly good accuracy.

Sometimes, minor events may precede major events. For example, in the case of earthquakes, minor tremors may be felt a few days before a major tremor. Soil may tumble down in small amounts and slowly before a major landslide. Heavy incessant rainfall is a forewarning to impending floods. A fall in the atmospheric pressure over a particular ocean indicates the building up of a cyclone. What is important in all these cases is analysing the minor phenomena correctly and passing on the information promptly without addition or deletion to the concerned authorities and the local people.

For predicting disasters, monitoring stations have been set up in different countries of the world. Round the clock weather monitoring is most essential in this regard. This is done very efficiently and effectively in the US. Though we have many weather forecasting centres in India, none are as well equipped or as effective as those in the US. The forecasting is so accurate in the developed countries that every day, even kindergarten students see the internet on their own to decide whether they should carry their rain coats to school on that day or not! The author has seen this happening in the US, and knows officials in Indian ships who find global weather forecast by the US more reliable than the forecasts by others. We Indians are in no way less intelligent than our counterparts anywhere in the world; what we need is more hard work, dedication and devotion to the cause.

Applications of Remote Sensing and GIS: *Remote sensing is the science of identification and classification of the surface features of the earth using electromagnetic radiation.* Every object reflects or scatters some electromagnetic radiations. The pattern of this radiation is different for different objects. These are understood and classified by the sensors of satellites left by different countries in the space. Indian Remote Sensing Satellite (IRS-1A and 1B) were launched successfully in 1988 and 1991, respectively. More such satellites have been launched by India later.

With the help of images from the remote sensing satellites and aerial photos, vegetation, water, geology, etc. can all be mapped. The scales of these maps can be from entire continents to details of a few square metres. Satellites can detect the events which occur before many disasters like floods, cyclones, etc. By remote sensing, the event can be monitored while it is occurring in full force. Thus, it is because the satellites are positioned in the space that we are able to understand the disaster in detail and plan management methods.

After the disaster also, remote sensing is of great help. It gives a clear picture of the damages and the area affected. From this, the quantum of relief needed can be assessed. After that, the new situation can be mapped and the data stored. Based on this data, reconstruction can be done. Finally, the whole procedure can help preventing repetition of such disasters in future.

Geographic Information System (GIS) is a very good tool in predicting and monitoring environmental disasters. GIS is a system for creating, storing, analyzing and managing geographical information. It can be used to analyze data from the space (spatial data) and non-spatial data also. The processed data can be used at different stages of disaster management. With the help of GIS, plans can be made about the emergency measures to be taken during natural disasters. Maps can be made of areas which are prone to earthquakes.

In any country, *information on environmental disasters must be regularly included in planning any development project and making plans for various types of investment.* Projects on development and investment should include a cost/benefit analysis, i.e. the cost of installing management measures should be compared with the losses that would occur if these measures were not introduced. Remote sensing and GIS can be used at many levels:

A] *National level*: At the national level, the government as well as the public can get useful information on many aspects of disaster management with the help of GIS. This will guide governments in formulating policies and create awareness among the public. This is done by two methods:

a) By making a map of all disaster prone zones of the country, and
b) By providing a list of all previous disasters

This information can be used for many purposes:

- Developing disaster free regions
- Avoiding most of the development in zones of high disaster
- Taking measures to minimise losses in disaster prone zones which have been already developed
- Investigating reasons for occurrence of disasters in different parts of the country, etc.

B] *Regional level*: GIS can be made use of at the regional level by urban and rural planners before giving final shape to various developmental activities. For example, engineers can decide the location of dams, the height of buildings to be constructed, etc. Information on the firmness of the ground, nature of rivers, weather patterns of the area and the like will be of great help in this.

C] *Medium level*: At inter-district or inter-municipal level, extending over a few hundred square kilometres, studies on how far the planned project is suitable to the geography of the place and the chance of various environmental disasters are most essential. These should be completed before starting the project. GIS analysis of intensity of possible disasters in the area is of great use in such situations.

D] *Local level*: A single district or municipality is the target area at the local level. Here, GIS is used for deciding the suitability of various developmental works. However, more than this, it is used to prepare local risk maps for existing settlements and cities and to plan how to be prepared for disasters and how to carry out relief measures. The data will include results of field tests of materials and field measurements. Mapping will be accurate to one hectare or less. This is of great help at a local level.

Thus remote sensing and GIS have become very valuable tools in all stages of disaster management. Before the disaster, there are two phases:

1. Disaster prevention
2. Disaster preparedness

After the disaster, work has to be done in three phases:

1. Relief
2. Rehabilitation
3. Reconstruction

In both the phases, i.e. before and after the disaster, and also during the occurrence of the disaster, as explained earlier, remote sensing and GIS are very useful:

3. *To educate the people on how to manage disasters*: Disasters occur on a very large scale. Naturally, *most people will not know how to tackle them*. Many of them will be so shocked by the destruction all around that they will be at a complete loss as to what is to be done to relieve their own as well as others' sufferings. In such a situation, there is urgent need of help from trained persons. *Specialists in disaster management should be rushed to the site of the event.*

However, the real process of educating the people starts much before the disaster strikes. *People must be always prepared for any disaster, especially those living in risk zones.* In addition to the topic taught in schools and colleges, this can be achieved by creating awareness among the general public. For this, the media must be employed, like newspapers, television, radio, documentaries, etc. People must be educated on the need to avoid risk zones for developmental activities, etc.

In general, it is the duty of the government to warn the people about the approaching disasters. Therefore, scientists and research workers must inform the concerned government departments all their findings promptly. So, the government can take steps in the following directions:

a) Making decision makers and the public aware of the gravity of the disasters
b) Training the decision makers in handling the situation
c) Informing public about the disaster well in advance
d) Providing the concerned people with the required maps
e) Explaining to the people how to reduce damages, like constructing quake proof houses, etc.

4. *To do research on improving the management methods available at present and find out better methods of management*: Many organizations in the world, like The International Council of Scientific Unions (ICSU), are doing research programmes in disaster management. This organization studies man induced environmental changes, the origin of natural disasters, etc. It has founded many sub-committees and established research programmes. Some of these are:

- The Scientific Committee On Problems of the Environment (SCOPE)
- The Global Change Programme (GCP)
- The International Geosphere-Biosphere Programme (IGBP), etc.

Social scientists also have started different research programmes like The Human Dimension of Global Change (HDGC). The UN has also launched such research programmes with the aim of improving early warning systems, relief and rehabilitation in different countries.

Research is done in many fields in disaster management. Some of these studies are:

- The factors which cause the disasters
- The mechanism of action
- Classifications based on various parameters
- Potential of the disasters
- Mapping of land and water
- Population density
- Population distribution
- Systems of transport
- Methods of communication

- Facilities available in clean water and food
- Break out of epidemics
- Best methods of administration
- Role of police and army
- Rehabilitation methods
- Medical care including treatment of trauma, etc.

In conclusion, *disaster management is a multidisciplinary area*. It involves forecast, warning, evacuation, rescue, medical care, relief, reconstruction, rehabilitation, etc. *It also involves persons from various fields* like administrators, scientists, decision makers, volunteers, doctors, nurses, engineers, financiers, etc. in addition to NGOs, communities, charitable organizations and the like. When so many agents are working together, there is every possibility of confusion and repetition of work. Therefore, co-ordination among all the agencies is a must for any successful disaster management/mitigation programme. Disaster mitigation is promoted all over the world these days. Disaster Mitigation means lessening the negative effects of the natural disasters.

We have governmental organizations like the Indian Space Research Organization (ISRO), the National Building Research Organization (NBRO), etc. which are doing commendable research and development. We also have the Indian Meteorological Department (IMD), The National Disaster Management Centre (NDMC), etc. as governmental institutional machineries engaged in useful work.

Yet, in general, in India, we have to learn a lot from our past experiences. It will not help if different agencies compete with each other or conduct parallel assessments, after the disaster. The efforts have to be cohesive for best results.

[Note: Disaster management discussed in this book is common for all disasters. Therefore, while preparing long answers individually, after describing each disaster, the same Disaster Management practices are to be written, leaving out one or two sentences specific to other disasters, if present. Time available to write the answer should be kept in mind.]

QUESTIONS

A] Very short answer type questions:

a) Multiple choice type: Tick the right answer

1. Radiation is a health hazard because it leads to:

   (a) Typhoid (b) Cancer

   (c) Colour blindness (d) Pneumonia

2. Which of the following gas causes air pollution?
   (a) Nitrogen (b) Hydrogen
   (c) Water vapour (d) Carbon monoxide
3. Kitchen wastes contain:
   (a) Non-biodegradable pollutants
   (b) Biodegradable pollutants
   (c) Radioactive pollutants
   (d) None of the above
4. Water pollution is caused by:
   (a) Sewage (b) Industrial effluents
   (c) Discharges from farms (d) All of the above
5. DDT pollutes:
   (a) Air (b) Water
   (c) Soil (d) All of the above
6. Which of the following is not a biodegradable pollutant?
   (a) Plastic
   (b) Skins of vegetables and fruits
   (c) Dry leaves
   (d) Paper
7. Minamata disease is caused by:
   (a) Cadmium (b) Strontium
   (c) Mercury (d) Uranium
8. BOD stands for:
   (a) Biochemical Oxygen Demand
   (b) Biological Oxygen Demand
   (c) Biogeochemical Oxygen Demand
   (d) None of the above
9. Greenhouse effect is caused by excess of
   (a) CO_2 (b) H_2
   (c) He (d) O_2
10. Photochemical smog is caused by:
   (a) PAN + O_2 (b) PAN + O_3
   (c) $O_3 + O_2$ (d) PAN + CO_2
11. Which of the following is not a greenhouse gas?
   (a) CO_2 (b) CH_4
   (c) CFC (d) H_2
12. In plants, PAN blocks:
   (a) Reactions of respiration
   (b) Reactions of hormone formation

(c) Absorption of water from soil
(d) Hill reactions of photosynthesis

13. Ionizing radiations of ecological reactions are:
    (a) α-rays (b) β-rays
    (c) γ-rays (d) All of the above
14. The proportion of methane in biogas is:
    (a) 20 to 30 % (b) 10 to 30%
    (c) 60 to 70% (d) 80 to 90%
15. Which is not a natural disaster?
    (a) Cyclone (b) Nuclear explosion
    (c) Earthquake (d) Volcano
16. Landslides are caused by:
    (a) Earthquakes (b) Dam building
    (c) Mining (d) All of the above

b) Define the following:

1. Pollution
2. Pollutant
3. BOD
4. Saprobic habitats
5. Saprophiles
6. Radioactive isotopes
7. Incineration
8. Vrmicomposting
9. Vermitechnology
10. Vermiwash
11. Earthquake
12. Seismometer
13. Flood
14. Environmental disasters
15. Tropical cyclones
16. Landslide
17. Remote sensing
18. GIS

c) Name the following:

1. Two major components of air
2. Five examples of pollutants
3. Five examples of non-biodegradable pollutants
4. The three main types of pollution based on the part of environment being polluted
5. Five types of pollution based on the type of pollutant
6. Three natural pollutants
7. Three man-made pollutants
8. The five primary pollutants in the air
9. Three harmful gases emitted from vehicles
10. Five greenhouse gases
11. Three waterborne diseases
12. Five industries polluting rivers
13. Three types of ionizing radiations of ecological importance

14. Five solid wastes
15. Three hazardous wastes
16. Two disasters which are natural and also man-made
17. Three most flood prone rivers in India
18. Three man-made causes of floods
19. Two natural causes of floods
20. Two factors which determine the degree of destruction of earthquakes
21. Two causes of earthquakes
22. Five benefits of vermicompost

d) Write very short notes on the following, in 3 or 4 sentences:

1. Biodegradable and non-biodegradable pollutants
2. Three ways in which cost of pollution can be measured
3. Methods of educating the public regarding the ill effects of pollution
4. Primary pollutants in air
5. Greenhouse effect and global warming
6. Effects of carbon monoxide on human health
7. Methods of control of air pollution by industries
8. Photochemical smog
9. Methods of control of air pollution by automobiles
10. Water bloom
11. Bioaccumulation
12. Biomagnification
13. Thermal pollution
14. Natural process of recovery in rivers
15. Soil pollution by industries and automobiles
16. Importance of acting on earthquake warnings
17. Concentration of radionuclides along food chain
18. Advantages of recycling
19. Biogas
20. Source reduction and segregation of waste
21. Causes of landslides
22. Recycle, recovery and reuse
23. Minamata disease
24. Hazardous wastes
25. Richter scale
26. Hypocentre and epicentre

B] Short answer type questions:

Write short notes on:

1. Harmful effects of environmental pollution
2. Methods that can be adapted to reduce existing pollution
3. Sewage pollution
4. Smog and photochemical smog
5. Sewage treatment plants
6. Nitrate pollution
7. Methods to control water pollution
8. Estimation of water pollution
9. Herbicide pollution
10. Harmful effects of noise pollution
11. Control of noise pollution
12. Bombing of Hiroshima and Nagasaki
13. Chernobyl disaster
14. Resource recovery of solid wastes
15. Biogas and its advantages
16. E-wastes
17. Flood control measures
18. Disastrous effects of earthquakes
19. Characteristics of cyclones
20. Applications of remote sensing
21. Fields of research in Disaster Management
22. Educating people on how to manage disaster

C] Long answer type questions:

a) Give an account of industrial and automobile pollution
b) Describe in detail air pollution by smoke and smog
c) Give a detailed account of water pollution, its harmful effects and control measures
d) Which are the methods by which soil pollution take place? How can soil pollution be minimised?
e) Describe in detail noise pollution, its effects and control measures.
f) Write an essay on nuclear hazards.
g) What do you understand by solid waste management? Elaborate.
h) Write long notes on any two:
    1. Marine pollution
    2. Soil pollution due to wrong agricultural practices
    3. Harmful effects of pesticides
i) Give an account of the major steps in municipal solid waste

management in cities.

j) What is vermicomposting? Explain in detail.

k) Give a detailed account of role of individuals in prevention of pollution

l) Write long notes on any two:
   1. Solid waste management in industry
   2. Hazardous wastes
   3. Earthquakes and their effects

m) Write an essay on floods, their control and management

n) Give an account of tropical cyclones.

o) Describe in detail the four aspects of management of environmental disasters.

p) Write long notes on any two:
   1. Providing immediate relief measures after disasters
   2. Prediction and forewarning of disasters
   3. GIS and its usefulness at various levels

Unit 6

Natural Resources

Ecology of natural resources is called Resource Ecology. A resource is a source of supply. *Natural resources are substances, different forms of energy, etc. that living organisms can take from nature for various purposes.* Forests, soil, grasslands, water bodies, minerals, sunlight, different gases in the atmosphere, etc. are all natural resources. They may be living or non-living. Man depends on all the natural resources for various purposes. In fact, there is no existence for him in the absence of these resources.

These natural resources on which we depend are provided by four 'spheres':

- The atmosphere – *The atmosphere is a protective covering around the earth, and is a reservoir of several elements essential for life. It is a mixture of different gases in different proportions.* Of all the gases, oxygen is the most important, as all living organisms depend on it for their survival. Carbon dioxide is needed by green plants to synthesize food. Without plants, there can be no life on earth. Water vapour is needed to form clouds. Ozone absorbs the harmful ultraviolet rays. Nitrogen and some other gases are also useful to us in various ways.
- The hydrosphere – *All the water on the earth together is called the hydrosphere. It includes the rivers, lakes, ponds, etc., the seas and oceans and the glaciers and the water stored underground in aquifers.* Only very little of the available water is fresh water. Water is essential for us and is used for drinking, cooking, bathing, washing, cultivation, industries, transportation, recreation and many other purposes. Plants and animals also cannot survive without water.
- The lithosphere – *The soil, the rocks and other parts of the earth's surface together is called the lithosphere.* We need soil for

cultivating plants. We also need minerals for various purposes. We get coal and petroleum from the lithosphere. Stones and sand, used for building construction, are obtained from the lithosphere. In this way, we also depend on the lithosphere for our survival.

- The biosphere – *The place on earth where life exists is the biosphere or the ecosphere. The biosphere includes all the living organisms on the earth and the different parts of the environment on which life depends. It is therefore the whole earth itself.* This is the largest and the most self-sufficient biological system that we know. In the biosphere, living and non-living components are constantly acting and reacting with each other. Therefore, any disturbance in one component of the system will affect the others.

TYPES OF NATURAL RESOURCES

Natural resources are of two types:

1. Inexhaustible resources
2. Exhaustible resources

1. *Inexhaustible resources are those which are in unlimited supply in nature.* Activities of human beings will not affect these resources, because they are present in nature in very large quantities, much in excess than required for the existence of human beings and also all other living organisms. Some of these are necessary for man, but some others may not be necessary directly for human life, though they can be made use of.

Important inexhaustible natural resources are:

- Solar energy
- Rain
- Wind
- Tidal power, etc.

2. *Exhaustible resources are in limited supply in nature and can therefore be exhausted if not used carefully. These are of two types – renewable and non-renewable.*

Renewable and Non-renewable Resources

Renewable resources are resources that can be restored and replaced. Examples are:

- Wildlife
- Forests
- Water

- Soil fertility
- Aquatic animals and plants, etc.

These can be renewed by us or they have the capacity to renew themselves by replacement, reproduction, or recycling. All this takes time, yet they can be renewed. However, even these may be lost, at least temporarily, if management is not proper.

Non-renewable resources are those which, once used, are exhausted for ever, and cannot be restored or replaced. Examples are:

- Fossil fuels like petroleum, coal, etc.
- Minerals
- Metals
- Salts
- Species of plants, animals and microorganisms.

There is an end to the reserves of coal, petroleum, metals, salts, etc. in nature. Similarly, the gene combination of a particular species can never be duplicated once it becomes extinct.

NATURAL RESOURCES AND ASSOCIATED PROBLEMS

The world is facing many problems associated with both of these two types of natural resources. Recklessness in the utilization of these resources without replacement is the main reason for the problems associated with them.

The important natural resources besides air in the atmosphere that are useful to man are listed below:

(A) Forests
(B) Water
(C) Minerals
(D) Food
(E) Energy, and
(F) Land

All these natural resources are being over exploited by human beings. Forgetting the fact that we are part of nature, and are using a lion's share of the natural resources. We are also not taking care to return to nature what we have taken out of it. This is creating many problems like:

- Imbalances in nature due to pollution
- Extinction of valuable species
- Deep division between the More Developed Countries (MDC's) and the Less Developed Countries (LDC's) because of:

a) More population growth in the LDCs
b) More poverty, hunger, diseases, illiteracy, etc. in the LDCs
c) More per capita income in the MDCs
d) More non-vegetarianism in the MDCs, leading to more land use for rearing animals
e) More energy being used by the MDCs
f) More wastage of food, electricity, etc. in the MDCs
g) More pollution due to industrialization in the MDCs, etc.

- Energy crisis affecting industries as well as day-to-day life
- Water scarcity, leading to social tensions
- Scarcity of available land for agriculture, resulting in food scarcity
- Depletion of minerals for industry
- Deforestation leading to desertification, soil erosion, floods, draughts, loss of biodiversity
- Global climatic changes
- Increase in man made environmental disasters
- Social conflicts and wars, etc.

In short, it is high time for all of us to pay more attention to the problems created by overexploitation of natural resources, discuss the problems associated with it and find out and implement the remedial measures on a case by case basis.

A) Forest Resources

Forests are the great national wealth. A dense forest is full of trees. The forest floor is densely populated by various shrubs (bushes), herbs (small plants), ferns and mosses. Many climbers and epiphytes (plants growing on the branches of trees, without absorbing any water or food from them) depend on the trees. The atmosphere in such a forest is cool and humid, and wind velocity is very low. Forests are the great tourist's attraction in any country.

Uses of Forest Resources

The uses or functions of forest resources can be classified into 3 categories– protective uses, regulatory uses and productive uses.

Protective uses:

- Trees provide excellent protection to the soil against soil erosion. The deep and spreading roots of trees and other plants in the forest hold the soil particles strongly and prevent them from being washed away in rain.

- Forest trees build up a thick layer of humus to prevent floods, droughts and soil erosion.
- Roots of trees and other plants serve as a natural dam, and hold rain water.
- Forests prevent early siltation of dams and lakes.

In the absence of trees, soil at the edges fall continuously into the water bodies making them shallow, reducing their water storage capacity. This is especially bad for dams because when a dam is built, its life is usually predicted. Siltation shortens the life span of dams. Siltation also increases floods.

Regulatory uses:

- Forests regulate various biogeochemical cycles, especially that of carbon. This is because forest trees absorb huge quantities of carbon dioxide from the air during photosynthesis.
- During photosynthesis, trees release huge quantities of life giving oxygen, which is one of the end products of the process.
- Trees help purify the atmosphere, thereby reducing pollution drastically.
- Acting as sponges, forests absorb and hold water which charges springs, streams and ground water. Thus they regulate the flow of water from mountain highlands to croplands and urban areas.
- Forests also influence local, regional and global climate.
- Trees absorb large quantities of water from the soil and release it into the atmosphere during the process of transpiration. About 50 to 80% of moisture in the air above tropical forests comes from trees in this manner. This moisture soon becomes rain clouds. Therefore, if large areas of forests are cleared, rainfall will be affected.
- Wild animals and plants are most essential to balance the nature itself. In their absence, ecosystem will be misbalanced.
- Forests buffer us against noise, absorb air pollutants and raise human spirits.

Productive uses:

Worldwide, about half the timber cut each year is used as fuel for heating and cooking, especially in the poor countries. Some is burnt directly as firewood, while the rest is converted to charcoal, which is used in urban areas in poor countries and also by many industries. One third of the world's annual timber is made into building materials. Forests are great providers. They provide us with:

- Timber for various construction purposes and furniture
- Firewood, one of the most popularly used fuels by the rural poor
- Wood for paper pulp
- Alkaloids
- Essential oils
- Gums
- Resins
- Perfumes
- Tannin
- Charcoal
- Dyes
- Wood alcohol
- Plywood
- Turpentine
- Drugs
- Honey
- Lac and many other things.

Many forest lands are also used for:

- Mining
- Grazing animals
- Recreation, etc.

The wild animals and plants are used by plant breeders, genetic engineers and biotechnologists as a reservoir of wild type of genes. This is also a productive function.

Forest Resources of India

As per the Survey Report 1999 of Forest Survey of India, an assessment of the forest cover of the country was presented. For preparing this report, satellite remote sensing technology was used. *13 states of India together comprise 63% of our forest cover. These states are:*

1. Andhra Pradesh
2. Arunachal Pradesh
3. Assam
4. Delhi
5. Himachal Pradesh
6. Madhya Pradesh
7. Maharashtra
8. Manipur
9. Meghalaya

10. Mizoram
11. Nagaland
12. Sikkim
13. Tripura

Forests cover of India is 678,333 sq km which constitutes 20.63% of its geopraphical area. The three types of forests which together make up this percentage are:

Type of forest	% area covered
Dense forest	11.48%
Open forest	7.76%
Mangrove forest	0.15%
Total	19.39%

Among the states, Madhya Pradesh tops the list with 24.79% of its area under forest cover. The 7 north-eastern states together cover 25.70% of India's forest cover. Of the mangrove forests, 85% are in West Bengal, Gujarat and Andaman Nicobar islands.

Over Exploitation of Forests

Most of the overexploitation of forests is taking place in the tropical countries. Tropical forests are of different types and cover about 6% of Earth's land area. All types of forests are affected, but the worst situation is in the rainforests. Forests are overexploited in different manners. The most important methods of overexploitation are deforestation, mining and dam building. It is often thought that tribal people destroy the forests in which they live. However, this is not true. The quantum of damage done by the forest dwellers is not much. They depend on the forests and their produce for livelihood. Why should they do anything to diminish their own sources of income? The real causes lie elsewhere.

Deforestation, Mining and Dams and their Effects on Forests and the Tribal People

The three main reasons for the dwindling forest resources in the world are deforestation, mining and dam building. When the area under forest cover is decreased due to any reason, it is the overall prosperity of the country that suffers. And the people most affected are the tribal people who depend on the forests for their livelihood. The tribal people live a simple life. They get most of their food in the form of leaves, fruits, roots, rhizomes, etc. and the small animals from the forests. They drink water from the streams of the forests. They collect firewood for cooking and also for selling in the nearby villages. They also sell flowers, fruits, leaves, twigs for brushing teeth, honey, lac, gum, fibres and other such forest produce to earn livelihood. They make huts using bamboos and other building materials available in the forests. They use the medicinal plants available in the forests for different types of ailments. In this way, *the life of the forest dwellers is dependent entirely on the forests.* Therefore, the direct effect of deforestation by any method – whether it be lumbering or mining or dam building – is felt by the tribal people the most, though everybody else is affected too in various ways.

Deforestation

Worldwide, most of the destruction happened since 1950. But as far as India is concerned, the greatest loss was about 150 years ago, when trees were cut on a huge scale by the British for making railway sleepers for England and other European countries. Forests degraded into open woodlands, savannas, or deserts. However, even after the British left, the process of deforestation continued. Rising population, developmental activities, industrialization and pollution have decreased the forest cover in India drastically. It is most unfortunate that this is happening in a country which gave to the world the earliest reference to forest conservation in the Mahabharata, the longest epic of the world. Even after the Mahabharata, more than 2,500 years ago, Lord Budha of India had pointed out strongly the necessity of forest conservation to the people of the country. Over the years, all the words of wisdom of the past have been forgotten.

Deforestation is occurring at a dangerously high rate in many parts of the world, like South-East Asia and South and Central America. Philippines have lost 97% of its original forest cover, while Madagascar has lost 84%. Of the total tropical deforestation, about 40% is taking place in the vast Amazon basin of South America alone. All these are important hotspots of the world. Thus there has been great loss to precious biodiversity due to deforestation alone. Madagascar is the world's 4th largest island which is located in the Indian Ocean 250 miles off the eastern coast of Africa. Most of its species are endemic. Half of the island's birds and all its reptiles and mammals are endemic. These are the most endangered species in the world.

Tropical forests contain 50 to 90% of the Earth's total species. So deforestation of these is the most serious problem. Half of the supply of hardwood, coffee, tea, cocoa, spices, nuts and tropical fruits are obtained from these forests. 25% of ingredients of medicines are obtained from medicinal plants growing in tropical forests. Most varieties of rice, wheat and corn used in the world have been developed from wild tropical plants. Thousands of plants in tropical forests have not been studied yet for their food or other values.

Causes of deforestation: Deforestation is the result of many interconnected factors. It may be due to natural causes like forest fires, diseases, climatic changes, etc. or due to human activities listed below:

- *Population growth and Poverty*: The landless poor keep on shifting to forests, where they try to do cultivation just to survive.
- *Road construction*: The process of degradation of a forest in fact starts with construction of a road passing through it. This forest road is usually made by the company/contractor for transporting timber. Once there is approach to the forest, many people enter it and start clearing the forest.
- *Unsustainable small scale farming*: People living at the margins of the forest move into the interior, build huts and plant crops in small cleared areas.
- *Shifting cultivation*: When soil fertility of areas once cultivated as farms decreases, these areas are abandoned. New areas are cleared and made into farms. This is called shifting cultivation. Valuable forest patches get degraded by this practice.
- *Cattle grazing*: Villagers living around the forests leave their cattle to graze in the forests. The surface soil becomes hard under the hooves of animals and water percolation decreases. Seeds cannot germinate on such hard and dry soils. Reproduction is hampered as the tops of plants are eaten too. Forest land thus gradually turns into barren land with thorny bushes.
- *Commercial logging*: Wood is cut from forests for various purposes by industries. It is made into doors and windows, furniture, packing cases, match boxes, paper, etc. The problem is not that forest trees are cut, but the cutting is at a much faster rate than the natural process of regeneration. Replacements are not done by the industries either. Moreover, when one tree falls, many others around it are injured. Many of them die. It has been estimated that many timber exporting countries in South-East Asia will not have any timer to export at all in a few years.

- *Forest fires*: Forest fires may be natural or man made. In nature, on very hot and dry days, forests may catch fire. Man made fires are also common. Large forest areas are burnt down for cultivation of cash crops like tea, coffee, sugarcane, bananas, pine apples, different types of spices, etc.
- *Faulty Governmental policies*: Without realizing the full ecological value of trees, Governments put subsidies on timber and make it comparatively cheap in the third world countries. This has speeded up deforestation.
- *Mining and dam building*: These factors also cause immense damage to the forests.

Forest Conservation: Forest Management includes afforestation, reforestation and forest conservation.

- *Afforestation* is planting trees in areas where there were no trees earlier.
- *Reforestation* is bringing areas, presently devoid of trees, back to the original form of forests.
- *Forest Conservation* is protecting the present forests and preventing them from degradation.

Conservation is not hoarding or rationed use of limited resources but is a constant effort to increase resources. All the natural resources need to be conserved, whether renewable or non-renewable.

Tree plantation is now done on a large scale in different parts of the world. In India, the festival of Van Mahotsav was introduced for this purpose. It is celebrated every year in July. Planting is done in wastelands, open lands, inside and around cities, etc. Thus new trees will come up to replace the old ones. This is a sound planning for future. However, Van Mahotsav should not become a mere ritual. Trees planted should be cared for and not forgotten after the festival. It should be made into a well planned country wide afforestation programme. The needs of the different zones should be looked into while selecting the saplings. For example, arid zones should be planted with drought resistant trees. The new trees should blend well with the surrounding trees.

To overcome the negative effects of the large scale deforestation in India, a plan of action should be followed:

- Degraded land, which is not fit for agriculture, must be converted to forest land with suitable species.
- Special types of salt tolerant trees can be planted in the saline soils.

- Plant trees on roadsides, near railway tracks, canal banks and farm boundaries.
- Encourage planting of trees around villages, school yards, public buildings, church yards, cemeteries, market squares, etc.
- Enforce the present laws against tree felling, strictly.
- Force industries using wood as a raw material to plant trees 20 to 50 times the number they cut.
- Develop fuel efficient stoves to decrease the quantity of fire wood used in rural households.
- Make alternative energy sources popular like solar energy, biogas, etc.
- Practise selective harvesting, i.e. marking trees according to age, cutting selectively and filling gaps with saplings.
- Popularise social forestry, i.e. involve local people in programmes.

Realising the importance of forest management, the Ministry of Environment and Forest (MoEF), Government of India formulated the Joint Forest Management (JFM) in 1988. Accordingly, the Forest Departments work with local communities. In 1990, Village Forest Communities (VFC's) were formed. New guidelines were given in 2000. At least 25% of the income from the villages which volunteer to manage the forests goes to the villagers. All the states in India now have JFM's and all are functioning well.

Mining and its effects

Mining degrades forests to a great extent. In many countries, large tracts of forest land are cleared for mining operations. *Mining disturbs the physical, chemical and biological characters of the soil.* The fertile topsoil of the forests goes deep inside the mines while the infertile subsoil devoid of humus and organic nutrients, necessary for plant growth, is brought to the top, which increases erosion. The whole area becomes useless, unless soil reclamation practices are not done promptly. Even then, it is not possible to revert the land back to the natural forest. But what can be done at the most is to grow some plantation forests, which lack the biodiversity and beauty of natural forests.

The extracted minerals like iron ore, bauxite, etc. are used in the native countries or are exported to the developed countries. Mines of Mica, Manganese, coal, limestone, iron ore, Copper and many other metals and minerals lead to large scale destruction of forests. Gold mining has been all the more damaging. This is because mercury, which is a toxic substance, is used to separate gold from sand and gravel. Once the mining is over, the mercury is left at the site, polluting the soil and the ground water. This has happened to very large gold mines in the Amazon Basin of Brazil.

Asbestos is a carcinogenic substance – it induces cancer. It has been shown to cause cancer of lungs and other organs as well as breathing disorders. The World Health Organisation (WHO) estimates that at least 90,000 people die every year of asbestos related diseases, yet it is used extensively for different purposes, mainly in building construction, mostly to add to cement, in the LDCs. It is also used as an insulating and fire proofing agent in buildings, ships, etc.

Mining of asbestos is done in many places. To point out the harmful effects of asbestos mining, a comparatively recent incident is described here. More than twenty years ago, Hyderabad Asbestos Cement Products Ltd. (now known as Hyderabad Industries Ltd.) used to mine asbestos from the Roro mine, situated about 20 kms away from Chaibasa in Jharkhand. This was one of the largest asbestos mines in India, producing about 4 to 5 lakh tons of asbestos annually. In 1983, when the company realized that the grade of the asbestos was no longer good, the mine was closed down and the assets taken away.

However, thousands of tons of asbestos waste were left on the Roro hills. This waste is poisoning land, water and vegetation around. River Roro, flowing down from the hills, joins Subarnarekha downstream, carrying tons of asbestos waste. A community of about 5,000 tribal people is living within a 5 km radius of the waste dump. The small paddy fields of the tribal people are covered with the waste. People cross the dump sites while grazing their cattle. When children slide down the slopes of the hill, clouds of the lethal dust rise in the air. It is high time to dispose of this carcinogenic waste so that the health of the poor tribal people is not affected any more.

In this case, mining has resulted in a double environmental destruction. On the one hand, it has resulted in deforestation of the Roro hills along with all the dangers associated with it, and on the other hand it is also causing serious pollution problems.

Iron ore is mined out from several mines in Noamundi, again in Jharkhand. Large scale afforestation programmes are being done continuously for regeneration of the forests cut down for mining. Yet plantation forests are very unlike natural forests in that they do not have the original biodiversity, either plants or animals. It is also possible that diverse species have been lost forever.

Aravallis are one of the oldest hill systems of Rajasthan and Haryana. These mountain ranges have a major role in the control of climate and the drainage system in the entire region. It also acts as a mini watershed as many seasonal streams originate from the Aravallis. A lot of mining goes on here. The mines are of marble, granite and other types of stones. In 2002, the Supreme Court of India put a blanket ban on all mining activities in the Aravallis. As this ban affected the lives of 65,00,000 lives in Rajasthan – workers in the 9,700 industrial units with their families – this had to be lifted after one week. However, during that 7-day period, the water level in the wells in the region went up. The people of Rajasthan realized that Magsaysay Award winner Rajendra Singh was right when he said that "Mindless mining is responsible for the depletion of water levels in Rajastan. In fact, mining affects *jal, jangal* and *jameen*(water, forest and land)".

Dams and their effects on forest and tribal people

There are more than 45,000 dams in the world. More than half of the large dams are in India and China. *Dams built across major rivers and streams serve many purposes*. These are:

- They store water to be used in irrigation during dry seasons
- They prevent soil erosion
- Provide drinking water
- Control floods
- Hydroelectric projects at some dams generate electricity

Dams like Bhakra Nangal, Damodar, etc. in India have turned large tracts of unproductive lands into prosperous fields. To conserve, retain and store surplus water, many projects have come up in India, like the Gandak and Kosi in Bihar, Thungabhadra in Andhra Pradesh and Karnataka, Hirakud in Orissa, Upper Krishna in Karnataka, etc. Thus, in general, dams have many uses. This is why the dams – major, medium and minor – are built across many small and large rivers all over the world. One example is the Aswan dam across the Nile in Egypt. It has given a boost to the economy of Egypt.

However, *dams do cause various types of miseries to the local people*. Dams built on large rivers flood vast areas of tropical forests. Electricity is generated using the water flowing through the dams in the developing countries. This electricity is used mostly in mining and smelting minerals which are exported to the developed countries. In the Amazon basin, many large dams have been built, displacing many indigenous people. The local people living near River Nile in Egypt are facing loss of soil fertility and a rise in diseases caused by snails, as predicted by environmentalists.

The report of the World Commission on Dams (WCD) was released by Nelson Mandela in November 2000. Though the Commission did not say that all big dams are bad or big dams should not be built, the bad experiences with such dams came to be known to the world. Many countries around the world decided to follow the recommendations of WCD. India, however, rejected the report and is still planning to get many big dams built. *The harmful effects of major dams can be listed as follows*:

- The energy generated from large hydroelectric projects has been continuously falling over the last 11 years in India.
- As per the projections made by the Planning Commission for future electricity needs of the country, annual capacity has to be increased by 5,040 MW for the next 25 years. This will need Rs. 30,000 crore each year.
- Approxinately 0.67 million people, mostly tribals, will be displaced each year.
- Nearly 100,000 ha of land, mostly forest land, will be submerged each year.
- Surrounding land will become water logged, and this will also lead to salinisation of soil.
- Habitats of animal populations will be disturbed; their migratory routes will be cut off.
- Dams built on seismic (earthquake prone) zones can be very dangerous. The huge quantity of water stored exerts a tremendous pressure on the ground below, which can lead to earthquakes.
- Fishing and water transportation will be affected.
- The cost of irrigating land through big dams is much more, compared to that by minor dams.
- Big dams are affected worst by siltation; this will lead to degradation of soil in the forests.
- It is easier to reduce siltation and to desilt minor dams.
- In spite of having so many major dams, water use efficiency in Indian agriculture is one of the lowest in the world as admitted by Finance Minister P. Chidambaram, in February 2005.
- Ground water recharge is done better by minor dams.
- The cost of building major dams escalates unimaginably high. When the ecological cost of deforestation is added to this, it

becomes all the more. Minor dams do not lead to deforestation to this high extent.

- Advocates of major dams say that we need more storage capacity in the country. But storage is a means, not an end. Storing huge amounts of water and then releasing it all of a sudden whenever there is a danger of breaching of the dam leads to flash floods bringing death and misery all around. This affects the poor people, mainly tribal people, as they live in the forests or in lowlands.
- Resettlement and Rehabilitation packages are never properly implemented. This is very evident in the 3,000 odd dams built on River Narmada. This river has the largest number of dams built on than any river in the world. Indira Sagar, India's largest dam, and Asia's second largest, holds one third of Narmada.

The total storage capacity of the many existing minor dams and small surface storages has never been calculated. This must be very large. The existing storage capacity of major dams themselves is not utilized fully.

Considering all this, it will be wise to go for other options. These include:

- Rainwater harvesting
- Watershed development
- Minor irrigation projects
- Improving the infrastructure of existing projects
- Raising end use efficiencies, etc.

Behind almost every major dam in India, there are stories of broken promises. This is because of the administration's mishandling the issues. Moreover most of the rural poor, mostly tribal people are not aware of their rights. They have no idea how much compensation they are entitled to get or how much land will be submerged. Many of them do not know that there is a Measurement Book. Those who know get annoyed when surveyors come, just measure the lengths and breadths of houses and put very low values for them. Actually, the purpose of the Measurement Book is to help engineers to evaluate houses properly so that compensation to build similar houses can be given.

The aim of the government is to evict people in minimal time with minimal expenditure. The residents are caught in the red tape and at each step, lose more and more. Sometimes, the houses of the affected people are demolished before alternate arrangements are provided. In many cases, tribal and other people have not got the land to land settlement even after the dams have been built. At other times, substitute land given is far away and barren or rocky, and absolutely uncultivable.

In Arunachal Pradesh, 89 large dams have been cleared for construction. People of many tribes (some of them as small as 3,500) will be badly affected by these dams socially and economically.

Tehri dam on River Bhagirathi is 260.5 m high – the 5th tallest dam in the world. Rehabilitation had started in 1979, yet, till now it is not completed. The dam has been built on earthquake prone Central Himalayas. Thus social and environmental issues have been ignored.

In short, though dams are most essential for the development of a country, many small dams will fully serve the purpose of one large dam. The problem of rehabilitation of tribal and other people too, if at all is there, can be tackled much more easily.

B) Water Resources

Water is one of the most valuable natural resources. *All the water on the surface of the earth together is called the hydrosphere*. It includes the seas and oceans and also the rivers, streams, springs, lakes, ponds, glaciers, etc. The hydrosphere covers 70% of the earth's surface.

A major part of the hydrosphere is the marine ecosystem comprising the seas and oceans, estuaries, backwaters, etc. Only very little is the fresh water. This includes ponds, tanks, lakes, rivers, springs, streams, glaciers, etc. The freshwater in the lakes, rivers, glaciers, etc. which is continuously being renewed by a process of evaporation and rainfall. Some of this fresh water is stored underground, called aquifers. This is also part of hydrosphere. Thus hydrosphere includes all types of water resources.

Of the total amount of water on the earth, 97% lies in our oceans and only 3% is the freshwater. Of this 3% of fresh water, 2% is locked in the polar ice caps while only 1% is available as drinking water. Water found below the surface of the earth is called *groundwater resources* while that above ground is called *surface water resources*.

This shows that water available for our use is very little. More than 2/3 of all countries have already started experiencing some degree of water shortage. Global population is increasing at a fast pace. Each individual needs at least 250 litres of water per day. This is in addition to the much larger hidden water demand required to produce food and other consumer items. A terrible water crisis is therefore not far. The prediction that "the Third World War will be fought for water" may really come true.

Table: Distribution of world water resources

S. No.	Location	Total water (%)
1.	Fresh water lakes	0.009%
2	Saline lakes	0.008%
3	Stream channels	0.0001%
4	Ground water (<0.8 km deep)	0.31%
5	Ground water (>0.8 km deep)	0.31%
6	Soil moisture	0.005%
7	Ice caps and glaciers	2.15%
8	Atmosphere	0.001%
9	Oceans and seas	97.2%
	Total	100%

Ours is a country of vast water resources. We have annual monsoons which regularly recharge the water storages maintained in:

- the great glaciers (ice blocks) of the Himalayas
- the rivers and streams of our plateaus
- the enormous multi layered underground water pools, etc.

Yet, the per capita availability of clean water is decreasing day by day due to many factors like population explosion, urbanization, industrialization, pollution, etc.

As such, available fresh water in the world is very less, in comparison to other forms of water. Therefore, every effort must be made to conserve this precious natural resource.

Use and Overexploitation of Water

Water is the most abundant ingredient of living beings. About 60 to 95% of the body is water. Man can live without food for a few days, but he can not live without water even for a day. Thus after oxygen in the air, water is the most important need of man and also other animals. The uses of water are many:

- Most chemicals can function only in the presence of water.
- Cytoplasm, the basis of life in all living organisms, can function only in a hydrated condition. If dehydrated beyond a certain level, living cells will die.
- Water is a raw material for photosynthesis (synthesis of food by plants) on which all organisms depend directly or indirectly for food.

- The circulation of minerals within the plants, absorbed from the soil, and also the food made during photosynthesis takes place through the medium of water.
- Water is necessary for the condensation and also for the synthesis of various compounds inside the living body, and outside in the soil.
- Water is the universal solvent in nature.
- In addition to all this, many aquatic organisms, plants and animals, live in water.

In short, without water, life on earth can not exist.

Man needs water for drinking, cooking, washing, taking bath, cultivating various crops and rearing domestic animals, gardening, etc. This water may be taken from underground sources like wells, or surface sources like rivers or lakes. Wells may be dug wells or tube wells. Both depend on underground water. Surface water present in rivers and other water bodies are used by us for different purposes, like:

- Drinking
- Irrigation
- Bathing
- Transport
- Fishing
- Hydroelectric power
- Aquatic sports and recreation, etc.

Overexploitation of water: Without thinking of the consequences, human beings are overexploiting water from different sources in various ways. The main culprit, one can say, is the population explosion itself. This has led to an increase in the amount of water used by human beings directly and indirectly. When the world population is more, agriculture is more, industries are more; everything is more. Thus, the use of water has increased 4 to 8% since 1950.

Maximum usage of water is in the field of agriculture. About 69% of the total water consumed in the world is for agricultural purposes. Only about 8% is used for domestic and municipal purposes. The rest, i.e. 23% is used for various types of industries like mining, textiles, cement, leather, pharmaceuticals, paper, sugar and for energy production (oil and gas production and power plant cooling) and for various construction works of buildings, bridges, etc.

At the time of independence, the population of our country was 40 crores. Now it is 120crores! Just imagine how much we would have developed in every field and how less problems we would have had to face, including the present shortages in water, electricity, and everything else if only our population had been stable at 40 crores!

Continuous overexploitation of water from nature has led to many problems.

- The worst problem is fresh water shortages. There are 4 reasons for this:
  1. Aridity, i.e. shortage due to dry climate
  2. Drought
  3. Desiccation, i.e. drying up of soil due to deforestation, overgrazing, etc.
  4. Water stress, i.e. low availability due to overpopulation
- When too much water is withdrawn from the surface resources, wetlands dry up.
- Ground water is also depleted due to overuse and lack of recharge done.
- Too much irrigation in dry areas leads to soil turning saline when the water evaporates. This results in the reduction of yield.
- If water from near sea coasts is overdrawn, sea water moves into these wells.

Floods

In many countries like India, there is enough rainfall, but most of it is for a few months only. There is no or very less rainfall in the rest of the year. The results are very disastrous.

- Soil gets waterlogged (too much water in the soil).
- Soil nutrients move deeper down in the earth.
- Topsoil is washed away.
- Flowing water overflows its normal channel leading to floods.

Causes of floods: There are many causes of floods. These causes act in various combinations and thus become very complex. It is because of the complex nature of their causes that floods are difficult to manage and control.

1) *Natural causes*:
   - *Heavy Rainfall*: Continuous heavy rainfall for long periods or melting of enormous quantities of snow lead to a huge increase in the quantity of water in a river. When this happens upstream, floods occur downstream. In arid (dry) and semi-arid areas, rivers cannot contain large quantities of water at

any given time. Therefore, in such areas, if a heavy rainfall occurs for a continuous period, rivers get flooded. Example of this is the flood happened in some parts of Rajasthan in the monsoon season of year 2006.

- *Winding course of rivers*: Some rivers take many twists and turns during their course of flow. In such cases, the speed of flow is reduced which results in accumulation of water a some places, which overflows beyond the banks.
- *Occurrence of vast floodplains*: On either side of some rivers like the Ganges, there are vast floodplains. Even comparatively small increase in the quantity of water in such rivers spread over large areas, causing floods.
- *Sudden change in slopes*: Rivers flow along slopes. If the slopes are uniform throughout the course of the river, the flow remains uniform. But if there are sudden changes in the gradient of slope in the river bed, floods result during heavy rains.
- *Landslides*: If there is a huge landslide upstream (in nature, these are caused by heavy rains and the resulting loosening of soil especially on the mountain sides), flash floods occur downstream.

2) *Anthropogenic causes:*

- *Deforestation*: The commonest and most important man-made cause of floods is deforestation. Trees in forests are felled by man for various purposes like logging, mining, dam building, cultivation, overgrazing, forest fires, etc. This removal of plant cover has many grave consequences:
  a) Infiltration capacity of the soil decreases, i.e. less water filters into the soil, resulting in more runoff.
  b) When there are trees and other plants, rainwater filters through the gaps in their leaves and branches in the form of tiny streams on the soft forest litter (fallen leaves, etc.). But in the absence of plants, rainwater strikes at the land with force and does not seep down and leads to rapid runoff.
  c) Roots of plants tightly hold soil particles and keep them in place. With the removal of plants, soil becomes loose and is washed off into rivers, making them shallower which increases floods.

d) Removal of the top layers of soil leads to a decrease in infiltration of rain water in the valleys. Therefore, rain falling in the valleys runs off into rivers, causing floods.

- *Urbanization*: When any city spreads sideways, there is an increase in the cemented and concreted areas of land, tarred roads, pavements (footpaths) laid with bricks, etc. All these reduce infiltration and increase runoff. The water flowing fast through the drains of the city reach the nearby river. Constructions of bridges across the river, dumping of garbage in the river, etc. also follow urbanization. All these factors increase floods.
- *Changing the course of rivers*: Sometimes, the course of flow of rivers is changed by man to suit his convenience. If this is not done keeping the topography of the place in mind, rivers get flooded.
- *Construction of dams*: One purpose of dams is to control floods. However, construction of dams results in the river flow obstruction. So, when there is a heavy rain, dam gates have to be opened for the safety purpose to prevent the dam from cracking under the enormous pressure of released water. This results in flash floods.
- *Some agricultural practices*: In many farmlands, after the winter crops (rabi crops) are harvested and the farms are left to dry. Rain washes off this soil into rivers, leading to siltation as a result rivers become shallow and floods become more severe.
- *Man-induced landslides*: Major dams store enormous quantities of water. The mere pressure of this water may lead to earthquakes. Mining activities also may lead to earthquakes. Such man-induced earthquakes result in landslides which block the water flow in rivers and cause floods.

Flash floods are caused by opening of dam gates when the stored water crosses the danger level, which happens in years of very high rainfall. This leads to untold misery to people living in the adjoining areas as dam gates are opened unannounced in many cases. Storing too much water in dams for the purpose of hydroelectric projects and irrigation may cause much harm, because in addition to frequent flash floods, the weight of the water in the dam itself may stress the earth, causing earthquakes. This is why environmentalists caution that dams should never be built on seismic zones.

People in poor countries have no choice for living. *They usually live in lowlands prone to floods.* Therefore, it is all the more necessary that government, non-government organizations and the people themselves understand the causes of floods and take corrective measures to prevent them.

It is a myth that the proposed interlinking of rivers in India will control floods. The average flood flow in the Ganga and the Brahmaputra is about 50,000 cumecs. The maximum flows that the project will be able to divert will be 1,700 and 3,000 cumecs, respectively which will not make any big difference. On the other hand, the 1000 km of canals planned may make the flood situation worse.

Worldwide, it is being understood that *instead of flood control, flood management must be done.* India's failure to control floods even after almost 60 years of independence and spending crores of rupees proves the necessity of flood management. The flood affected areas have been increasing steadily in India, even after all this expenditure.

Methods of flood control: Following are some flood control methods:

- Channelization, i.e. deepening, widening or straightening a suitable section of a stream
- Lining the stream with concrete
- Building artificial embankments (stream bank protection)
- Building check dams across the streams or rivers

Methods of flood management: Some flood management measures practised globally are given below.

- Prepare a flood frequency curve (FFC) based on the history of floods in the area
- Based on this FFC, prohibit certain types of buildings/activities in the flood prone zones, and estimate the height of buildings
- Construct flood proof buildings
- Construct floodways which allow the flood water to flow through with minimal damage
- Make a rule that construction projects proposed on flood prone areas will not be approved by the Government.
- Arrange for flood insurances for the people living in such risk areas

Droughts

Drought is a period when rainfall is nil and evaporation from the soil is higher than normal. There are at least 80 dry or semi-dry countries in the world. 40% of the world's population live in these countries. They experience droughts throughout the year. Drought kills thousands of people every year in the world. Millions of poor people are forced to live on drought prone land in the developing and the underdeveloped countries. Droughts are experienced to different degrees in many countries of Africa, Gulf countries of the Middle East, Northern China, much of the former Soviet Union, Mexico, parts of India, etc. (See *Biogeographical Regions of India* in Unit 7)

Land degradation and deforestation which induce droughts are also closely related to uncontrolled population growth. When there are more people in a country, more trees are cut, more cattle are grazed, land is cultivated more, leading to loss of fertility. Moreover, in some countries, though huge rivers flow through forests but their water is not available to the people, who face drought in other parts.

Poor natural resources and frequent droughts result in highly variable production. Because of this, farmers often do not have enough money, or sometimes they have no money at all, to invest for the future. This is one pf the reasons that farmers remain poor for ever in the dry lands.

Reasons for drought:

1. Actual deficit of rainfall
2. Uneven distribution of rain
3. Not managing the monsoon water by proper methods
4. Deforestation
5. Overgrazing
6. Huge wastage of water by wrong methods of agriculture, etc.
7. Global climatic changes (In *El Nino* years when the ocean currents reverse their direction (this happens periodically), there are storms in America and drought in Australia).

One third of India is either facing drought or is under the threat of drought. Therefore, it is essential to adopt suitable measures to counter the effects of drought.

Methods to reduce drought and its effects:

1. Minimize wastage of water. About 70% of total water available is used for agriculture. Therefore, there is an urgent necessity to reduce this amount. Many farmers practise flood irrigation even today. Here, the fields are flooded with water; while the plants

absorb the amount they need but the huge quantity is lost by evaporation from soil. This is a highly wasteful method. Use of sprinklers, drip irrigation, etc. must therefore be promoted among Indian farmers who are not aware of such systems.

2. Plan water resources development and management properly. Also, water harvesting techniques must be adopted wherever possible
3. Crops suited to dry land must be cultivated in drought prone zones. This is called dry land farming.
4. Cultivate drought resistant varieties of crops. Plant breeders have already developed some such varieties. Biotechnologists are also trying their best to develop such varieties, with some successful achievement. For example, the International Crops Research Institute for the Semi-Arid Tropics (ICRISAT), Headquartered at Hyderabad, is using transgenic technology (transferring genes from one species/variety to another) to develop drought resistant chickpea and groundnuts.
5. Do not overuse rivers and wells for irrigation.
6. Do not clear forests, because this leads to droughts. When trees are cut down, in dry years, the soil dries out more quickly. The result is a man-made drought. It was such a draught that caused a devastating forest fire in Indonesia in 1997-98 — the biggest forest fire recorded in the world. Tropical forests are usually moist and thus do not burn fast; but due to felling of trees, the forest had dried so much that the fire set to clear land spread uncontrollably.
7. Community based rain water harvesting has the potential to drought proof the entire country. India receives most of its rainfall in only 100 hours out of the 8760 hours in a year. If this water is not captured and stored, there will be no water in rest of the year. The solution is rain water harvesting.

Methods to manage deserts and prevent further desertification: Deserts are dry zones due to actual deficit of rainfall. Here, soil is sandy. Pore spaces between sand particles are large and so water holding capacity is very low. Soil erosion takes place by wind. Therefore, some practices must be adopted to make the soil cultivable.

1. Making furrows for holding water – this helps in keeping the soil moist. A moist soil is not loose; it can hold water and it is not blown away by sand.

2. Plant the sand binders (plant species which can bind the sand).
3. Later, as the soil improves, do strip cropping (planting in rows) across the direction of the wind.
4. Plant trees in rows, across the direction of the wind – such trees are called wind breaks.
5. Prevent overgrazing by animals and deforestation near the desert as both these will lead to desertification of normal land.

Drought is a natural phenomenon in most cases, therefore we can not prevent it fully; but we surely can reduce the severity of droughts by suitable water conservation measures and prevent desertification.

Conflicts Over Water

International scene: Water is necessary for all living beings. Water resources are becoming lesser and lesser while population is increasing fast in many parts of the world. Agriculture based on irrigation has increased greatly and so has withdrawal of water from underground aquifers. Thus there is no wonder that war-like situations have arisen in many parts of the world over water. The next war in the Middle East may be over water, not oil. Most water here comes from 3 rivers – Nile, Jordan and Euphrates-Tigris. At 3% growth rate, population is expected to double in 20 years. Conflicts over water from Nile among Ethiopia, Sudan and Egypt are spreading fast.

Nearer home, the sharing of the waters of Ganga and Brahmaputra has always been a tricky problem between India and Bangladesh. When floods bring misery to the people of Bihar, on most occasions, it is because Nepal opens its dam gates.

The national scene: Within India, there are many conflicts over water.

- *Karnataka - Tamilnadu*: Sharing of water from Cauvery is a big conflict even now. The case is in the Supreme Court at present. The Cauvery Water Authority has been set up under court order.
- *Punjab – Haryana*: The Supreme Court ordered the Punjab Govt. to release part of the water from Satluj to Yamuna many times. But Punjab has not constructed a canal for the same till now.
- *Orissa – Chatisgarh*: These two states are supposed to share the waters from River Indravati as per a water sharing agreement drawn up in 1975. But Chatisgarh is complaining that Orissa is not releasing the amount of water as per the agreement.
- *Tamil Nadu – Kerala*: The bone of contention here is the inter-

state tributary of Cauvery and River Bhavani. If Kerala diverts water in its territory, Tamil Nadu will not get water.

Similar water sharing conflicts exist between Goa and Karnataka; Maharashtra, Goa and Karnataka, etc. The most recent conflict, widely reported in the media, has been between Gujarat and Maharashtra. This conflict is over raising the height of a dam on River Narmada. Environmental activist Medha Patekar is still fighting her *Narmada Bachao Andolan*. Other important personalities have also joined her and thousands of her followers demanding proper rehabilitation of the displaced tribals and others.

Water Harvesting

Water was worshipped by all civilizations in the past as a divine gift of nature. As early as the 3rd millennium BC, farmers in Baluchistan used stored rain water for irrigation. Dams built of stone have been found in Baluchistan and Kutch. References to canals and embankments are found in the Vedas, Puranas, Mahabharata, Budhist and Jain scriptures and archaeological remains. Inscriptions in many South Indian temples prove existence of water harvesting systems. Wells were probably a Harappan invention. Every 3rd house in the Indus Valley Civilization had a well.

As society progressed, water was brought by pipes to homes and then people gradually forgot the importance of water. Today, the water shortage is due to different reasons.

1. Rising population
2. Urbanization
3. Industrialization
4. Deforestation
5. Overexploitation
6. Lack of proper water management policies
7. Drying of wetlands
8. Neglecting excellent traditional water conservation methods
9. Overgrazing and desertification
10. Siltation of water bodies due to increased soil erosion
11. Wastage and lack of water reuse
12. Pollution, etc.

From this list, it is clear that *scarcity of water is mainly a man made problem*. Therefore, it is for us to take corrective measures. If we do not take action now, it may be too late. Thus there is an urgent need to practice as many water conservation methods as possible.

Methods of water conservation

There are many methods of water conservation.

1. Increase efficiency of irrigation
2. Protect watersheds and wetlands
3. Afforestation, Reforestation and Forest Conservation
4. Preventing wastage in:
   - Industries
   - Leaking in dams and reservoirs
   - Leaking in municipal pipelines
   - Households
5. Recycling of waste water by setting up waste water treatment plants
6. Constructing a large number of small ponds, tanks, etc. in low lying areas
7. Building, check dams at appropriate places in all rivers
8. Provide drainage facilities in canals and use drainage water for irrigation
9. Rainwater harvesting in rural and urban areas
10. Consider water as an economic good and levy reasonable charges to discourage its wastage

Rain is the primary source of water in the hydrological (water) cycle whereas rivers, lakes, ponds and ground water are all secondary. However, forgetting the primary source, we depend completely on the secondary sources. This trend must be changed. The potential of rain is tremendous. It is most unwise to let rain water go waste and then cry hoarse over water scarcity.

Water harvesting in villages: Some of the common water harvesting systems used in villages are as follows.

1. *Tanks* (*Tankas*): These are common in arid zones. *Tankas* have been constructed for centuries in Indian households, near temples, schools and public places. Such structures have been reported from many countries like Botswana, Ghana, Yemen, Kenya, Thailand, Indonesia, Sri Lanka, etc. A *tanka* is a covered underground cistern – a hole, 3 to 4 meters in diameter, with plastered sides and a lid of thorns or dried grass. In the Barmer District of Rajastan alone, more than 10,000 tankas are functioning successfully at present.
2. *Nadis*: These are excavated or embanked village ponds, 2 to 12 meters deep. Different linings are used to reduce seepage. To

reduce evaporation, rubber or polythene sheets or bamboos are used.

3. *Check dams*: These are small dams built with stones or earth and boulders across rivers or streams. Capacity is 1000 to 50,000 cuubic meter.
4. *Subsurface water harvesting*: This is usually done in riverbeds and is common in India, Kenya, Tanzania, etc. The riverbed must be of sand. Bunds are made below the riverbeds. They recharge all the surrounding wells with a large quantity of water.

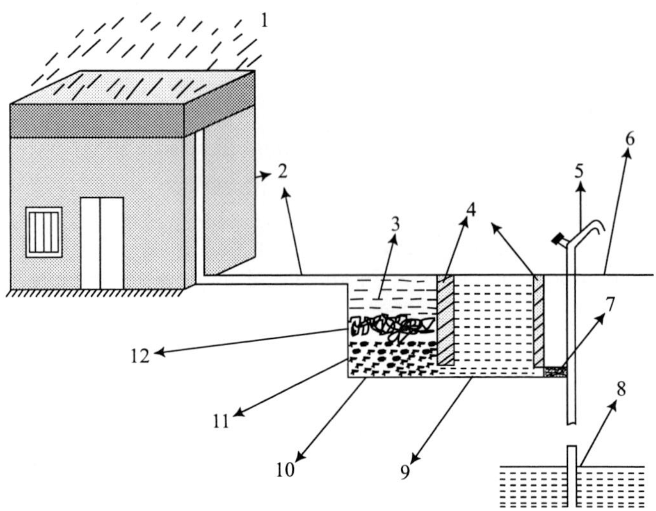

Basic Plan of Roof top Rain water Harvestin: 1. Rain 2. Drain Pipes 3. Collected rain water 4. Brick walls 5. Hand pump 6. Ground level 7. Sand filter 8. Water table 9. 2nd Tank 10. 1st Tank 11.Gravel 12. Sand

5. *Floodplain reservoirs*: These are ponds dug in the floodplains. They get filled with water during floods and remain full after the flood is over.
6. *Quarries*: These are pits left after rocks or minerals are extracted for building materials. These are used as storages.
7. *Baolis*: These are stepwells. Baolis were very common in Delhi during the periods of Mughal and other dynasties.

Water harvesting in urban areas: Water harvesting systems are gradually becoming popular in our urban areas. The common ones are:

1. *On channel storage*: During the rainy season, storm water flows very fast along the sides of the roads and other paths. This water can be stored in the channel itself by constructing regulators at suitable places. Deepening and widening the channels wherever possible will increase the storage capacity.
2. *Off channel storage*: Channels cannot store more than 20% of water flowing through it, as their capacity is less. Therefore off channel storage systems (reservoirs away from the channels) can be made, wherever possible.
3. *Lakes and Ponds*: In cities, storage capacity of lakes and ponds can be increased by desilting and widening them.
4. *Rooftop Rainwater Harvesting*: Rainwater falling on roofs can be collected and stored for direct use or can be recharged into underground aquifers.

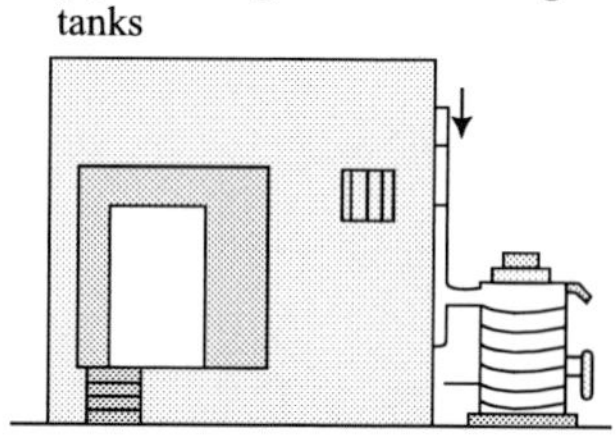

Storage of rainwater in over-ground containers

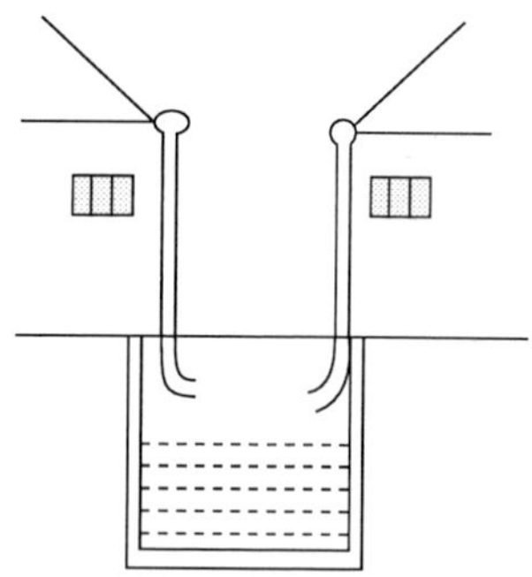

Storage pf rain water in underground tanks

The method to be adopted for water harvesting depends on the rainfall of the place. In states like Kerala, Mizoram, etc. rain falls almost throughout the year small domestic sized water tanks can be used to store water. But in dry zones like Delhi, Rajastan, Gujarat, etc. where rain falls only for 2 – 3 months in a year, huge containers will be needed to collect water and thus it is better to recharge aquifers.

- *Storage for direct use:* Rainwater is directed from rooftops to storage tanks through drain pipes, after filtration. Sand filter must be used and chlorination must be done if water is to be used for drinking.
- *Artificial recharge of rainwater into aquifers*: There are different

methods to recharge rainwater into aquifers.

a) *Recharge Pits*: These are constructed for recharging shallow aquifers. They are excavated pits, 1-3 meters wide and 2-3

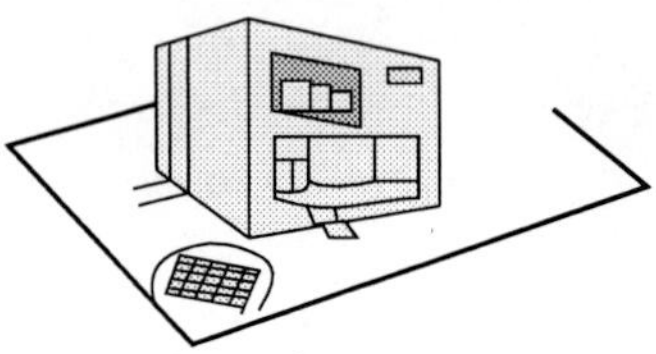

Recharge pit

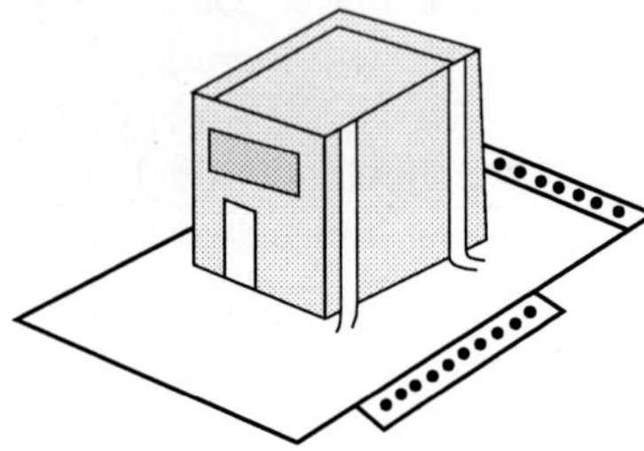

Recharge trench

meters deep. The pit is lined with brick or stone wall with openings called weep holes at regular intervals for sidewise infiltration of water. The top can be covered with a perforated lid. Filter material should be arranged in layers: Pieces of bricks or stones at the bottom, gravel in between and coarse sand on the top.

b) *Recharge trenches*: These are long continuous trenches dug in the ground. They are filled with porous media like pebbles,

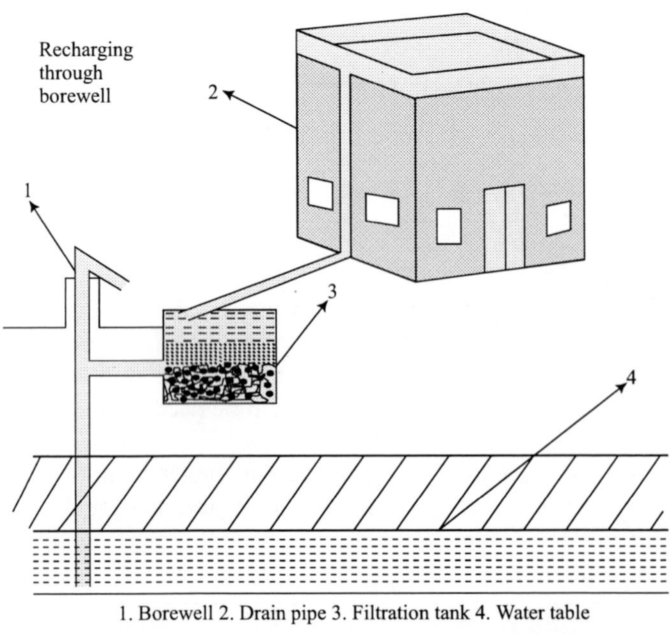

1. Borewell 2. Drain pipe 3. Filtration tank 4. Water table

boulders or brickbats. These are also constructed where rain water is fed to shallow aquifers. They are made near the outer walls of buildings. Dimensions can be 0.5 to 1 m wide and 1 to 1.5 m deep. Recharge trenches should be cleaned regularly to maintain their capacity to take in rain water flowing towards them from all sides of the compound.

c) *Bore wells or dug wells*: Rainwater from the rooftop is diverted through drainpipes to a settlement or filtration tank from which water flows into the well – bore well or dug well, working or abandoned. If a borewell is used for recharge, the casing of the bore should be a perforated pipe. This is for sidewise infiltration. Developing the bore well is beneficial. In this process water or air is forced into the well under pressure to make the soil around the bore loose. This helps to increase infiltration.

   If a dug well is used, its lining should have weep holes at regular intervals. The bottom of the dug well should be desilted every year to maintain its intake capacity.

Roofs of houses, flat complexes, public buildings like schools, shopping centres, hospitals, companies, etc. can all be used for rooftop rainwater harvesting. Many schools in India have installed such systems. Some companies have also installed such systems.

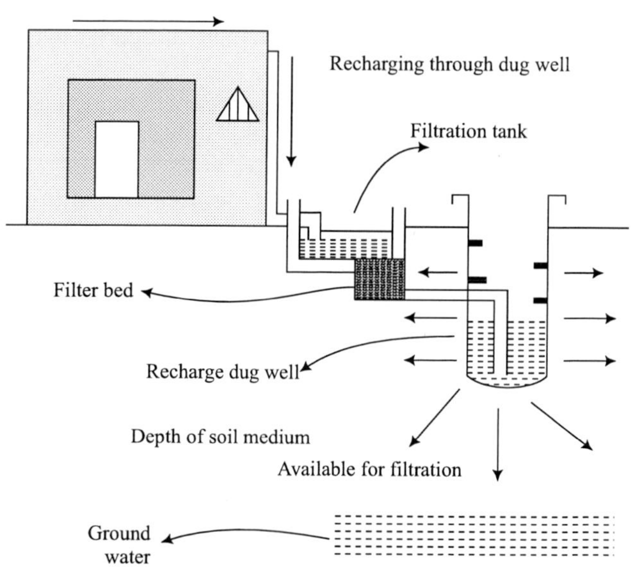

Some states and districts have decided to make rainwater harvesting systems mandatory for all future buildings. This is a welcome sign but community awareness and participation are most essential for any such programme to succeed.

(For information on the potential of rain water and the size of settling tanks, see Appendix – 1).

C) Mineral Resources

A mineral is an element or an inorganic compound that occurs naturally and is solid. Some minerals are made of a single element, like gold, silver, diamond (carbon) and sulphur. However, most of the 2000 minerals identified till now are inorganic compounds, e.g., salt, mica, quartz, etc. All these minerals are of economic importance to man while other minerals form rocks.

A mineral resource is a concentration of a naturally occurring solid, liquid or gaseous material inside or on the earth's crust and its extraction is profitable.

Man requires different types of minerals for industrial, technological and cultural purposes. Mineral resources are of three types:

a) Energy resources – examples are coal, oil, natural gas, uranium, etc.
b) Metallic mineral resources – examples are iron (Fe), copper (Cu), aluminum (Al), silver (Ag), gold (Au), lead (Pb), zinc (Zn), etc.
c) Nonmetallic mineral resources – examples are clay, sand, gypsum, phosphates, salt, sulphates, etc.

Earlier it was thought that mineral resources are unlimited. So nobody thought that conservation was necessary and anyway, it was impossible to preserve them (Odum, 1971). But now we know that these beliefs are wrong. As population is increasing, per capita consumption is also increasing. But total resources available remain constant. The mineral resources are non-renewable. The only hope is that new deposits may be found out or new technologies are discovered to extract existing minerals profitably.

Any solid natural material containing compounds of metals in the combined state along with earthly impurities and found below the earth's surface is a mineral. Those minerals from which metals are extracted commercially at a comparatively low cost and with minimum effort are called ores of metals. Hence all ores of metals are minerals, but all minerals are not ores.

Formation of minerals: Minerals are formed in many ways. In some cases, erosion of rocks leads to concentration of ores. Seawater contains such ores brought from land, for example, Magnesium and Bromine. Sometimes, materials from rocks dissolve in circulating ground water, and get deposited somewhere else as veins in the cracks of rocks. Examples – Gold, Copper, Tin, Lead, Zinc and Molybdenum. Sometimes, dissolved materials precipitate out of solutions and form minerals. Examples – sandstone, banded iron ores, etc. When water in marine conditions evaporates, salt, Potassium minerals, gypsum, etc. are formed. Some minerals are formed from organisms, for example, Phosphate rock deposits are phosphates from the bones and teeth of fish and other aquatic animals. Petroleum is formed from remains of organisms; coal is formed from dead trees under heat and pressure of the earth.

Mining of minerals from the earth: Mining is a method for the extraction of minerals. There are three stages in any mining operation:

1. *Prospecting and Exploration*: First of all, the areas where minerals are hidden have to be located. This process is called prospecting. In the early days, spots where veins of minerals were exposed on the surface were mined. But now there are sophisticated methods of prospecting. Areas where mineral ores are located are now detected with remote sensing devices. Geographical Information System (GIS) is extensively used for this purpose. The size, shape and extent of spread of the underground minerals can also be assessed using these methods. (For GIS and remote sensing, see *Disaster Management* in Unit 5).
2. *Development*: Once the locality is known, transportation to the target area has to be assured. For this, roads must be constructed, and all other necessary infrastructure developed. This is the Development stage.
3. *Exploitation*: Minerals may be superficial or deep seated in position. Accordingly, mining methods differ. Surface or open cast mining methods are different from deep or shaft mining. The mined mineral is first processed and then transported to various industries as per requirement. In all mining operations, safety of the miners is given due importance.

Mineral Resources of India

Minerals show uneven distribution in the world. Some areas are very rich in mineral resources such as Canada, South Africa, Some countries formed from the erstwhile USSR, etc. while large areas like the Amazon basin, Netherland, Denmark, etc. are comparatively poor. In India, mining for minerals is mainly done in Uttar Pradesh, Bihar, Jharkhand, Orissa, Madhya Pradesh and Andhra Pradesh.

Some minerals are found in great abundance only in some parts of the world, for example, more than 69% of the world's supply of gold comes from South Africa. Similarly, there are vast reserves of petroleum in the Middle East.

Coal is a very important solid fuel needed for many industries. It is also the commonest domestic fuel. It is the ultimate product of slow decomposition of large tree forests buried underground and subjected to high heat and pressure. The amount of carbon in coal increases as time passes. Thus different varieties of coal are present in nature. These are, starting from the poorest quality coal based on carbon content, peat, lignite, bituminous and anthracite. Peat has only 60 – 70% carbon whereas anthracite is the most superior coal, with 90-95% carbon and burns with maximum heat and minimum smoke.

Coal occurs in sedimentary rocks in layers of varying thickneses. It is mainly found in Europe, North America and Asia. China is the world's largest producer of coal which accounts for 32.09% followed by the United States that accounts for about 23% of world's annual production. In Asia, China has a large reserve, followed by India. In India, the biggest coal deposit is in the Damodar Basin centred round Ranigunj. If exposed on the surface, open cast mining is done; if deep below, underground mining is necessary, which is expensive and also dangerous.

Petroleum is a mineral oil, occurring in the pore spaces of sedimentary rocks. Crude oil is used as fuel for steam ships and other locomotives. Petrol is used in automobiles and aeroplanes. Diesel is used in automobiles and trains. Huge amounts are burnt in industries. Kerosene is used as oil for lamps, for cooking, lighting and heating. Hundreds of other byproducts are made from it like lubricants, vaseline, tar, wax, ink, medicines, soap, terylene, etc. Another product mined with petroleum is asphalt (bitumen or pitch). It is in fact very thick oil. It is an excellent material for road surfacing.

Iron is the basis of the modern industrial complex. About 90% of the world's iron ore is used in making steel. Many industries are based on iron and steel. Important iron ores are:

- Magnetite – 55 to 72% iron. Seen in igneous rocks. Black in colour. World's best magnetite is found in Sweden.
- Hematite – 50 to 70% iron. Seen in sedimentary rocks. Reddish in colour. Most of the iron ore in Lake Superior area is hematite.
- Limonite – Commonest of the commercial iron ore. Iron content up to 50%. Seen in sedimentary rock. Impurities are high. Brown in colour. Found mainly in Austria and England.
- Siderite – A low grade ore, which has only less than 45% iron. Seen in sedimentary rocks. Brown in colour. Found in large quantities in England and France.

USA and countries of the earlier USSR are leading producers of iron and steel. India produces 4% of the world's iron ore, mainly haematite from Jharkhand, Bihar and Orissa. Production is about 50 million tones annually. The area also has India's large scale iron and steel works at Jamshedpur, Rourkela, Bhilai and Durgapur.

Manganese is used in making good quality steel. Addition of Manganese makes steel tough and hard and it does not rust easily. It is also used in many industries. Largest producer is earlier USSR. India produces 8%. India is the 3rd largest producer in the world and exports large quantities to the iron and steel producers of the world. About 80 million tones of Manganese deposits are found in India. Orissa is the leading producer state.

Aluminum is the most abundant metal accounting for about 80% of the earth's crust. It is obtained from an ore called bauxite which contains about 25% of the metal. Aluminum is a light-weight metal and does not rust. It has a wide variety of uses. Leading producers of Aluminum are USA, Canada, France, earlier USSR, Germany, Japan, China and India. India has about 1250 million tones of bauxite deposits located in Jharkhand, Bihar, Orissa, Madhya Pradesh and Gujarat.

Mica is a transparent mineral found in most igneous rocks. Some varieties are coloured. Mica splits easily into thin plates. It is chiefly used as an insulating material in electrical goods. USA is the largest producer with 72%. India produces 20% of the mica in the world. Indian micas are the finest in the world because of the large size and perfect plates. Most important mica producing states are Bihar, Jharkhand, Andhra Pradesh and Rajasthan.

Gold differs from other metals as it is not combined with any metal. It occurs as veins running through quartz or other hard rocks. This can be exploited only by using costly machines for mining and crushing rocks.

Gold is also found as grains in sand on riverbeds as in the case of River Subarnarekha flowing through Jharkhand and Orissa. The name Subarnarekha itself means a line of gold. The greatest producer of gold is South Africa (60%). Erstwhile USSR accounts for 21% while Canada accounts for 6% of the world's annual production. India has scanty deposits in the Kolar Gold Fields of Karnataka.

Copper was probably the first metal used by man. It a constituent of different alloys like bronze, brass, German silver, etc. Copper is used mostly in electric appliances. The largest producer of copper is USA (17%). There are scanty deposits of the metal found in Jharkhand, Bihar and Rajasthan.

Tin is a very useful metal for coating cans used to preserve food materials. It is also used for soldering and making different alloys. Malayasia accounts for 40% of the world's production.

Diamonds of superior quality are used in jewellery. Poor quality diamonds have many industrial uses. Zaire (Congo) produces 32%, earlier USSR produces 17% and South Africa accounts for 16% of the world's diamonds.

Uranium and Thorium deposits in India are seen in Jharkhand, Bihar and Rajasthan. These are used for generation of atomic energy.

India is rich in about 35 minerals. The coal reserves of India amount to about 86,000 tonnes. Annual production is more than 150 million tonnes. Annual production of petroleum is more than 9 million tones. Reserves are seen in the valleys of the Brahmaputra, deltas of Cauvery, Krishna, Godavari and Mahanadi. Coastal areas of Kerala, Andhra Pradesh, Orissa and offshore areas near Mumbai also have deposits of petroleum.

Minerals of Jharkhand

Jharkhand is a mineral rich state. A major part of this young state's revenue comes from minerals. The commonest minerals of Jharkhand are:

- *Iron ore*: The 2,000 MT haematite reserve in Chiria near Manoharpur in Singhbhum district is one of the single largest deposits in the world. The average iron content in the deposit being 62-63%, it has the potential to feed a 10 MT capacity mine.
- *Coal*: Almost 100% of prime cooking coal, 93% of medium cooking coal and about 30% of the semi-cooking coal or blendable coal reserves are available in Jharkhand. The Jharia coalfield is the only source of prime cooking coal in the country.
- *Limestone*: There are 22 mines of limestone spread across Palamu,

Hazaribagh, Singhbhum and Ranchi districts. All grades of limestones are available in these mines.

- *Graphite*: The Graphite reserves of Jharkhand are located mainly in the Palamu district. Important deposits are found in Sikra, Satbarba, and Barwadih areas where both flaky and amorphous graphite occur in lumpy form.
- *Granite*: Extensive deposits of granite and ornamental stones such as dolomites, peridotites, banded hematite, migmatite, etc. are also available in Jharkhand.
- *Mica*: Jharkhand has 46% of the country's reserves of mica. The deposits are seen in Koderma, Giridih and Hazaribagh districts.
- *Fireclay*: Deposits of fireclay occur in the coal bearing areas of Dhanbad, Hazaribagh, Giridih, Ranchi, Latehar and Palamu districts.
- *Copper ore*: Rich deposits of copper ore occur in East Singhbhum and Ghatsila.
- *Base metals*: Many small but rich chunks of lead and zinc co-exist with copper deposits in Singhbhum, Deoghar and Hazaribagh districts.
- *Other reserves in Jharkhand*: Include uranium, kyanite, quartz, silica, china clay, bentonite, talc-soapstone, etc. Rich placer deposits of gold have been traced to Singhbhum and Ranchi districts. Explorations should be done in this regard. Moreover, five coal bed methane (CBM) blocks have been identified in Jharkhand.

Jharkhand is thus an ideal place for setting up mineral based industries. However, it is most essential to ensure that the miners are maintained in good health. Underground mining is more dangerous than surface mining. Accidents occur more often in underground mines mainly due to lack of ventilation, falling of mine roofs, flooding of mines, explosions, inhalation of toxic fumes, etc. Adequate precautionary measures are taken in mines of radioactive substances, like Uranium mines in Jadugoda. If such measures are not taken, there will be the risk of radiation sickness or even cancer, in addition to mutations (changes in the genes) of reproductive cells leading to birth of deformed children.

D) Food Resources

It has been estimated that the Earth has about 30,000 plant species with parts that people can eat, but *90% of our food comes from just 15 plant and 8 animal species*. Wheat, rice, corn and potato together contribute most of the world's food production than all the other crops taken together. Plants are our main food providers. Meat of animals and animal products like milk, cheese, butter, etc., some types of birds and their eggs, and fish form another part of our food. Thus we get our food from agriculture, animal husbandry and fishing. However, all animals ultimately depend on plants for food. Thus we need to increase agricultural production greatly to keep pace with the ever increasing population.

In India and many other developing countries, several methods used to improve agriculture show harmful effects on the environment. These are against the policy of sustainable agriculture. *In sustainable agriculture, air, water and land are not polluted by any means and the natural gene pool is also not disturbed.* If the aim of agriculture becomes just to increase in yield without any thought of the environment, the ecosystems in nature will be harmed beyond repair.

The fact that only sustainable agriculture will ultimately pay has been now realized by all. Production of all crops, including cereals, oil seeds, fruits, vegetables, etc. have been increased in India considerably in the recent years. After a very serious drought of 2002-03, we have made an impressive recovery in all areas of agricultural production.

Agriculture can be of following types.

- *Industrialised agriculture* uses large amounts of fossil fuel energy, water, fertilizers and pesticides to produce large quantities of single crops or animals for sale.
- *Plantation agriculture* is a cultivation of cash crops like banana, coffee, etc. in the developing countries mainly for their sale in developed countries.
- *Traditional subsistence agriculture* is done by farmers on a small scale – they produce only what they need in the family; a small quantity may be sold off.
- *Traditional intensive agriculture* uses a lot of inputs including human and animal labour.

In *traditional agriculture*, intercropping (2 or more different crops grown in the same field), poly varietal cultivation (many varieties of the same crop in one field), agroforestry (crops and trees planted together), etc. are done. All these practices are good for the soil and are being practised in India.

In the *fisheries* sector, coastal fish stocks are being exploited in India while resources in the deeper seas are not exploited. This is again an offshoot of poverty as most of our fishermen cannot afford to buy trawlers.

As for freshwaters, as the state governments have the ownership rights of ponds and lakes, proper development is not taking place. In general, Indian *aquaculture* (rearing and using fish and other water animals and plants) needs to improve a lot to be able to compete with global markets.

On the other hand, *milk production* in India shows great promise. A growth rate of 4.5% has been achieved by our dairy sector during the last decade. We have already overtaken the USA in milk production. We are moving further ahead and the target to double the milk production by 2015 seems to be achieved.

India's ranking is good in *poultry* market too. The output of eggs in our country in 2004-05 was 48 billion. Our rank is among the top 6 egg producing countries in the world.

Livestock population in India is large and widely diversified. We possess the largest cattle and buffalo population in the world. Goat population is second only to China. Total production of livestock products has also increased greatly over the past 3 decades.

With all these increases in production of food resources, the reason for poverty in India is mainly due to population explosion.

World Food Problems

In the begining of year 2000, it was estimated that 1.1 billion people were undernourished and underweight and the estimation of World Bank shows that 1.3 billion people were living in poverty. Hunger and poverty go together.

Today, the world is facing a food problem because of following reasons:

- *Population explosion* – This is one of the major problems, especially in the Asian and African continents. In India, with a population of more than one billion, 53% of all children are undernourished, whereas Bangladesh alone accounts for 56%. In all developing countries most of the poor population lives in villages and the population growth rate is alarming in such countries.
- *Subdivision of farms* - When a farm is transferred from one generation to the next, it is subdivided among the siblings. In India, in the last 30 years, the number of farms with less than 2 hectares of land have been doubled. In Bangladesh, the average farm size has already fallen below 1 hectare. Thus in addition to

the real landless, many others are 'effectively landless' as their plots are extremely small.

- *Scarcity of water* - Another major setback in increasing world food production is scarcity of water. Of the nearly 3 billion people to be added to the world population in the next 50 years, almost all will be added in countries already facing water shortage, like India, Pakistan and many countries of Africa. In India, water tables are falling in large areas. Water is taken for granted and is wasted in many ways in agriculture and other fields.
- *Increasing demand for animal food* - Over the last century, the world's demand for animal food has increased. There is great economic disparity in different countries and also in different parts of the same country. Thus the number of rich people has increased greatly in the world even when the number of poor has increased. The rich demand animal protein more. To rear animals, more and more land is converted to grazing pastures; great quantities of world's grains are also used. This adds to the shortage of grains for human consumption.
- *Falling of fish production* - Water pollution has increased in many water bodies. Added to this is the fact that in many countries, fish are caught even in the breeding season. Due to both these, fish production is falling in many places. As fish is an important source of protein, fisheries must be given due importance. Overexploitation using modern trawlers must be discouraged.
- *Overdependence on some countries* - Many developing countries import food grains from developed countries. Only 5 countries – United States, Canada, France, Australia and Argentina together account for 88% of world's exports. Thailand, China, Vietnam and the US account for 68% of all rice exports. In the case of corn, US alone contributes to 78% of the world's exports. In such a situation, importing countries face great risks. If the production of grains in the US falls due to any natural calamity, the whole world will suffer.
- *Lack of taste for new foods* - There are so many plant species on this earth of which human beings use only very few as food. There is a need to develop a taste for food made from plants which have not become popular yet. Similarly, the sea has abundant food resources in the sea weeds. Yet these have not found a popular market till now, except in some countries like Japan.

- *Gender inequality* - A social problem can also be cited here. In many countries, women are not getting their due share in anything. Food is no exception. Whenever there is a shortage of food in the family, it is the women who sacrifice it for the sake of the other members of the family. This is one reason why anaemia and other such effects of malnutrition are more prevalent among women and girls. Considering that women form 50% of the people in the world, this is not a good sign. To correct such anomalies, social changes are essential.

Changes Caused by Agriculture and Overgrazing

Wrong policies used in agriculture and also overgrazing are having disastrous effects on air, water and land. These are:

- Monoculture (cultivation of single crop in a field) is very bad for land. It depletes nutrients of soil.
- When a field is irrigated too much, the salts in the deeper layers of the soil are brought up. This makes the soil saline and also waterlogged.
- In many places, marshes are drained and made into fields. This disturbs the whole ecosystem.
- Trees in forests are burnt down to convert them to croplands. This also has many serious consequences.
- Shifting cultivation is bad for the land. Here, when a plot becomes infertile due to agriculture, the farmer abandons that plot and starts cultivation in another plot.
- By depending on hybrid seeds only, farmers are neglecting completely the traditional seeds. This leads to loss of genetic diversity.
- When grains and other crop produce from the developed world are imported at cheap price in the developing countries, the farmers in the latter countries adopt any method, sustainable or unsustainable, to compete with the imported stuff. This degrades the quality of the land forever.
- Too much fossil fuel is burnt to run the machinery and pump water.
- Use of too much fertilizers and pesticides in agriculture pollute the environment.
- Overgrazing is very bad for the soil. The soil becomes hard as heavy animals walk on it for long. Water cannot go down in such

a hard soil. Seeds cannot germinate in hard soils. Soon, the area becomes barren as the tops of the grasses and other plants are munched off by cattle and other animals. Thus overgrazing results into the spread of deserts.

Effects of Modern Agriculture

The main solution to the world food problem is raising cropland productivity. All countries, including India, are trying their best to do this. Many effects of modern agricultural practices are seen all over the world these days.

- *The most popular and common practice is to try to increase the yield per unit area. This process is called Green Revolution.* For this, many methods are being used.
    1. Plant breeding has given rise to many varieties better than the usually cultivated ones in many aspects like increased yield, disease resistance, draught resistance, etc.
    2. Irrigation is being done on a much larger scale.
    3. More chemical fertilizers and biofertilizers are being used
    4. Better weather forecasts help to reduce losses
    5. Better awareness regarding modern practices of soil testing, disease control, seed treatment, use of hormones, biopesticides, etc. have helped to increase the yield
    6. Biotechnology and genetic engineering are also of help. Tissue culture plants have become very common. Transgenic plants (Genetically Modified Crops) have also made their appearance.
- Another method is to increase the number of crops in each field by *multiple cropping*. The farms where monoculture was the practice are being converted into multiple cropping fields. This has lead to increasing world food production greatly.
- The third method is to *utilize different crop residues for various purposes*. Thus now we have rice bran oil, pellets made of crop residues for fish ponds, biogas plants, etc. Such measures increase the overall efficiency of *integrated farming* (farming in which agriculture, animal husbandry, poultry, fishing, etc. are all integrated).

Green Revolution involves many steps:

1. Developing monocultures of high yielding varieties
2. Using fertilizers, pesticides and water in large quantities
3. Multiple cropping

The First Green Revolution was started in 1950. Where as the Second Green Revolution has been taking place since 1967. It was at this time that fast growing dwarf varieties of rice and wheat specially bred for tropical countries were released. Between 1970 and 1992, India doubled its food production.

However, the harmful effects of Green Revolution are being felt more and more. An abstract from M.S. Swaminathan's address to the Indian Science Congress held at Varanasi in January 1968 is being reproduced here:

"Intensive cultivation of land without conservation of soil fertility and soil structure would lead ultimately to springing up of deserts. Irrigation without arrangements for drainage would result in alkaline or saline soil. Indiscriminate use of pesticides, fungicides and herbicides could cause adverse changes in biological balance as well as lead to an increase in the incidence of cancer and other diseases, through the toxic residues present in the grains or other edible parts. Unscientific tapping of underground water would lead to the rapid exhaustion of this wonderful capital resource left to us through ages of natural farming. The rapid replacement of numerous locally adapted varieties with one or two high yielding strains in large contiguous areas would result in the spread of serious diseases capable of wiping out entire crops … ."

Thus it is most important to think of the consequences before starting new programmes in agriculture or any scheme in increasing world food production.

Fertilizer – Pesticide Problems

Use of inorganic fertilizers in agriculture to increase production is nothing new. These have been and in fact are being used worldwide. As far as India is concerned, fertilizers have played an important role in making the country self-sufficient in food grain production. *Urea is the most widely used fertilizer. It contributes about 80% of the nitrogen needs of the country.* Other common fertilizers used are Ammonium sulphate, Potassium nitrate, Superphosphate, etc.

Some agricultural scientists argue that *fertilizer use is a must in India for following two reasons.*

1. As more land can not be brought under agriculture in the already overpopulated India, productivity has to be increased per unit area.

2. The main grains used here are wheat and rice and both these give higher yield on nitrogen addition.

The farmers even feel that the present prices of fertilizers should be brought down by the Government as most of them are small or marginal farmers. They believe that there is a need to increase both fertilizer consumption and productivity of agriculture.

On the other hand, some other agriculturists as well as all environmentalists are against the use of fertilizers in more quantity. They argue that by reducing the price of nitrogen fertilizers, the Government is promoting inefficient fertilizer use. Even after a 10% reduction in the cost of nitrogenous fertilizers, production did not increase proportionately. This is because soils in India suffer from deficiency in many nutrients.

Moreover, the *gap between nutrient addition and nutrient removal is very high*. This shows that soil fertility is being depleted. Therefore, the solution is not in adding more fertilizers, but an integrated nutrient supply. Along with fertilizers, organic manures like cow dung, green manure, crop residue, compost, vermicompost, biofertilizers, etc. must be given.

Use of too many inorganic fertilizers pollute not only soil, but also water and air. From the soil they reach surface water along with the runaway water. From the surface waters, they leach into aquifers also. Fertilizers, which are used in greater quantities than pesticides, are bad for many aquatic organisms. It has been observed that some amphibians are very sensitive to the nitrogen compounds leaching out of fields. For example, the Oregon spotted frogs are poisoned by even very low nitrates or nitrites – so low that it passes the drinking water quality standards set by the US Environmental Protection Agency (EPA).

One main problem is that *farmers apply too much nitrogen to the soil* for better yields. It has been proved that in China, crops use only 40% of the nitrogen given. The rest get dissolved in water and moves down to join the aquifers. In many parts of Europe, US and India also, farm runoff is polluting surface and ground water.

Some of the extra nitrogen from the fertilizers is released into the air also in the form of nitrogen oxides. These are converted to nitric acid (HNO_3) that returns to the earth along with rain water. This acid rain makes the soil acidic.

Solutions to fertilizer problem:

1. Reduce use of inorganic fertilizers in farms; use them in combination with organic manure like compost, vermicompost, etc.

2. Use biofertilizers like blue-green algae, Azolla (a water fern), etc. in fields.
3. Do not use fertilizers on slopes at all.
4. Use slow release fertilizers if possible (such fertlisers have been introduced in the market – they release chemical only slowly into the soil).
5. Grow alternate rows of crops with pulses – the bacteria present in the roots of pulses fix atmospheric nitrogen, so there is no need to give extra nitrogen to the plants.
6. Plant some permanent plants between cultivated fields and nearby water bodies.

Pesticides are chemicals that kill pests, i.e. undesirable organisms like harmful insects. In a broad sense of the term, insecticides (chemicals that kill insects), fungicides (chemicals that kill fungi), weedicides (chemicals that kill weeds), etc. are all pesticides. However in general, pesticides are considered to be the same as insecticides.

To save plants from pests of different types, pesticides like DDT, dieldrin, aldrine, etc. are used. DDT is a non-biodegradable compound. It does not decompose but accumulates in the soil. Different types of weedicides like 2,4-D, 2,3,5-T, etc. are sprayed in croplands to kill unwanted weeds.

Harmful effects of pesticides:

- Pesticides pollute soil.
- They are washed off from the crop plants during rain or irrigation and reach water bodies too.
- The remains of pesticides on the leaves or fruits make them poisonous. Those who eat unwashed fruits and vegetables have a chance of getting food poisoning.
- Pesticides remain on the grains also.
- Along with the crop residue, they enter into the bodies of farm animals and make them ill.
- Dioxin is a by-product formed during the synthesis of herbicides. This compound was used extensively in the Vietnam War to kill crops in large areas. It can be seen many kilometers around the site of application. Dioxin is extremely toxic to mammals, causing liver disorders, nerve problems and even cancer.

Methods to control pesticide pollution:

1. Apply pesticides only when needed.
2. Reduce the need of pesticides by using biological control.

3. Use biopesticides like neem seed oil, neem oil cakes, etc. in place of chemical pesticides.
4. Reduce non-farm use of pesticides in lawns, public lands, golf courses, etc.
5. In animal farms, control animal density.
6. Use runoff from animal farms on croplands or forests.

These measures are usually not taken seriously by many. According to a survey made by the WHO (World Health Organization), more than 50,000 people are poisoned and more than 5,000 die in developing countries as a result of toxic chemicals used in agriculture. Samples of fruits, vegetables, eggs and milk collected from many cities in India contain pesticide residues much higher than human tolerance limits by WHO. Pesticide content in soft drinks like Coco Cola and Pepsi has become a big controversy in the recent days.

E) Energy Resources

All forms of life need different types of resources to live. Energy is one of them. *Life on earth depends on the sun.* 99% of the energy used by living organisms comes directly or indirectly from the sun. Thus solar energy is the most important source of energy for us.

Green plants are the primary producers, i.e. it is the green plants that trap solar energy and use it in making food for themselves and all other organisms on earth. (This process by which plants make food using solar energy is called photosynthesis, i.e. synthesis or manufacture in presence of light). In addition to man's primary need of food, most of his other needs are also fulfilled ultimately by plants.

It is the oxygen given out mainly by plants during photosynthesis that we breathe in. Similarly, the carbon dioxide that we breathe out is used by plants for photosynthesis. This shows that the whole cycle of exchange between oxygen and carbon dioxide so very essential for life to exist on earth is running smoothly because of the solar energy.

Solar energy also helps directly or indirectly in the circulation of material in this world. For example, it is the heat of the sun that evaporates water in the water bodies and soil into water vapour which forms clouds in the sky. Rain formed from the clouds falls into the water bodies and on the soil. This is the main part of the water cycle or the hydrological cycle.

It is because of this water cycle that rivers do not dry up. The electricity that we get in our homes comes mainly from hydroelectric projects which use the force of running water of the rivers. Electricity is most essential for industries and to run agricultural machineries. In fact, we cannot imagine life without electricity.

Electricity is also made using thermal power. For this generation of heat, coal is the fuel used. Coal is nothing but trees that lived many years ago and buried under the ground, which have been subjected to high pressure and high temperature. Thus *the energy that we get while burning coal is actually the solar energy* trapped by those trees thousands of years ago in photosynthesis.

Petrol and diesel used for transport and the kerosene used for lighting by the rural/poor people are obtained from petroleum which is an another fossil fuel. It occurs as crude oil in underground oil fields in different parts of the world. *Petroleum is formed by bacterial decomposition of marine microscopic plant and animal fossils.* The energy in these organisms that lived in the past was of course solar energy converted into food energy. This energy is used in transport of people, raw materials and manufactured goods from place to place and plays a most significant role in day to day life.

Thus *almost all the energy that living organisms use comes ultimately from the sun*. Other forms of energy are atomic energy, wind energy, tidal energy, etc.

Growing Energy Needs

People need energy for many things – for agriculture, domestic appliances, industry, transport, etc. All these are much more in developed countries. Agriculture is almost fully mechanized there; industries and transport facilities are more; and people use a variety of electrical appliances. In short, the energy requirement of the developed countries far exceeds the energy requirement of the developing countries.

Some of the most important minerals are used as fuel for power – like petroleum and coal in all countries – developed or developing. These and natural gases are all fossil fuels. These form a major chunk of the fuel energy used all over the world. These resources evolved millions of years ago as plant and animal matter decomposed and was converted under high temperature and pressure under the earth's surface into hydrocarbon compounds that we now call fossils fuels. Since the beginning of the machine age, industrial societies have become increasingly dependent on fossil fuels. This must cause us concern. There are two main reasons for this.

1. *Fossil fuel resource is limited in amount.* These were produced

by solar energy in the distant past and when they finish off they cannot be replenished. Fossils are still being formed, but at a very, very negligible rate compared to the rate at which we are consuming them. We first began consuming fossil fuels at an appreciable rate only about 150 years ago. How long will they last? On a global scale, we still have some coal for a few centuries more, but natural gas and oil will be in short supply in a few next decades.

In the United States, the situation is worse than the global average because they are depleting their resources at a faster rate than in other fossil fuel rich countries. In fact, the entire stock of fossil fuel available has been held in storage under the earth's surface for more than a hundred million years; and now it is being completely exploited in only a few centuries.

2. *Unexpected environmental consequences result from the extensive use of fossil fuels* for everything from cooking in our kitchens to powering our automobiles. When we burn coal, natural gas or oil to obtain energy, gaseous compounds are formed and dumped into the atmosphere. This is causing problems. For many years, it was felt that the emitted gases were not significant as the earth's atmosphere is very vast. But now with increasing world population and industrialization, this is no longer true. Atmospheric pollution is causing health problems and even death. Carbon dioxide emissions are threatening to produce climatic changes over the entire globe. It is thus high time we find solutions to these problems of resource depletion and environmental pollution.

Coal, petroleum and natural gas give conventional forms of energy. Wood which is used mainly in the developing countries also provides a conventional form of energy. In these countries, *large quantities of wood from forests are used as fuel*. This leads to deforestation which causes soil erosion, floods, desertification and many other grave consequences. Burning of firewood leads to atmospheric pollution. Thus *all conventional forms of energy cause pollution*.

The energy that is sold in the market is called commercial energy. This is what the developed countries use most. In the developing countries, most people use non-commercial energy like firewood, dung and crop wastes that they collect on their own. These together can be called biomass. Thus, while the quantity of biomass used as energy source is just 3% in the developed countries, it is as much as 35% in the developing countries.

Similarly, there are differences in the other forms of energy used too. The percentage of natural gas and nuclear energy used by developed countries is much higher.

The developed countries are sure to face shortage of energy resources in a few decades, but the developing countries are already facing it. Cutting wood for fuel has lead to deforestation in vast areas in these countries.

The USA uses maximum energy resources in the world. The Americans use 25% of the world's commercial energy even though they account for only 4.7% of the world's population. They also waste a large quantity of energy. At the same time, though India contributes to 16% of the world's population it consumes only 3% of the world's commercial energy.

Yet, a critical analysis will show that *energy needs are growing everywhere*. The developing countries are striving hard to become developed, and for this, they will have to spend more energy. At the same time, the developed countries are also making material advances fast. Thus energy needs are increasing at a fast pace all over the world. It is estimated that the *world's energy needs have been increased four times during the period from 1950 to 1990*. The world's demand for electricity has doubled in the past 22 years. Added to this is the fact that electrification is far from complete in all settlements of the developing world. A majority of the rural areas in the Less Developed Countries (LDCs) do not have electricity at present. Even where there are electric connections, supply is irregular due to real shortage in generation or loss in transmission. In such a situation, wasting electric power is a crime.

It is because of non-availability of conventional energy that man is now trying to tap non-conventional energy resources more and more for his various needs.

Renewable and Non-renewable Energy Resources

Energy resources are of 2 types – non-renewable and renewable. The renewable energy sources pollute the environment very little while the non-renewable forms pollute environment on a large scale.

There are basic differences between the use of commercial energy in the developed or More Developed Countries (MDCs) and the developing or Less Developed Countries (LDCs). *While the MDCs use more of non- renewable energy, the LDCs use more of renewable energy.*

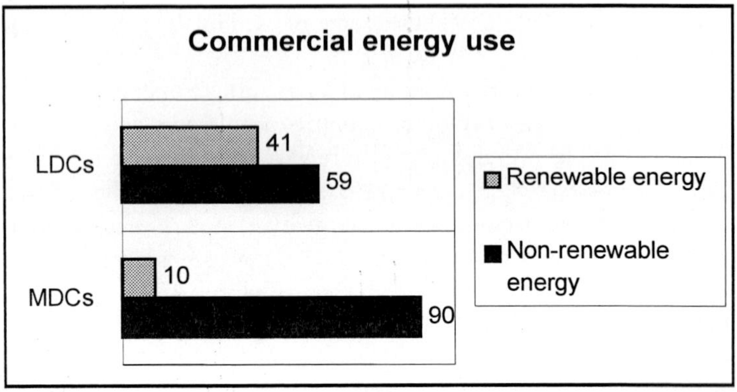

Use of renewable energy

	Biomass	Hydropower, Geothermal, Solar	Total ***c***
MDCs	33	7	10
LDCs	35	6	41

Use of non-renewable energy

	Coal	Oil	Natural gas	Nuclear power	Total
MDCs	25	37	23	5	90
LDCs	25	26	7	1	59

From this data reproduced from Tyler Miller Jr., it is clear that the MDCs are using more non-renewable sources of energy available in the world. But this does not mean that we can relax. We should also try to reduce our use of non-renewable energy, especially coal and oil as they are highly polluting also.

Renewable Energy Resources

If a resource can be renewed in a short time, such as a few hours to a few decades through natural processes, it is called a renewable resource. Examples of renewable energy resources are solar energy, hydropower, wind energy, geothermal energy, tidal energy, energy from burning biomass (firewood, dried dung, municipal solid waste, agrowaste, etc.) and the like.

The non-renewable energy sources have been in use for centuries now. These are therefore also called conventional energy sources. The renewable sources, also called non-conventional or alternate energy sources are becoming more and more popular these days as the scare of the underground deposits of coal, petroleum and natural gas getting exhausted has become a major concern. On the whole, we should turn to alternative energy sources as soon as possible and as much as possible. There is no fear that they will be exhausted; they are environment friendly also.

Non-renewable Energy Resources

Energy resources that are available only in limited amounts in nature, as it takes a very long time, tens of thousands of years or more, to be renewed are called non-renewable energy resources. So if continuously used, their supply will end one day. Examples of non-renewable energy resources are coal, mineral oil, natural gas and nuclear power. Of these, coal, oil and natural gas are organic in origin and can be considered as fossils. So they are also called *fossil fuels*.

Coal: Coal is one of the oldest forms of energy. Location of industries was earlier decided by nearness to coal deposits. However, discovery of oil and invention of electricity have reduced the importance of coal. Its low heat output, untidy appearance and great volume which increases transport costs are negative factors. Still, it is used widely in many countries as domestic fuel, as a source of industrial power and as a raw material in some chemical industries.

Coal is a very important solid fuel needed for many industries. It is also the commonest domestic fuel. There are different types of coal containing different amounts of carbon. The more the carbon content, the higher is the quality of the coal.

S. No.	Type of coal	% of carbon	Qualities as a fuel
1	Peat	50-60%	Lowest grade coal, producing less heat and more smoke and ash
2	Lignite or Brown Coal	60-70%	Burns easily with some smoke
3	Bituminous coal (Black hard coal)	70-90%	Burns with lesser smoke, used as domestic fuel
4	Anthracite	90-95%	The best quality coal. Very hard, shining, producing intense heat and minimum smoke

- *Anthracite* comprises only 5% of the available coal. It is used mainly for making steam in industries and for home heating in cold countries. Due to scarcity and higher prices, anthracite is used only when other types of coal are less suitable.
- *Bituminous coal* is the most abundant. It is of 3 types. Steam coal is the superior grade used in steam ships, locomotives and engines. Coking coal softens on burning to form coke. Coke is used in blast furnaces to extract iron from iron ore. Gas coal is the lowest rank of bituminous coal. When burnt, it gives off coal gas and other byproducts.
- *Lignite* is a low category coal. It is widely used in domestic heating to generate thermal electricity and in chemical industry.
- *Peat* is formed from dead plant remains that settle at the bottom of ponds. It is taken out, dried, cut and burnt in homes. It has only local importance.

Petroleum: Petroleum is also called mineral oil. Oil fields are situated near coastal regions. An oil field contains a layer of petroleum and natural gases above a layer of salt water. On drilling a hole into the oil well, which may be a few hundred to a few thousands of feet deep, crude oil and natural gases gush out. The crude oil is then carried by pipelines to oil refineries where it is refined into petrol, diesel, kerosene, etc. A large number of by-products are also obtained from the crude oil.

Uses of Petroleum:

- It is the indispensable motor fuel. Without petrol and diesel, the modern transport system will simply break down.
- It is a vital source of power. Vast amounts of mineral oil are burnt daily in heaters, boilers and furnaces to provide power in factories and to generate thermal electricity.
- By-products are many, like grease used as lubricant for machines, paraffin and wax used to make candles, etc.

Natural Gas: Natural gas occurs along with petroleum. Some wells yield oil but not much gas while some other wells yield only gas. Natural gas is mainly methane and ethane. The gas has to be properly purified before using as a source of power. It is used in industries and also for domestic purposes, for heating and lighting. It is often cheaper than electricity. In America, Europe, Russia and New Zealand, natural gas is pumped through pipes for industrial and household purposes.

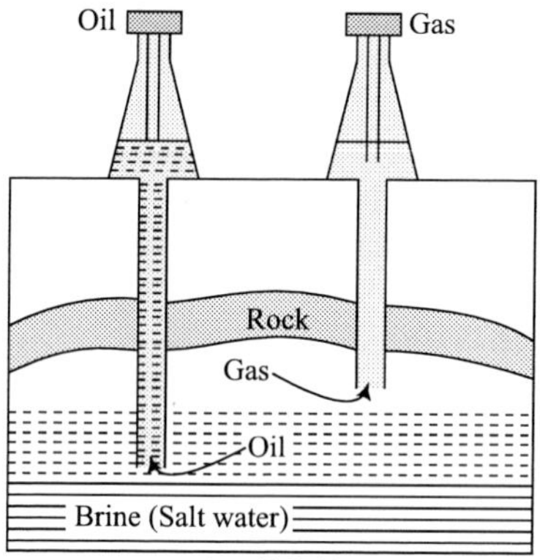

An oil field

Nuclear Energy: Strictly speaking, only very minute quantities of Uranium are needed to make nuclear power and so in effect, the supply is limitless. However, as Uranium deposits underground are limited, nuclear energy is included in non-renewable resources.

Nuclear energy was discovered in 1938-39 in Germany. To produce nuclear energy, the nuclei of uranium atoms are split with the help of neutrons. When the nuclei split, mass is converted to energy. The amount of this energy is very large. The heat generated during fission (splitting of nuclei of atoms) is used to heat water. One kg of uranium can generate energy equivalent to that obtained from 3,000 tons of coal.

Initially nuclear energy was used for producing nuclear weapons for the world war. The first nuclear reactor was assembled at the University of Chicago in 1942. After the war, efforts were made to utilize nuclear reactors for the generation of commercial electricity, for propelling submarines and other ships of the navy.

Electricity for civilians was made first by a reactor at Pennsylvania in 1957. This was just the beginning. Many countries set up nuclear reactors. In India, Dr. Homi Bhabha is the father of nuclear power development. Bhabha Atomic Research Centre (BARC) conduct research in this line. India has 10 nuclear reactors and produces 2% of our electricity. India has uranium deposits in Jadugoda in Jharkhand. Thorium deposits are found in Southern Kerala. Our power plants are located in Maharashtra, Rajasthan, Tamil Nadu, Uttar Pradesh and Gujarat.

Nuclear energy is an inexpensive and environment friendly source of energy. However, it has some serious disadvantages:

1. Even when maximum care is taken, some radioactive gases may escape through microscopic cracks in the reactor.
2. Serious accidents can occur with nuclear reactors as happened in Chernobyl.
3. The biggest problem is that of safe nuclear waste disposal.

Due to these drawbacks, many countries have stopped tapping this source. For example, Sweden has decided to become nuclear free by 2010.

Use of Alternate Energy Resources

Solar energy: The energy from the sun is a renewable resource as the sun will continue to give energy to the earth for more than about 6 billion years. So we can consider solar energy as inexhaustible. The energy that the earth receives from the sun in the form of electromagnetic radiation is solar energy. It includes infrared, ultraviolet, etc. in addition to visible light.

To put solar energy to use, different types of solar cells have been designed. These are used for different purposes ranging from heating water to cooking to converting solar energy directly to electrical energy. Various technologies have been developed for these.

Solar energy is used most commonly for water heating. Such systems have two parts – a solar collector and a solar tank. The collector kept on the terrace heats the water; the heated water flows into a well insulated storage tank. Collectors are of many types. A flat-plate collector is an insulated rectangular box. It has a transparent cover that faces the sun. Inside the box, is a metallic plate painted black. Many tubes are attached to the metallic plate. Through these plates, water flows. As the collector gets hot, the water is heated too. Pipes carrying this hot water to the tank and from there to wherever it has to be used are well insulated too. Such solar water heaters have become very common. These can be seen in many hostels, restaurants, etc.

Solar energy is nonpolluting and safe. However variations in solar intensity over day and night, summer and winter and also changing cloud cover complicate the use of solar energy. There must be a storage system for cloudy and rainy days.

Solar cells, also called solar photovoltaic cells have become common in some countries. A solar cell has two layers of silicon. Each layer contains an electric charge. When sunlight strikes the cells, the charges move between the two silicon layers and electricity is produced. Calculators, watches, etc. running on solar cells have become very common.

In India, in more than 300 villages of Rajasthan, Gujrat, UP and Haryana solar energy is used for street lighting. Solar furnaces at Pune, Rohtak, Hissar, etc. are functioning very well. As such, we dry our clothes in the sun, which is also in fact usage of solar energy. In developed countries, imagine how much electrical energy is used in the washing machines with dryers.

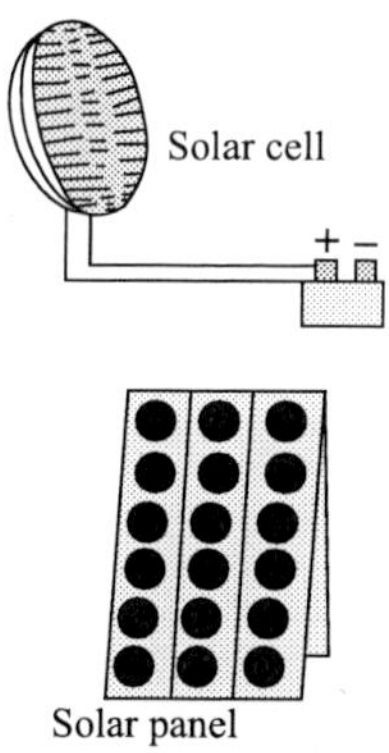

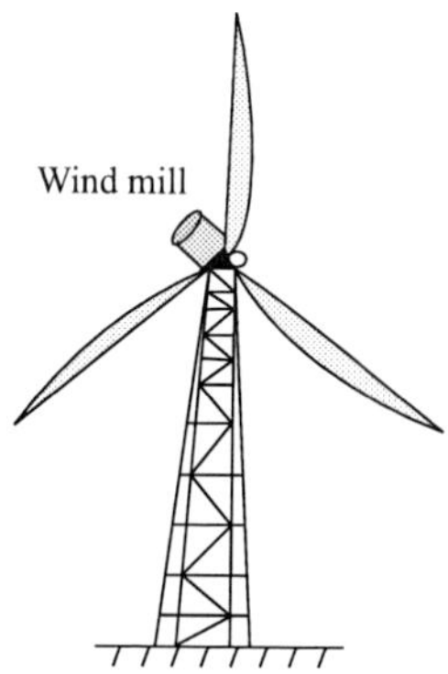

Wind energy: Even in ancient times, sailors used wind blowing on the masts to steer their boats. The large ships of the Nineteenth century could extract as much as 10,000 hp (horsepower) energy from wind. Wind mills in thousands have been operating in different countries from the nineteenth century onwards. Wind energy is used to run mills, pumps, etc. and also to generate electricity.

In Denmark, 13% of the country's electricity is made using wind energy. The Government of Denmark is encouraging people to establish wind cooperatives. In these Cooperatives, established by the Government, people can invest their money. Thousands of Danish families have shares in such wind cooperatives. Wind mills produce noise and obstruct view; so the government has made it compulsory for the investors who benefit from wind energy to live within 3 km of the site. This ensures that no movements of opposition arise from any quarter. On the other hand, the non-polluting nature of the energy is making it more popular.

In India, Gujarat is the state where wind energy is being used extensively. Tamil Nadu produces 850 mW of electricity from using wind energy. India is the third largest producer of wind energy in the world at present.

Wind energy has two disadvantages. Firstly, as wind velocities are not constant, energy produced will fluctuate. Secondly, large areas are needed to set up wind mills. Yet advantages outweigh disadvantages.

Geothermal energy: *The naturally occurring heat from the interior of the earth flowing to the cooler surface is called geothermal energy.* Where heat flows are greater in volcanoes, hot springs, etc. Underground hot water has been used to produce electricity. Steam from geysers is used to drive turbines. Deep wells can be dug into the underground hot water reservoirs and the hot water is brought to the surface. Technologies for this have been developed which has become necessary as natural geysers are very rare.

No widespread commercially beneficial technologies have been developed thus far to use geothermal energy of hot rocks, molten magma etc. However, people have used geothermal energy from hot springs for bathing, aquaculture, green-housing, etc. In India, about 350 geothermal springs have been located. Good sites for power generation exist in Ladakh, MP, UP and HP.

Tidal energy: Tidal energy is extracted from the potential and kinetic energy of the earth–sun–moon system. Ocean tides result from this interaction. Water level fluctuates twice daily, alternately filling and emptying along the shore line. These currents can be used to drive water turbines connected to generators to produce electricity. Two systems are in practice. In the one way system, the incoming tide is blocked and the water collected is used to turn turbines when it goes back. In the two way system, both incoming and outgoing tides are used to generate electricity. The technology is very similar to hydro power. Tidal power stations are operating in Russia, France, Canada, etc. In India tidal power potential has been identified in the Gulf of Kutch, Gulf of Cambay, and the Sunderbans.

Disadvantages of tidal energy: The times of the tides keep changing. Storms may bring irregularities. Power stations disturb aquatic life and block navigation.

Hydroelectric power: *Hydel power or hydroelectric power is the most important source of electricity for us*. In large projects, water stored in dams built across large rivers is made to flow through huge pipes. This spins turbines to produce electricity. This is called hydropower. 50% of electricity in the developing countries like India is made by this method.

In Norway, 95% of electricity is hydropower. The hydroelectric plants have long durability. The source of energy is renewable, cost of operation is low, and it is a non-polluting energy. These are the advantages of hydroelectric power. However, large dams are not built anymore in any country as their disadvantages are more. Therefore only small projects are approved that use low dams on small streams, with or without reservoirs.

Energy from Hydrogen—the gaseous fuel: Hydrogen fuelled automobiles are set to hit the market on an experimental basis within months. BMW, the automobile manufacturers of Germany, have designed such cars. These cars will release water vapour instead of the poisonous carbon monoxide, nitrous oxides, etc. into the air! If such cars do become popular gradually, much of our air pollution due to automobiles will be reduced.

Energy from Biomass: This is of different types. Agro wastes like waste wood, green plant matter, straw, paddy husk, saw dust, etc. can be converted into fuel. In 1989, a rice straw fired thermal plant was set up in Punjab. Paddy husk can be made into smokeless solid fuel briquettes. These can be used for cooking at home, hotels, kilns and boilers. Such briquettes are made at the Alternate Hydro Energy Centre at Roorkee. Rice bran is used as a source of oil which is later converted to a methyl ester for use of fuel. Saw dust is converted into low calorific producer gas for thermal power generation. At the Central Mechanical Engineering and Research Institute at Durgapur, since 1988, solid garbage (leafy waste) is used to produce electricity. The garbage is digested anaerobically to produce biogas which is in turn used to generate electricity.

Biogas plants: It is true that India lives in its villages. For the villages, the best form of non-polluting alternative energy is biogas. Biogas plants are very useful in converting solid waste into energy. *Bio gas is a mixture of methane (CH_4), hydrogen (H_2), carbon dioxide (CO_2) and nitrogen (N_2).* It is produced by degradation of biological substances. 60 – 70 % of biogas is CH_4; 25 – 30% is CO_2 and the rest is composed of H_2 and N_2. About 600 litres of biogas can be produced per kg of dry organic matter.

In India, in many states, small and cheap biogas plants have been setup in villages. Biogas is commonly called gobar gas in India as the main raw material used is dung. Biogas is the best solution for the energy problem in rural areas. The technology involved is very environment friendly. Cow dung is available in our country in large quantities. Biogas made from cow dung can be used in place of firewood and kerosene in our villages. In the process, tons of wood can be saved every year. A by-product of the process called slurry can be used as manure.

There are different designs for biogas plants but in all modern ones the digesters (fermenters) are almost entirely underground. In a biogas plant, dung and water are mixed in the ratio 1:1. Dung is fed at a constant rate to the fermenter. Other than dung, the input may include recycled slurry, night soil, kitchen waste, etc. These help in the growth of anaerobic bacteria. pH is maintained at 7.

The cow dung mixing tank is placed on an earthen platform. It is fed with a continuous supply of cow dung mixture. From this tank, cow dung moves through a cow dung inlet pipe into the fermenter tank. There is a platform at the base of the tank.

The fermenter tank and other units are made with brick and cement or concrete. The lid is usually of tin. The biogas produced in the underground fermenter tank is kept under pressure below the lid held in place by a heavy iron rod. Biogas collects above the slurry level. From here, it is directed into gas outlet pipes. The flow of gas is controlled by the gas cock. The gas outlet pipe is connected to main pipes carrying biogas to gas stoves in homes. Biogas burns without smoke and is an excellent fuel.

The used slurry is directed to drying beds where it is converted into organic manure. Biogas can also be used as fuel in generator to produce electricity. Using 1m^3 of biogas, 160 watt bulb can give light for six to seven hours, or three meals for six persons can be cooked, or one HP motor can be run for two hours.

Benefits of biogas plants: Biogas plants are beneficial in different ways.

1. Prevents pollution – a lot of smoke is emitted on burning *uplas* (goyttas). Moreover, making *uplas* is difficult in the rainy season.
2. Increases fertility – wastes from biogas plants are used as manure.
3. Deforestation is prevented – no need to fall small trees and shrubs for firewood.
4. Improves family health – by avoiding smoke and dust, soot and coal.
5. Improves economic condition – villagers can lead a more prosperous life as biogas is cheap, convenient and healthy.

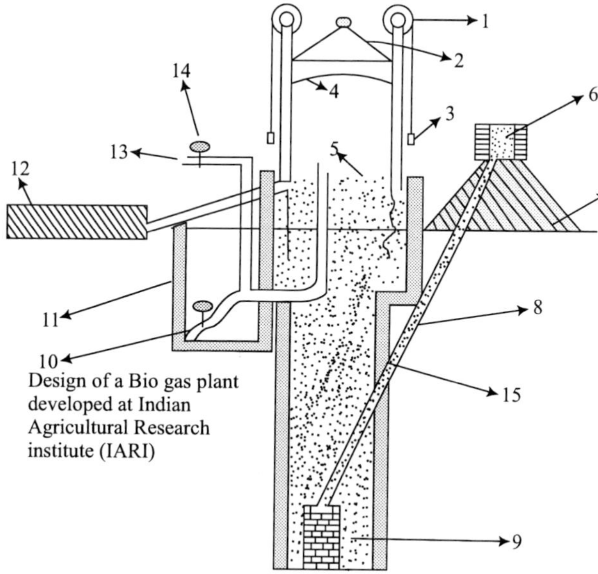

Fig. Design of a Bio gas Plant developed at India Agricultural Research Institute (IARI) 1. Pulley 2. Iron rod 3. Counter weight 4. Gas holder 5. Slurry level 6. Cow dung mixing tank 7. Earthen platform 8. Cow dung inlet pipe 9. Platform 10. Gas moisture exit trap 11. Brick wall 12. Drying bed 13. Gas outlet pipe 14. Gas cock 15. Fermenter tank

In India, biogas produced in villages is being used as a fuel in homes. We have a few million families using biogas as a fuel. In China also, biogas is very popular. In many European countries like Germany, France, England, Denmark, etc, biogas is used for making electricity on a very large scale. In India, some cities have begun such large scale projects. There is need to set up more such plants.

(F) Land Resources

Land is another very important natural resource. The forests which maintain the whole balance of nature and which conserve valuable wildlife, the croplands on which we grow all our crops, the pastures on which our cattle graze, the land on which rural and urban settlements are made and industries are set up, etc. are all land resources. Thus it is most essential to maintain the quality of land all over the world. Areas under shallow water are called wetlands. These occur between dry land and water bodies. Protection of wetlands is also very essential to maintain the equilibrium of the ecosystems of the earth.

One-fifth of the earth's surface is land. Of this, 51% is not usable as it consists of ice or snow, or is deserts or mountains. Forests, croplands, grazing land and arid land all are included in the rest 41%.

In India, however, these percentages are very different. First of all, cultivated land is much more here; it is the maximum in the world. This is because ours is an agricultural country and also grazing lands (pastures) are very less here. Moreover, forest land is also less here, compared to many countries in the world.

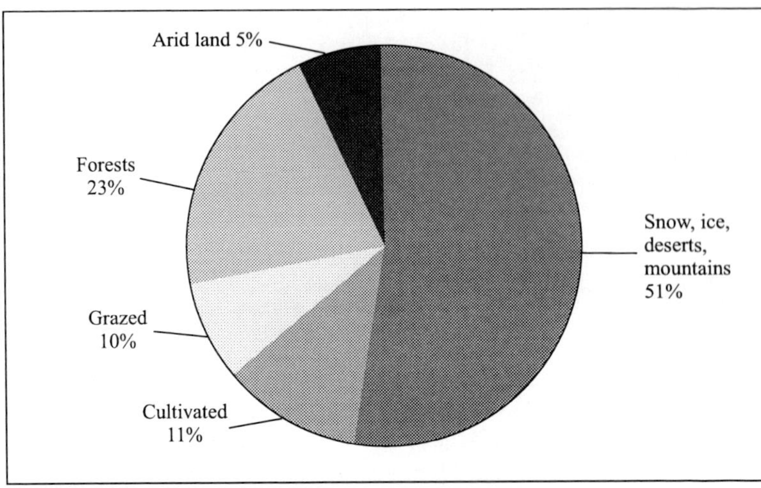

Land use percentages in world

Wastelands include the alkaline, saline, arid, rocky and sandy deserts and the mountains and the highly uneven land. Part of the wasteland is used as residential land.

Land is also needed for transport, industries, commerce, recreation, etc. As the total land area is fixed, towns and cities will have to grow

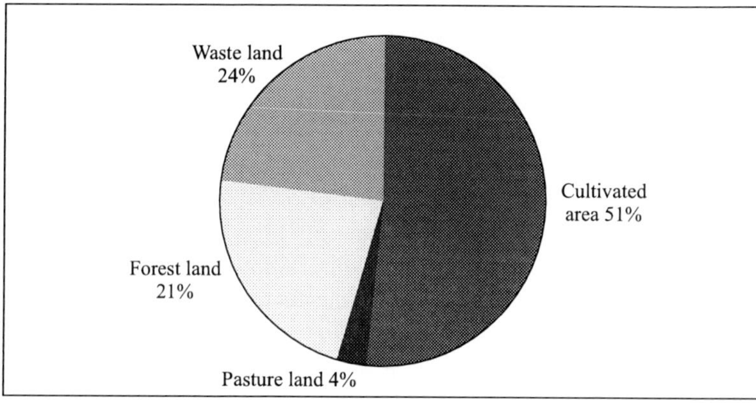

Land use percentages in India

upward now. Sidewise growth will hamper land being used for other purposes.

Land use planning is not properly done in India. For example, fertile agricultural land in the Yamuna basin is being used for spread of industry. Because of our high population, per capita land availability is very less in India. Man to land ratio is very low, 0.48 ha/person. The Government must take steps to survey the whole country properly, formulate the best land use plans and implement them strictly.

As 76% of Indians live in villages, it is all the more important to have proper plans because this vast population depends on land for their livelihood. If forests are depleted or pastures are overgrazed or toxic industrial wastes are dumped on it, valuable land will be converted to waste land. On the other hand, if utilized carefully and intelligently, land becomes a renewable natural resource. Maintaining some natural areas as protected areas is also essential to conserve our vast biodiversity.

Land Degradation

There has been a great increase in agricultural land in India since 1950. This was necessitated by the ever increasing population. In addition to more and more land being converted to croplands, the existing croplands are also over utilized. Increased irrigation has resulted in water logging and the bringing salts in the deeper layers of soil over ground. Use of agrochemicals (fertilizers and pesticides) has polluted land. These and many other factors have been responsible for the fast pace at which land is being degraded in many parts of the world. It has been estimated that every year, about 5 to 7 million hectares of land is getting degraded in the world. Thus the area of usable land is dwindling year after year.

Reasons for land degradation: Land is degraded in several ways.

1. *Deforestation*: Deforestation is a main cause of land degradation. Unmindful of the consequences, large areas of forest land is being degraded all over the world. Since agriculture began about 10,000 years ago, world's forest areas have decreased from 34% to 26% of the world land area. Forests are not only disappearing, they are being fragmented also due to roads, rails, human settlements, dams, etc.

In India, the figures available are not much reliable because of two reasons. Firstly, many non-forest areas like scrublands are included as forest land. Secondly, contradictory reports from different sources and agencies are circulated and one cannot know which one is true. Moreover, area of plantation forests should not be added to that of natural forests. In India, the forest area is 23% of our land area but the forest cover is only 19%. This difference is because plantation forests are considered as forests, but in real sense they are not true forests. What we see today in India is the sight of natural forests depleting and plantation forest increasing in area.

Dense forest cover which is undisturbed is called pristine forest cover. When the area of the pristine forest cover decreases, the population of wild animals like tiger also falls. Forest cover depletion is not only due to tree felling, but also due to human activities like converting forest land into croplands or fruit orchards.

In Himachal Pradesh, large areas of natural forests have been cleared for apple cultivation. This is causing land degradation on a large scale due to many factors:

- Apple starts yielding in 5 to 7 years. During this period, potato is cultivated as intercrop. Potato plants cannot bind soil at all; thus soil erosion increases greatly.
- These leads to death of many wild plants, animals and microorganisms.
- To pack apples cultivated in 1 hectare of cleared forest land, wood from 7 hectares is needed for the packing cases. This leads to further deforestation.
- Huge amounts of fungicides sprayed on the apple trees to save the apples from the scab disease pollute streams and land equally.

2. *Urbanization*: As urbanization goes on unchecked in the world,

more and more land is degraded due to many reasons. As the population of cities increases, there is a corresponding increase in many factors. The total concreted (pucca) surface increases greatly. The quantity of wastes also increases in amount. Moreover, there is a rise in the number of:

- Slums
- Buildings
- Roads
- Streets
- Sewage drains
- Vehicles
- Factories, etc.

All these lead to deterioration in the quality of land. Soil becomes hard and loses the capacity to hold and retain water. Pollution kills many useful soil microbes.

3. *Industrialization:* Industrial development is good to boost the economy of a country. But it should not be at the cost of environment. The main problem arises due to industrialization is pollution. The effluents, sometimes even hazardous wastes dumped outside the industries or even in the dumping yard are injurious to land. Substantial land area has been degraded due to dumping of industrial wastes.

4. *Mining:* During mining operations, large quantities of wastes are produced. For example, while mining metallic ores, large quantities of rocks, earthen materials and other wastes are produced. While mining for coal, surface materials from the earth's crust are removed to expose the coal seams. These wastes are heaped at other places. These may contain toxic substances too. The wastes gradually become mountains which occupy valuable land surface. The mined areas also, if not reclaimed, become degraded land.

5. *Water logging*: Excess irrigation leads to water logging. Waterlogged soils cannot support plant life. When the interspaces in the soil become filled with water, the air is driven away. In the absence of soil air, plants and also many microbes die. Moreover, the salts in the soil dissolve in the irrigation water and rise to the surface. When the water evaporates, the salts are left on the soil surface. Thus water logging leads to land degradation. World Bank survey (1995) reveals that India loses 1.2 to 6 million tones of food grains every year due to water logging.

6. *Desertification*: Desertification is a process by which the productive potential of a land reduces by 10% or more. It is caused mainly by human activities. Overgrazing is the biggest reason for desertification. Forest

margins are very commonly overgrazed. Deforestation is another reason. The spread of the desert robs the land of fertility and moisture. Pastures are also degraded if more animals graze on it than the area can support.

7. *Acid Rain:* Industries like thermal power plants and crude oil refineries and also automobiles release large amounts of sulphur dioxide into the air. It dissolves in the water films on the suspended particles in the air forming sulphuric acid. This corrosive acid falls down to the earth along with rainwater. This is called acid rain. Nitrogen oxides released by some industries form nitric acid; also causes acid rain. Acid rain makes soils acidic thus degrading the quality of the soil.

8. *Nutrient disorder*: When a land is cultivated intensely and the crops absorb the nutrients they need from the soil, the land becomes degraded due to mineral deficiency. The opposite of this also happens in croplands. Too much fertilizer is added to the soil. Both these are included in nutrient disorder and both lead to land degradation.

9. *Extension of cultivation to marginal land*: In many developing countries, when the poor have no land to cultivate, they try to cultivate crops on marginal lands like slopes, sandy soils, etc. Such lands are less fertile and get degraded fast.

10. *Soil erosion*: This is a very major cause of land degradation. *Loss of top soil or disturbance of the soil structure is called soil erosion.* Since all the nutrients needed by plants are present only in the top soil, its removal is a great loss resulting in land degradation.

Methods of Land Reclamation: *Bringing degraded land back to the normal condition is called land reclamation.* There are several methods of land reclamation. These may be mechanical, chemical or biological.

Mechanical method:

- To reclaim alkaline and saline soils, this method can be used. Make a network of deep cuts in the land to lower the water table. Salts leach deep down. Then irrigate by fresh water. Remove alkali on the surface by scrapers. Plant salt or alkali tolerant species at first.

Chemical methods:

- Reclaim alkali soils by addition of gypsum (Calcium sulphate)
- Reclaim acidic soils by adding lime.

Biological methods:

- Do reforestation through micropropagation. For this, select tree species which are adapted to grow in degraded land. Multiply in thousands by tissue culture and then plant on the degraded land.

- To fix sand dunes and prevent desertification, plant *Casuarina* trees. They grow fast and improve soil fertility by nitrogen fixation.
- Varieties of many plant species tolerant to salts have been developed by plant breeding and genetic engineering. Plant such varieties on land degraded by excessive salts.
- On degraded land, only a brief period is favourable for growth. Within this period, if saplings do not establish, no vegetation cover is possible. Use those available which when injected into the roots of saplings, promote growth.
- Use microbes for improving soil fertility of degraded land. Plant trees which have nitrogen fixing bacteria in their roots, like Shisham (*Dalbergia sisoo*), Siris (*Albizzia lebbek*), etc. on degraded land.
- Microbes are available which can grow on land polluted by toxic heavy metals released by industries. These can reclaim land degraded by industrial pollution.
- Use blue-green algae to reclaim alkaline and water logged land. Grow them in summer. Later, remove them and grow crops like rice there.
- Make vermicompost in the degraded soil. The earthworms will make the soil fertile.
- Do afforestation programme on land degraded by mining. Select suitable species. Take care to plant as many species as possible for biodiversity and protect them well. Make water holes for animals.

Soil Erosion

Soil is the top cover of the earth in which plants can grow. 4′ of soil is necessary for growing any economically important plant – plants for food, clothing or timber, etc. We completely depend on plants directly or indirectly for all our basic needs. And plants depend on soil. This shows how much soil is important for us. Destruction of top soil has dangerous consequences. One inch of top soil takes 400 to 1000 years to form. Yet, a single dust storm can blow it away or a single flood can wash it away.

Removal of the upper part of soil is called soil erosion. In soil erosion, the soil particles of the upper layers are removed from their original places and deposited in some other place.

Soil erosion is one of the most serious problems faced by the world today. Every year, about 2,500 million tons of soil are lost in the world.

Due to unchecked soil erosion, large areas in Africa, China and India and parts of America have been converted into deserts. In the USA alone, 3 million acres of land has been damaged by soil erosion. In India, thousands of tons of top soil, containing nutrients worth thousands of crores of rupees, is washed away every year. Soil erosion results in siltation of lakes and reservoirs on which we have invested thousands of crores. The Bhakra Nangal dam was designed to last for 400 years; but due to siltation caused by soil erosion, it may not last even for 150 years. Silt is choking our harbours. Eroded soil is deposited on riverbeds, raising them. This leads to floods, killing people and damaging property worth crores.

Soil erosion by wind is also severe in many parts of the world. In India, it affects about 50 million hectares of land, mostly in Rajasthan. This area is spreading due to deforestation, overgrazing and desertification. Fine dust is always suspended over the desert and so moisture laden clouds passing over Rajasthan do not precipitate into rain.

Soil erosion can be broadly classified into two types, based on the rate of loss. These are:

[I] Normal (Geological) erosion: This is a natural, slow process. Here, equilibrium between loss and build up is never lost. Soil is lost only as slowly as it is formed.

[II] Accelerated erosion: This is very rapid. It is much faster than soil formation.

1. *Erosion by water:* There are 5 types of erosion takes place by water.
    - (*a*) *Sheet erosion*: *Removal of soil in small but uniform amounts all over like a sheet is called sheet erosion*. The removed soil can be seen as silt in some other place.
    - (*b*) *Rill erosion*: Here, the running water moves rapidly and cuts small streams like channels on the land.
    - (*c*) *Gully erosion*: Many rills converge along slopes and join to form broad channels called gullies. Surface becomes uneven.
    - (*d*) *Riparian or Stream bank erosion*: This happens on river banks. The surface of fast moving rivers cuts the bank laterally. When a lot of soil is cut off in this way from beneath, the soil above tumbles down into the river with a big splash.
    - (*e*) *Slip erosion* (*Landslide*): In the rainy season, rocks or

masses of soil slip down. This is called slip erosion or landslide.

2. *Erosion by wind*: The soil erosion by wind are of following types.
   (a) *Saltation*: Salty soil carried by wind in small quantities is called saltation.
   (b) *Suspension*: In suspension, fine dust is lifted in the air and remains suspended there.
   (c) *Surface creep*: This is pushing of heavy soil particles by strong wind.

Classification of land into Capability Zones: For proper management, land is divided into 8 capability zones. Class 1 and 2 are level lands with no erosion. These are recommended for intensive agriculture. In Classes 3 to 8, restricted agriculture or no agriculture is permitted, these must be maintained as grazing lands or made into natural forests or national parks.

Prevention of soil erosion: Soil conservation by different means must be done to prevent soil erosion. The method adopted should be suitable to the area concerned, and the topography (form and surface behaviour) of the land concerned.

Methods to prevent soil erosion by water: There are different methods which can be broadly classified into two – Biological methods and Mechanical methods.

I. *Biological methods*: These are of four types:

A] *Agronomic practices*: Economic agriculture is called Agronomy. Agronomic methods of preventing soil erosion are of four types.

1. *Increasing soil fertility* – Plants do not grow well on infertile soils. Such a soil with poor vegetation is more exposed to water and wind erosion. So it is essential that soil is made fertile. Fertility depends on mineral content and humus content. In a fertile soil, plants grow luxuriantly and leave more residual material, which in turn, add to the humus. To increase soil fertility, many practices are followed. The 3 most important ones are:

   i) *Mulching*: Where cereals like wheat, maize, oats, barley, etc. and also cotton, tobacco, potato, etc. are cultivated, the basal parts of the plants called *stubble* are left in the fields after harvest. These form a protective layer, holding the soil together and reducing evaporation from soil and rate and amount of runoff. All these prevent soil erosion. Later, the stubble is ploughed under the soil. It is soon converted into humus and provides nutrients to the next crop. Organic matter increases soil aeration and stimulates activity of microorganisms. *The stubble is thus used as mulch and the process is called mulching.*

ii) *Adding fertilizers*: Plants absorb many minerals from the soil. So the soil becomes poor in nutrients. If in such a soil, the next crop is planted, the plant cover becomes poor and water and wind result into soil erosion. To avoid this, soil fertility should be maintained. For this, farmers add different inorganic commercial fertilizers like urea, ammonium sulphate, potassium nitrate, superphosphate, ammonium diphosphate, etc. and also organic fertlisers like cow dung, green manure, compost, vermicompost, etc. to the soil.

iii) *Ploughing*: Other than absorption by plants, minerals are lost by leaching also. *Leaching is a process by which soluble minerals dissolve in rain water and move deep down into the soil.* The bad effects of leaching can be overcome up to some extent by ploughing. By deep ploughing, some subsoil minerals are brought to the surface. Ploughing also breaks big lumps of soil and increases aeration and the water absorbing capacity of the soil.

2. *Crop rotation*: Crop rotation is also an important agronomic practice. It is a very old practice of growing different crops in regular order on the same area during successive years. In crop rotation, in the 1st year, maize or cotton, potato, etc. are planted and in the 2nd year, oats or other grain crops are planted. In the 3rd year, pulses like any gram, beans, groundnut, etc. are planted. Crop rotation decreases soil loss and preserves productivity of the land, controls erosion, weeds and diseases. The same crop year after year depletes soil of

minerals. This is overcome by planting pulses as they have nitrogen fixing bacteria in their root nodules. So when they are ploughed under, valuable Nitrogen is added to the soil.

3. *Contour planting or contour farming*: This is one of the oldest methods of cultivation, used in areas of low rainfall. The fields are prepared with alternate ridges and furrows. *Ridges at the same level are called contours.* Water is caught and stored in the furrows. This reduces runoff and erosion.
4. *Strip cropping*: Strip cropping involves planting of crops in rows or strips to check flow of water. This is of 2 types.
   i) *Contour strip cropping*: If entire slopes are ploughed and planted as one unit, rainwater will flow down and collect in lowlands. If the slope is long, flow of rainwater becomes faster and it cuts the sides more and carries more soil away. So long slopes are broken into many strips laid across the slope. Strips of dense growing and soil covering species are alternated with other crops which are not dense, and are cultivated leaving bare soil between rows. In this method, the force of running water is reduced. More water filters into the soil. Planting is done at 90° to the direction of the slope.

   ii) *Wind strip cropping*: Here also, strips are made, but in parallel rows at 90° to the direction of wind.

B] *Dry Farming*: This is the second biological method, after Agronomic Practices. This is useful for crops growing in low and moderate rainfall areas. In such areas, ordinary farming will be a loss. So such areas are put to different uses.

1. *Developing grasslands*: Grasslands are very effective in preventing soil erosion. India is not very rich in natural pastures (grazing grounds). So grasslands have to be grown. Wherever possible, vacant lands should be developed into grasslands. Open places in educational institutions, public offices and private concerns, hospitals and residential areas in the cities, wastelands around villages and towns, banks of canals, etc. should all be utilized. Bare soil between trees in

orchards and open forests should be covered with grasses. Grasses provide food to our cattle, sheep, goats and horses. They also help to absorb rainwater and increase the stability and fertility of the soil.

2. *Land fallowing*: This method is expensive. Here, the fields are ploughed and then allowed to lie idle for one season. This method builds up a good supply of valuable minerals, especially Nitrogen in the soil.

C] *Agrostological Methods*: All practices here involve use of grasses in different ways. Grasses like the 'Doob grass' (*Cynodon dactylon*) is commonly used as erosion resisting plants. This species have a very high Conservation Value of 90 to 97%. *Conservation value is the percentage of soil that would have been eroded in the absence of the species*. Thus, if *Cynodon dactylon* plants were not covering an area, 90 to 97% of the top soil of that area would have been washed away in a heavy rain. There are 3 main agrostological methods of soil conservation.

1. *Lay Farming*: Here, grass is grown in rotation with field crops. This helps to build up the soil structure, improves its fertility and prevents soil erosion.
2. *Retiring land to grass*: Where a lot of top soil has eroded, only grasses are grown.
3. *Strip growing of grass*: Here, grasses are grown in strips between crops. The grasses act as stabilizers.

D] *Control of Grazing*: Control of grazing is also included as a biological method of soil conservation. As cattle and sheep eat the green parts of the plants, the resulting loss of leaves leads to starvation of not only the shoots, but also the roots which hold down the soil. The surface soil is trampled under the hoofs of heavy animals. This makes the soil hard so that rainwater does not soak into it readily. The ground remains drier in comparison to undisturbed soil. As the tops of the plants are chopped off, re-seeding reduces. After many years, the soil becomes weak and useless. It is therefore necessary to prevent excess grazing. Fodder trees should be grown in semi-arid overgrazed areas. Trees like kubabul supplies high quality food to the animals. Village should be asked to grow such plants. Also, goats and sheep can be kept in enclosures and fed on tree fodder. These animals waste large quantities of energy roaming about pastures. This energy will be converted to milk and wool, respectively, if they are kept in healthy and well ventilated enclosures.

II. *Mechanical Methods*: These are used to supplement biological methods.

A] *Contour Ttrracing*: Here, the slopes are divided into a series of flat fields called *terraces*. In each terrace, a broad bottomed channel is made, so that there are ridges on both sides of the channel. These channels break up long slopes into a number of small watersheds. This method is called *broad based ridge terracing*. Another method is *channel terracing*. Here, channels are made at suitable intervals and the dug up soil is deposited as a wide, low ridge along the lower side of the channel. These serve as bunds along the contours. This helps to slow down the flow of water and leads it to the sides of the fields where it flows down slowly without eroding soil.

B] *Basin listing*: Here small basins are constructed along the contours to retain water. This reduces the velocity of the water flow and thereby erosion.

C] *Gully control*: Fast flowing water cuts channels in the hills. Such channels are called gullies. To prevent the formation of gullies, some measures can be taken, such as:

- Building dams or bunds across the gullies to catch silt
- Planting grass, trees or shrubs in gullies
- Blocking the flow of water by boards at regular intervals across gullies
- Constructing drains or diversions through which excess water can run off, etc.

D] *Stream bank protection*: To check the cutting and caving of stream or river banks, stream bank protection must be done. For this, the following measures can be taken:

- Vegetation can be grown alongside river banks
- Drains can be constructed
- Concrete or stone pitching can be done

Methods to prevent soil erosion by wind: This is also called control of wind erosion. Many methods are used for this:

- Covering the soil with vegetation
- Keeping the soil rough rather than smooth
- Strip cropping across the direction of wind

- Making furrows across slopes to hold water
- Planting rows of trees across the wind (These are called wind breaks)

These methods help to check the blowing away of valuable soil and its drifting on to crops, roads and farms. If the land is left to itself, sand storms blowing at high speeds will cover neighbouring vegetation and residential areas with sand. Furrows hold water badly needed by crops. Making furrows thus help in water conservation also along with preventing soil erosion by keeping the soil moist for longer periods. This produces more plant growth and provides a better vegetation cover which protects the soil.

Effects of wind erosion are clearly seen in Rajasthan. In west of the Aravalli Mountain Ranges, there is a vast sandy waste. Gigantic efforts are being made in Rajasthan to arrest the spread of the desert and to make the land productive. The Central Arid Zone Research Institute located at Jodhpur is doing a lot of research work in prevention of wind erosion. Control of wind erosion is an expensive process. So to avoid increase in costs, it is essential to implement projects with full vigour and complete them in time.

The Indira Gandhi Canal has been a boon to the people of Rajasthan. It is fed by both the Bhakra Nangal Dam and the Beas Dam. Work done by Rajendra Singh, the Magsaysay Award winner, is an example of how a dedicated man forming an NGO with some followers can work wonders. He made a river that had gone dry for the past 15 years flow again! He had adopted rain water harvesting methods to make this seemingly impossible task possible.

QUESTIONS

A] Very short answer type questions:

a) Multiple choice – Tick the right answer:

1. Coal is a
   (a) Renewable source of energy
   (b) Inexhaustible source of energy
   (c) Non-renewable source of energy
   (d) None of the above
2. Which of the following is a non-polluting source of energy?
   (a) Coal (b) Petroleum
   (c) Natural gas (d) Solar energy

3. For the formation of one inch of top soil, it takes
   (a) 10 to 20 years (b) 100 to 200 years
   (c) 300 to 500 years (d) 400 to 1000 years
4. Percentage of carbon dioxide in clean atmosphere is
   (a) 3% (b) 0.3%
   (c) 0.03% (d) 0.003%
5. DDT is a
   (a) Fungicide (b) Pesticide
   (c) Fertilizer (d) Disinfectant
6. Deforestation is caused by
   (a) Mining (b) Overgrazing
   (c) Dam building (d) All of the above
7. Major (Big) dams may cause
   (a) Earthquakes (b) Hurricanes
   (c) Storms (d) Cyclones
8. The percentage of water accounted by oceans and seas is
   (a) 90% (b) 87%
   (c) 97% (d) 99%
9. Maximum usage of water is in the field of
   (a) Industries (b) Homes
   (c) Agriculture (d) None of the above
10. Medha Patkar is famous for
    (a) Chipko Movement
    (b) Narmada Bachao Andolan
    (c) Green Revolution
    (d) Silent Valley Project
11. Rain water harvesting can be done using
    (a) Recharge pits (b) Recharge trenches
    (c) Abandoned bore wells (d) All of the above
12. Which of the following is a non-metallic mineral?
    (a) Gypsum (b) Iron
    (c) Zinc (d) Copper
13. Father of Nuclear Power Development in India is
    (a) Sunderlal Bahuguna (b) Homi Bhabha
    (c) M. S. Swaminathan (d) Rajendra Singh
14. Examples of renewable resources are
    (a) Wind energy (b) Solar energy
    (c) Tidal energy (d) all of the above

15. Biogas is made up mostly of
    (a) Hydrogen (b) Carbon dioxide
    (c) Ethane (d) Methane

b) Define:

1. Resource ecology
2. Natural resources
3. Atmosphere
4. Hydrosphere
5. Lithosphere
6. Biosphere
7. Inexhaustible resources
8. Exhaustible resources
9. Renewable resources
10. Non-renewable resources
11. Afforestation
12. Reforestation
13. Forest conservation
14. Drought
15. Minerals
16. Ores of minerals
17. Mining
18. Green revolution
19. Pesticides
20. Commercial energy
21. Pristine forest cover
22. Soil erosion
23. Soil
24. Soil degradation
25. Soil (land) reclamation
26. Sheet erosion
27. Agronomy
28. Leaching
29. Contours
30. Conservation value

c) Explain briefly:

1. Atmosphere
2. Hydrosphere
3. Lithosphere
4. Biosphere
5. Van Mahotsav
6. Joint Forest Movement (JFM)
7. Surface water and ground water
8. Flash floods
9. Reasons for drought
10. Three types of mineral resources
11. Minerals and mineral ores
12. Prospecting
13. Exploitation
14. Gold mining
15. Uranium mining
16. Types of agriculture
17. Traditional agriculture
18. Food resources in India

19. Integrated farming
20. Solutions to fertilizer problem
21. Methods to control pesticide pollution
22. Uses of petroleum
23. Natural gas
24. Disadvantages of nuclear energy
25. Disadvantages of wind energy
26. Geothermal energy
27. Tidal energy
28. Hydroelectric power
29. Energy from biomass
30. Benefits of biogas plants
31. Land degradation due to mining
32. Land degradation due to water logging
33. Desertification and land degradation
34. Land degradation due to acid rain
35. Role of nutrient disorder in land degradation
36. Types of soil erosion by water
37. Wind erosion
38. Leaching
39. Dry farming
40. Terraced cultivation
41. Contour planting
42. Stream bank protection

d) Name the following:
   1. Three main reasons for deforestation
   2. Five renewable energy resources

e) Differentiate between the following pairs:
   1. Contour strip cropping and Wind strip cropping
   2. Geological (normal) erosion and accelerated erosion

B] Short answer type questions:

Write short notes on:

1. Renewable and non-renewable resources
2. Natural resources and associated problems
3. Human activities leading to deforestation
4. Plan of action for forest conservation
5. Advantages of dams
6. Uses of water

7. Overexploitation of water
8. Methods to prevent desertification
9. Methods of flood control and flood management
10. Reasons for today's water shortage
11. Methods of water conservation
12. Water harvesting in villages
13. Water harvesting in urban areas
14. Stages in mining operations
15. Coal
16. Petroleum
17. Sustainable agriculture
18. Food resources in India
19. Changes caused by agriculture and overgrazing to environment
20. Green Revolution
21. Fertiliser pollution
22. Harmful effects of pesticides
23. Renewable and non-renewable energy resources
24. Solar energy
25. Wind energy
26. Land resources
27. Land degradation due to deforestation
28. Land degradation due to urbanization and industrialization
29. Consequences of soil erosion
30. Increasing soil fertility
31. Crop rotation
32. Control of grazing
33. Wind erosion
34. Biogas plants

C] Long answer type questions:

a) Answer the following in detail:

1. Give an account of the uses of forest resources.
2. What are the harmful effects of major dams?
3. Compare and contrast the effects of dams on forests and tribal people.
4. Explain in detail the causes of floods.
5. What do you understand by Rainwater Harvesting? Explain in detail.

b) Write long notes on any two:
   1. Harmful effects of major dams
   2. Conflicts over water
   3. Minerals of Jharkhand

c) Almost all energy that man uses comes from the sun – discuss.

d) Describe in detail the methods of prevention of soil erosion.

e) Write long notes on any two:
   1. World food problems
   2. Nuclear energy
   3. Bio gas plants
   4. Methods of land reclamation

Unit 7

Ecosystems

CONCEPT OF ECOSYSTEM

Plant and animal communities of any region influence each other and interact with one another. They together form a *biome* or a *biotic community*. In a biotic community, plants, animals, microbes and the non-living environment are constantly interacting and exchanging material. Trees provide shelter to birds. Birds help in pollination and dissemination of fruits and seeds. Plants purify the atmosphere. Animals provide carbon dioxide to the plants for photosynthesis. Microorganisms decay dead plants and animals thereby returning the minerals to the soil for plants to absorb. These minerals enter animals through plants. These are just a few examples. In short, the living and non-living components in nature together form a stable system. In this system, different components, like the well oiled parts of a machine, are co-coordinated to form a single unit. This is an ecosystem. *A natural unit of living communities and the non-living environment interacting and exchanging materials between themselves is called an ecological system or an ecosystem.*

Ecosystem is the basic functional unit of ecology. The study of ecosystems in fact covers the major part of the study of environmental science. This is because now-a-days, most of ecological studies are done at the level of ecosystem. This is called the *bioenergetic approach to ecology*.

The term 'ecosystem' was proposed by A.G. Tansley in 1935 in England. He defined it as the system resulting from the interaction of all living and non-living factors of the environment. Different terms were used earlier to refer ecosystem, like biocoenosis, geobiocoenosis, biogeocoenosis, microcosm, biosystem, ecocosm, etc. However, the term ecosystem is generally preferred and so it has become most commonly used term. *Eco* means environment; and *system* means an interacting, interdependent complex.

Size of ecosystems: *There is no fixed size for ecosystems. They may be very small or large or very large.* An ecosystem may be as large as a grassland or a river or a desert or an ocean. It may be as small as water in a ditch, a small lump of soil or a single fallen leaf. A forest is an ecosystem; a single tree is also an ecosystem; a single fallen leaf is an ecosystem too.

A small roadside ditch is an ecosystem which may harbour some bacteria or some blue-green or green algae. The water in the ditch has a certain temperature and pH. It is influenced by light, wind, atmospheric humidity, etc. It gets mineral from the underneath soil. It contains organic and inorganic compounds formed by the decomposition of the living organisms in it. Thus a roadside puddle is an ecosystem.

A tree is an ecosystem. It produces food by photosynthesis. It gives shelter to different types of birds, and sometimes also to monkeys and snakes. Many bugs, ticks, fleas, etc. live on the birds' bodies. There may be parasitic fungi and bacteria on or inside the bodies of these birds. The tree gives shade to the plants that grow below it. It reduces the light intensity and heat but increases the humidity available to these plants. The plants growing under the tree are protected from strong winds. The light they get is of a different quality as the tree reflects off most of the green rays in the incident light from its leaves. Many red and blue rays are used by the leaves for photosynthesis. Raindrops do not fall forcefully on the plants underneath. The soil under the tree contains more humus formed by the decomposition of the fallen leaves. It is more fertile. As the fallen leaves form a soil cover, there is less evaporation from the soil. There is no erosion under the tree as the soil is held in position by the roots of the tree. There are many insects and worms in the soil. There may be burrows of rats or snakes or rabbits under the tree. There are trillions of microorganisms in the soil. All these are performing different chemical reactions like nitrification, denitrification, ammonification, nitrogen fixation, etc. Some microbes are parasitic on other animals or plants. Altogether, there are so many actions, reactions and interactions in and around a single tree. It is one complete ecosystem. (Thus when one tree is cut, it is not one single tree, but an entire ecosystem that is destroyed. Understand this truth, and tell the same to all your friends.)

A single fallen leaf is also an ecosystem. It is being decomposed by thousands of microorganisms. Environmental factors are of different types. Climatic factors like light, temperature, wind, rain; edaphic factors like soil water, soil temperature, soil pH, soil acidity and alkalinity, and all other factors related to soil; topographic factors like exposure to sun and wind and rain; biotic factors like competition among the microbes, interrelationships among them, etc. are all working on the dead leaf fallen on the soil. Thus it is also an ecosystem.

Forests, grasslands, deserts, croplands, ponds, lakes, streams, rivers, seas, oceans, estuaries, marshes, etc. are all large ecosystems. The largest ecosystem is the whole earth. This is called the biosphere. This is the biggest ecosystem possible on earth.

It is for convenience in studying that we take up different ecosystems differently as forests, grasslands, deserts, rivers, etc. However in fact, all of them are linked with each other. There are no functional boundaries between them. For example, molecules of different gases and minerals circulate throughout the earth, in and out of living beings. Ecosystems of different places are all thus connected to each other. They are connected through time also. Ecosystems present at any one place at a given time are indeed related to those present at those spots years ago and also to those that will appear there in future through soil, water, gases, etc. Energy is flowing through all these ecosystems. Energy is transformed, circulated and accumulated in different parts of the ecosystems. The source of energy responsible for maintaining this giant system is the sun.

Green plants absorb sunlight and use the energy contained in it in making food. This process is called *photosynthesis.* The food produced by the green plants is used by themselves and also the animals which do not have the ability to produce food. When plants and animals die, their dead bodies are decomposed by microorganisms. The complex molecules in the dead bodies are broken down into simpler molecules and are finally released into the environment, to be used by living beings again. This circulation of materials taking place continuously in nature is responsible for the maintenance of the ecosystems.

Ecological niche: *Each organism, whether plant, animal, or microorganism, plays a definite role and occupies a definite position in the structure and operation of an ecosystem. This status or role of an organism within an ecosystem is called the ecological niche.* It includes all the physical, chemical and biological factors that an organism requires to live. It also describes the food source, the organisms for which it serves as food, and its effects on other organisms and the environment.

Biomass and standing crop: *The total amount of living organisms present in an ecosystem is called biomass*. It may be expressed as fresh weight, dry weight, carbon content, nitrogen content, or calorie content per unit area (square metre, acre, hectare, etc.) This varies at different times of the day and in different seasons. *The total amount of living material in an ecosystem at any given time is called the standing crop.*

STRUCTURE AND FUNCTIONS OF ECOSYSTEMS

To study any ecosystem, its structure and functions must be understood.

Structure includes three aspects:

1. The composition of the living beings (the biotic communities) of the ecosystem, like the species, their numbers, life history, spatial distribution, etc. The living beings may be producers, consumers or decomposers.
2. The quality and distribution of the non-living materials like water, inorganic and organic nutrients, etc.
3. The range of climatic conditions like temperature, light, etc.

The functional aspects of ecosystems include the following:

1. Food chains
2. Food webs
3. Ecological pyramids
4. Biogeochemical cycles
5. Energy flow, etc.

Components of an Ecosystem

The two main components of an ecosystem are the *abiotic component* and the *biotic component*.

I] Abiotic component: This is the non-living component of the ecosystem. The abiotic component influences the biotic component and is also influenced by it.

The abiotic component comprises of *three parts – inorganic, organic and physical.*

A) *Inorganic*: The inorganic component includes water, minerals, gases, etc.

These form the nutrients or raw materials for the green plants. Chlorophyll — the green pigment that functions in photosynthesis — is also an important inorganic chemical. *The amount of inorganic substances like Phosphorus, Sulphur, Carbon, Nitrogen, Hydrogen, etc. present in the environment of an ecosystem at any given time is called the standing state or standing quality.*

B) *Organic*: The organic component includes compounds like carbohydrates, fats and proteins belonging to the dead animals and plants and the excreta of the animals in the ecosystem. The intermediates and end products of the decomposition of these substances are also part of the organic component. Decomposition by bacteria and fungi ultimately converts these organic compounds into inorganic form, which are reutilized by the plants. Bacteria and fungi thus link the biotic and the abiotic components of the ecosystem.

C) *Physical*: The physical component includes climatic factors like light, temperature, wind, etc. In water bodies, in addition to these, there is also the pH of water, wave action, etc. The most important physical component is the solar energy which flows through the entire biotic community.

II] Biotic component: The biotic component includes living plants, animals and microorganisms. They all need food to live. Ultimately, all energy comes from the sun. Organisms in an ecosystem can be divided into **producers, consumers and decomposers.** All these organisms are placed in different energy levels (food levels). *Each energy level in an ecosystem is called a trophic level; different trophic levels differ from one another in the relative distance which separates them from the primary source of energy, i.e. the sun.*

Based on the nutritional level, the biotic component is again divided into autotrophic and hetaotrophic components.

A. *The autotrophic component*: This component includes the autotrophs. *Photosynthetic plants which can manufacture their own food are called autotrophs.* This includes all plants that contain the green pigment chlorophyll and also the photosynthetic bacteria containing bacteriochlorophyll. *Chemosynthetic bacteria are also autotrophic.* All these autotrophs fix energy and build up complex substances from simple substances.

   In an ecosystem, the autotrophic plants are the producers or primary producers (P or PP) and these form the first trophic level (T1).

B. *The heterotrophic component*: *Heterotrophs are organisms that do not have chlorophyll and therefore can not prepare their own food.* Organisms that cannot make their own food and get their food from the green plants directly or indirectly are the consumers

C. Animals, fungi and a majority of bacteria are consumers. They take in complex food materials and break them down into simple forms of food. Conversion of one compound into another by rearrangement also takes place inside the bodies of consumers. The Consumers are of two types – macroconsumers and microconsumers.

1. *Macroconsumers*: *Macroconsumers are large sized consumers. They include the herbivores, carnivores and top carnivores.*

i) *Herbivores*: *Animals and parasitic plants which feed directly on the green plants are called herbivores. They are consumers of the first order (C1) and belong to the second trophic level (T2).* They consume only food manufactured by the producers.
As herbivores are eaten by organisms of the next trophic level, they are sometimes called secondary producers. But it is better to avoid this usage as herbivores do not produce any food. Their bodies are eaten completely by the animals of the next level.
Examples of herbivorous insects are grasshoppers, bees, butterflies, etc. Examples of herbivorous animals are cows, buffaloes, goats, rabbits, meadow mice, deer, elephants, camels, monkeys, etc. Parrots, koels, bulbuls etc. are herbivorous birds.

ii) *Carnivores*: *Animals eating herbivores are called carnivores. They are consumers of the second order (C2) and constitute the third trophic level (T3).*
Insects like praying mantis, beetles and crickets; animals like small fishes, frogs, lizards, snakes, etc. and birds like sparrows, crows, woodpeckers, mynahs, etc. are carnivorous.

iii) *Top Carnivores*: *Top carnivores are animals that are not killed and eaten by other animals. They are tertiary consumers (C3) and constitute the fourth trophic level (T4).* Top carnivores eat carnivores and herbivores while carnivores eat herbivores only. This is another difference between carnivores and top carnivores. Examples of top carnivores are animals like tigers, lions, sharks, crocodiles, etc. and birds like eagles, hawks, kites, owls, kingfishers, gulls, etc. Animals of T3 and T4 are also called *predators*.

The trophic levels of some animals are not clear. Best example is man. He uses both plants and all types of animals as food and is an *omnivore.*

2. *Microconsumers*: *Microconsumers are the decomposers*. These are the saprophytic fungi and bacteria that live on or near the soil level. They obtain food from dead organisms and animal excreta. *On the land, fallen leaves and twigs and in water, dead algae and parts of other water plants*

that fall to the bottom of the water body form the detritus. The detritus is eaten by animals like protozoa, nematodes, snails, etc. and these are called detritus eaters.

The function of microbes like bacteria, fungi, etc. and the detritus eaters is similar; they are all decomposers.

Decomposers (D) are the top consumers (C4) and occupy the 5th trophic level (T5). They have a very important role in the functioning of the ecosystem. Minerals necessary for plants and animals will remain trapped in the dead bodies if there are no decomposers. This will lead to mineral deficiency of the soil which will lead ultimately to the death and extinction of all plants and animals. This calamity is prevented by the decomposers. Reutilisation of minerals is the basis of an active ecosystem. Some decomposers also fix atmospheric nitrogen, thereby increasing soil fertility.

Moreover, in the absence of decomposers, also called scavengers, land and ocean will be filled with dead bodies. In such a situation, there will be no place for the living!

TYPES OF ECOSYSTEMS

Ecosystems operate as self-sufficient interacting systems in the biosphere. Though all ecosystems have a basic plan, they differ in species composition, rate of production, etc.

Ecosystems may be natural or artificial. *Natural ecosystems* like undisturbed forests, deserts, etc. function under natural conditions without human interference. In *artificial or man-managed ecosystems*, there is a constant human interference. For example, in a cropland ecosystem, he adds manures and fertilizers, provides irrigation, sprays pesticides periodically, etc. In a zoo, he provides food to the animals and looks after their health and safety.

Based on habitat, ecosystems can be broadly classified into terrestrial (on land) and aquatic (in water). Examples of *terrestrial ecosystems* are forests, deserts, grasslands, croplands, etc. *Aquatic ecosystems* may be freshwater or marine. Freshwater ecosystems may be *lotic* (in running water like rivers, streams, springs, etc.) or *lentic* (in standing water like lakes, ponds, marshes, puddles, etc.). Marine ecosystems are the seas, oceans, estuaries, backwaters, etc.

Terrestrial Ecosystem

Main types of terrestrial ecosystems are forests, deserts, grasslands, croplands, savannas, tundras, etc.

Forest ecosystem: Forests are of different types. They may be tropical, temperate, alpine, etc. based on the type of geographical zones they are seen. They may be coniferous, deciduous, evergreen, thorn, mangrove, etc. based on the type of plants grow there. They may be rain forests, dry thorny forests, etc. based on the climate of the region. There are many more types of forests. In any country, one third of the land should be under forests. But in India, only about 8 to 10% of the land is under forests.

Abiotic components: Light, humidity, humus in the soil, etc. are important abiotic components. Light conditions are different in different strata. The composition of the organic and inorganic materials in the soil differs greatly depending on the tree species living in the forest.

Biotic components: Producers in forests are trees. Species differ in different types of forests. Ground vegetation is formed of various shrubs, ferns, herbs and terrestrial orchids. In tropical deciduous forests, teak (*Tectona*), palash (*Butea*), shisham (*Dalbergia*), etc. are seen. In temperate coniferous forests, conifers of different types are prominent. Examples are pine (*Pinus*), deodar (*Cedrus*), etc. Ground vegetation is very less in temperate forests.

Primary consumers: Primary consumers are the herbivores like ants, flies, beetles, leafhoppers, bugs, etc. and animals like elephants, monkeys, deer, pigs, squirrels, etc. and birds like koel, bulbul, parrots, etc.

Secondary consumers: These are carnivores like snakes, lizards, fox, bear, etc.

Tertiary consumers: Lion, tiger, wolf, etc. are tertiary consumers of the forest.

Decomposers: Detritus eaters are millipedes, mites, etc. Different types of bacteria and fungi are very active. Rate of decomposition in tropical and subtropical forests is greater than that in temperate forests.

Grassland Ecosystem

About 20% of the land area is occupied by grasslands. Grasslands are formed where rainfall is low and soil is not deep and fertile. Grassland ecosystems are very common in the temperate countries, but they are not very common in the tropical countries.

Abiotic components: Water, minerals and organic materials in the soil, gases in the air, light, temperature, etc.

Biotic components:

Producers: Producers are the grasses of different types. Some herbs and shrubs are also present.

Primary consumers: Cattle, deer, rabbit, mice, etc. are primary consumers. Different herbivorous insects, termites and millipedes feeding on grass leaves and some birds are also primary consumers.

Secondary consumers: Fox, jackal, snakes, frogs, lizards and many birds are secondary consumers. Hawks, kites, eagles, etc. may eat primary or secondary consumers. So they may be secondary or tertiary consumers.

Decomposers: Different species of fungi, bacteria and protozoa decompose the dead organic matter.

Desert Ecosystem

About 17% of land area is desert. Here, annual rainfall is less than 25 cm. Deserts are hot; winds blow at high speed. Soil is sandy.

Abiotic comoponents: Organic matter in the soil is very less as vegetation is scanty. Water is also very little in the soil. Temperatures are high; wind velocity and exposure are also very high.

Biotic Components

Producers: The primary producers are mainly shrubs and also some herbs. Shrubs have wide spread, branched root systems with variously branched stems and branches. Thorny plants and succulent (water storing and fleshy) cacti are common. Lichens and xerophytic mosses also occur.

Consumers: Reptiles and insects which can tolerate high temperatures are the consumers. Different types of birds, snakes, rodents, etc. are also seen. Camel is the commonest herbivore.

Decomposers: There are very few decomposers in the sandy soil of the deserts. This is because there is very little organic matter in the soil. Some thermophilic (heat loving) fungi and bacteria live in the soil.

Cropland Ecosystem

These are man made (artificial) ecosystems, which have replaced natural ecosystems. Here, man has changed the environment by providing irrigational water, spraying insecticides, adding fertilizers, using chemicals for disease control, etc. In a cropland, a single species is grown, like wheat, paddy, sugarcane, vegetables, etc. and all efforts are made to increase production by that species. Unwanted plants are removed. The aim is to increase the biomass of the usable part only and not the whole plants, as is done by nature. Man tries to increase net production, and diverts most of the production into the part – grain, fruit or tuber as the case may be – which is important to him. There is no need to increase the weight of the whole plant. In nature, emphasis is on increasing gross production, and not net production.

Abiotic component: These are as usual; in addition, many factors are controlled by man. Fertilisers are added to the soil and water relations of the soil are changed.

Biotic Component:

Producers: Crop plants are the primary producers. Weeds like grasses also produce food in the cropland.

Primary consumers: Most important is man himself. Insects, rats, birds, rabbits, cattle also consume part of the produce.

Secondary consumers: Frogs, some birds, etc. are secondary consumers.

Tertiary consumers: Tertiary consumers are snakes, hawks, etc.

Decomposers: Different types of microbes are active in the soil as decomposers.

There are some more terrestrial ecosystems like those of savannas, tundras, etc. Savannas are grasslands with some shrubs and trees scattered in them. Tundra is an Arctic plain with permanently frozen subsoil. Some types of lichens, mosses and dwarfed vegetation live in tundras.

Aquatic Ecosystems

Aquatic ecosystems may be freshwater (those of rivers, lakes, ponds, etc.) or marine (those of seas, oceans, backwaters, etc.). A pond ecosystem can be taken as a very good example of an aquatic ecosystem.

Pond Ecosystem

This is a self-sufficient, self-regulating system. The living and the non-living exist in harmony in a pond ecosystem. The nature of the pond influences the living organisms in it; the living beings in the pond also have an influence on the pond.

Abiotic components: Include all the non-living components present in the pond. These are physical, inorganic and organic.

Physical: Important physical components in a pond are light, temperature, pH of water, wind etc. If the pond is very deep, the bottom may be dark as there is no light penetration. But in shallow ponds, there is some light even at the bottom. Based on water penetration and the associated plants and animals, a pond can be divided into three zones:

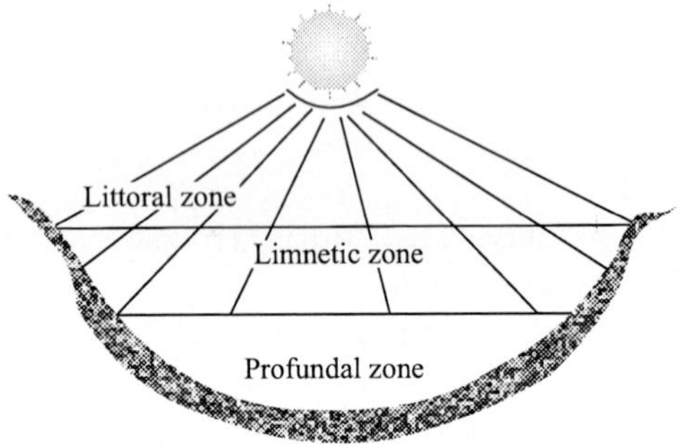

Zonation in a deep pond

a) Littoral zone – This is the shallow water zone where there is sufficient light and where free floating plants grow. This zone is rich in phytoplankton as well as zooplankton.
b) Limnetc zone – In this zone light is sufficient less but is enough for photosynthesis. Many rooted floating and submerged flowering plants and algae, crustaceans, insects and insect larvae grow in this zone.
c) Profundal zone – This is the deep water where there is no effective light penetration. Saprophytic (living on dead matter) bacteria and fungi and also different types of worms, snails, crabs, etc. live in the profundal zone.

The particles suspended in water reduce light penetration too as they absorb some light. The quality of light also changes depending on the depth of the pond. White light is made of seven colours – violet, indigo, blue, green, yellow, orange and red. Deeper waters get light of some colours more than other colours. Depending on all this, the types of plants that grow at different depths in a pond differ. The types of animals change too, as they depend on plants for food.

Temperature has a great influence on plants, animals and also microorganisms.

Based on variations in temperature, the water in a pond can be divided into three zones or strata. This is called thermal stratification. The three strata are known as hypolimnion, epilimnion and metalimnion. *The deeper part of the pond with cooler water is called hypolimnion. The upper part with warm water is called epilimnion. The middle part where these two layers meet is the metalimnion or the thermocline.* The presence of these layers based on temperature is called thermal stratification. pH means the

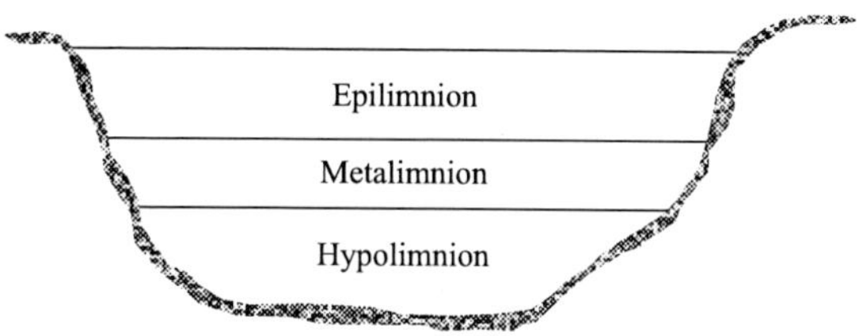

Thermal stratification in a pond

hydrogen ion concentration. The pH value of pure water is 7. This is the neutral value. Values below 7 indicate acidity and values above 7 indicate alkalinity. Many plants and animals, especially the phytoplanktons and the zooplanktons are affected greatly by the pH value of water. Some microbes prefer acidic waters while some others prefer alkaline waters. Some prefer neutral value also. Wind is another factor affecting life in a pond. It changes the temperature of the pond water. When fast winds blow, the hot and cool waters of pond get mixed up. Wind also causes wave action in water. The number, size and duration of waves in the pond are also important physical factors.

Inorganic: Water itself is an inorganic component. The different minerals dissolved in water are important nutrients for the plants. Animals get these minerals from plants. Nitrates, phosphates, etc. are always present in pond water. Many gases like carbon dioxide, oxygen, etc. remain dissolved in water. The solubility of CO_2 is more than that of O_2 in water. However, exchange of gases between the plants and water is not much as the air stored inside the plants form a reservoir of gases for them. The oxygen dissolved in water is used by the aquatic animals for respiration.

Chlorophyll, the green pigment present inside chloroplasts in all green cells of plants is essential for photosynthesis. This is an inorganic component too. *The amount of inorganic matter at any given time is called the standing state or the standing quality.*

Organic: The organic component in a pond includes organic compounds like carbohydrates, fats and proteins belonging to the dead plants and animals in the pond. The intermediates of their decomposition are also organic. Continued decomposition of these dead matter by bacteria and fungi ultimately convert them into basic inorganic form. This is reutilized by the plants. Bacteria and fungi thus form a connecting link between the abiotic and the biotic components of the ecosystem.

Biotic components: The biotic component includes the producers, consumers of different levels and the decomposers.

A] *Producers*: The primary producers of a pond are the autotrophic green plants and the photosynthetic bacteria. The large sized plants or the macrophytes may be flowering plants or non-flowering plants. These may be submerged fully in water or with leaves floating and other parts submerged or free floating. Of the flowering plants, *Hydrilla*, the insect catching plant Bladderwort or *Utricularia,* etc. are submerged; Lotus (*Nelumbium*), water lily (*Nymphaea*), etc. have floating leaves but they are rooted in the mud at the bottom of the pond while duckweed (*Lemna*), Water hyacinth (*Eichhornia*), etc. are free floating. Non-flowering plants are water ferns like *Salvinia*, *Azolla,* etc. and green algae like *Chara*,

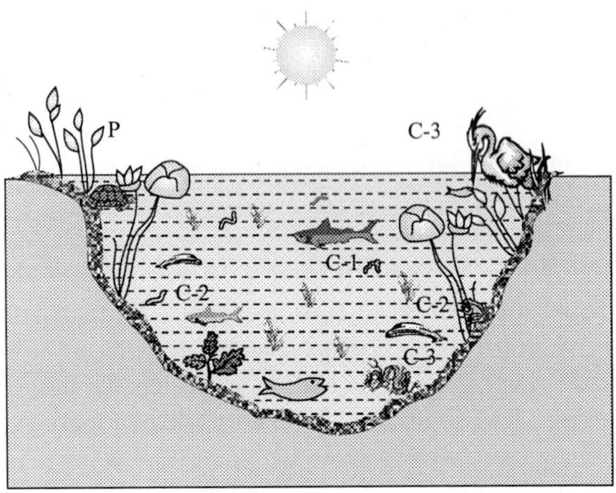

Pond Ecosystem
P – Producer; C1, C2, C3 – Consumers of first, second and third Orders;
D –Decomposer

Nitella, etc. All these are macrophytes. The commonest microphytes functioning as primary producers in the pond ecosystem are the phytoplanktons. These include, most commonly, green algae like *Spirogyr*a, *Zygnema*, *Scenedesmus, Ulothrix,* etc. and the unicellular golden brown floating algae, diatoms.

Blue-green algae like *Anabaena*, *Oscillatoria, Microcystis* etc. also are very commonly seen in ponds.

B] *Consumers*: The consumers depend on the producers for food. In a pond, consumers are of 3 types.

Primary consumers: These are herbivores. The zooplanktons which feed directly on the phytoplankton are the first level or primary consumers. Examples of zooplanktons are the rotifers. Some protozoans like *Cyclops*, *Euglena,* etc. are also primary consumers. The insects and animals that live at the bottom of the pond and eat plants and plant remains are called *detrivores*. Some other insects, some small fish, etc. eat phytoplankton and parts of other plants living in the upper waters of the pond. These, along with the detrivores, form the *benthos.* Benthic consumers are insects and insect larvae, beetles, crustaceans, molluscs, etc.

Large animals like cows, buffaloes, etc. and some birds visit the pond to eat the macrophytes at the edge of the pond. These are also primary consumers.

3. *Secondary consumers:* Insects and small fish which feed on primary consumers like zooplankton are the secondary consumers.

3. *Tertiary consumers:* Large fish and some birds which eat small fish are the tertiary consumers in a pond.

Thus in a pond, fish occupy all levels of consumers.

C] *Decomposers*: These are sometimes called microconsumers to differentiate them from detritus eaters which also decompose dead organic matter. They decompose only part of the dead matter. They break down the complex organic content of dead bodies and convert it into simple inorganic substances.

The commonest decomposers are different types of bacteria and fungi. As there is a deficiency of oxygen under water due to its low solubility in water, some dead debris is not decomposed fully. It accumulates as what is known as peat at the bottom of the pond.

Lake Ecosystems

A lake is a very large permanent pond. Its ecosystem has the same characters, nature and functions as those of a pond. Ponds and lakes are standing water (still water) ecosystems. In the other types of freshwater ecosystems of streams and rivers, water is flowing.

Stream and River Ecosystems

Here, the water is always flowing and so the different plants and animals are adapted to different rates of flow of water. Some can live in fast moving water while some others can live only in slow moving water. Some fish swim to the source of the stream or river during breeding times as they lay eggs only in clear water. The types of animals and plants in streams and rivers depend on the oxygen content also in addition to the rate of flow and clarity of water. Another factor is the nature of the bottom or the bed of the stream or river.

Marine ecosystems

These are saltwater ecosystems. Abiotic components are different aspects of water, waves, minerals, temperature, wave action, etc. Primary producers are mainly phytoplanktons like diatoms, unicellular and colonial algae, and the large sized red and brown algae. Some brown algae are very large, measuring a few metres in length. These are the sea weeds. All these producers show zonation, i.e. they live at different depths.

The primary consumers are the marine zooplankton. Some fish, crustaceans and mollusks are also primary consumers. Sea fish are different from river fish. The variety of fish in the seas and oceans are much more than in ponds or rivers. Other sea animals like crustaceans, starfish, jelly fish, etc. also live in large populations in the marine environment. Fish occupy different trophic levels. While some small fish are herbivorous and occupy the first trophic level, some others occupy higher trophic levels. Thus, while fish like Herring, Mackerel, etc. are secondary consumers, some other fish like Cod, Halibut, etc. are tertiary consumers. Decomposers are the various types of fungi and bacteria.

The flora and fauna (plants and animals) growing near the sea coast in shallow water is different from those that live in deep sea. Many types of animals live in the coral reefs under water in the shallow seas. The plants and animals that live on the seashore differ based on whether the shore is sandy, rocky, muddy, etc.

ECOLOGICAL PYRAMIDS

Producers are most closely related to the primary source of energy, the sun, as they get energy directly from the sun. So they occupy the first trophic level. Herbivores eat green plants and are therefore included in the next trophic level. Carnivores constitute the third trophic level and so on. *The different trophic levels in any ecosystem can be represented graphically and these graphic representations are called ecological pyramids.*

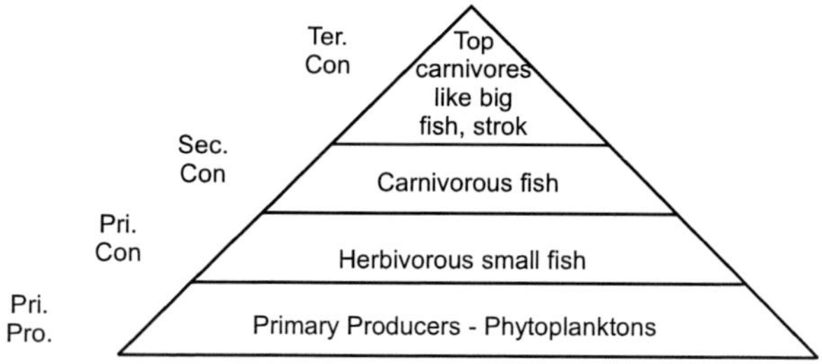

Upright Pyramid of numbers of the freshwater ecosystem of a pond

Each ecosystem has its own characteristic ecological pyramid. *Ecological pyramids are basically of 3 types – pyramids of number, pyramids of biomass and pyramids of energy.* At the base of the pyramid are the producers, the green plants. The apex is occupied by the top carnivores (as the decomposers get food from all trophic levels). Herbivores and carnivores are placed in between.

Pyramids of Numbers

In pyramids of numbers, the number of individuals in the different trophic levels at any given time is shown.

The shape of the pyramid depends upon the size of the individuals of the primary producers. If primary producers are small sized compared to the consumers, the pyramid is erect. Pyramid of numbers of freshwater ecosystems like that of a pond is erect.

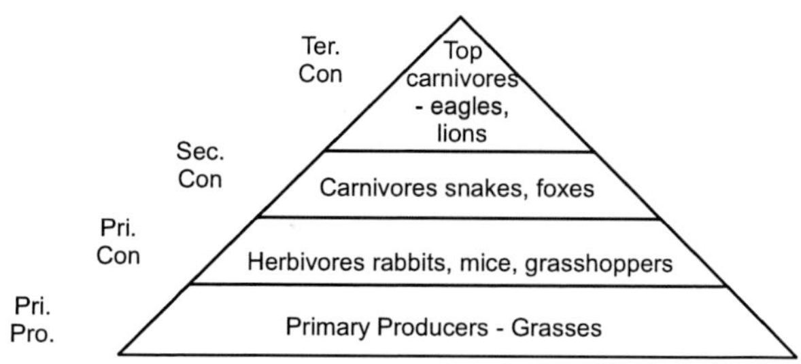

Upright Pyramid of numbers of a grassland ecosystem

The primary producers in a pond are the phytoplanktons (microscopic algae). Though these are extremely small, their numbers are very large.

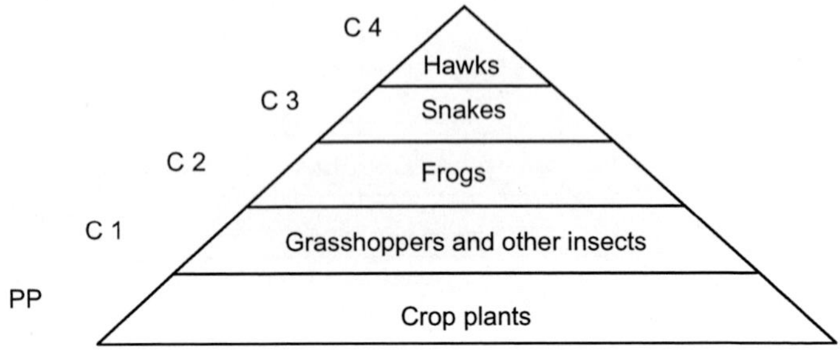

Upright pyramid of numbers of a cropland ecosystem

So they form the broad base of the pyramid. Above this are the herbivorous fish followed by carnivorous fish. The top carnivores at the top may be large fish or birds like storks. Here, the number is maximum at the base, decreasing gradually towards the tip of the pyramid. Pyramid of numbers of grasslands and croplands are also erect.

If the primary producers are large in size, and the size goes on decreasing in the consumer levels, the number of organisms goes on increasing in the successive trophic levels. This results in an inverted pyramid. A very good example is a tree ecosystem. A single tree is an ecosystem too, as different types of birds live on the tree. Moreover, monkeys, snakes, squirrels, etc. may depend on the tree for various purposes.

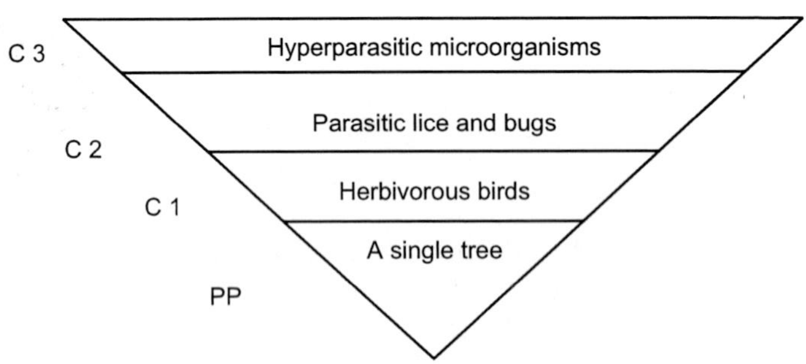

Inverted pyramid of numbers of a tree ecosytem

Here, the single tree is the producer. On the tree, live many herbivorous birds. These are consumers of the first order, C1. On the body of each bird, many parasites like lice, bugs, etc. consumers of the second order, i.e. C2 live. On each louse or bug, many bacteria, fungi etc. live. These hyperparasites constitute C3. Here, the number of organisms gradually increases from the producer to the consumers.

Pyramids of Biomass

Pyramids of biomass indicate the relationships between the different trophic levels of an ecosystem on the basis of living weight – fresh or dry.

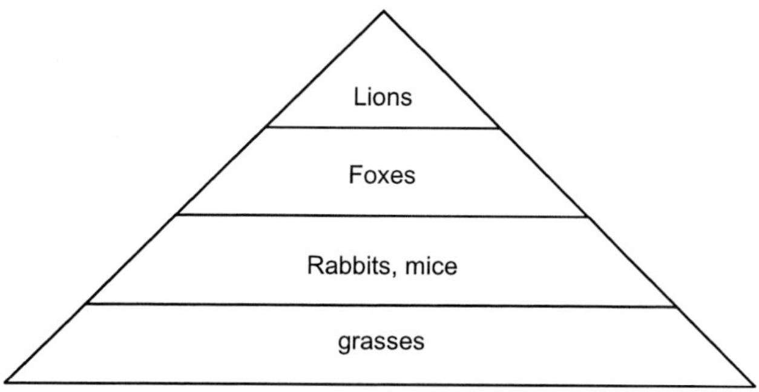

Upright pyramid of biomass of a grassland

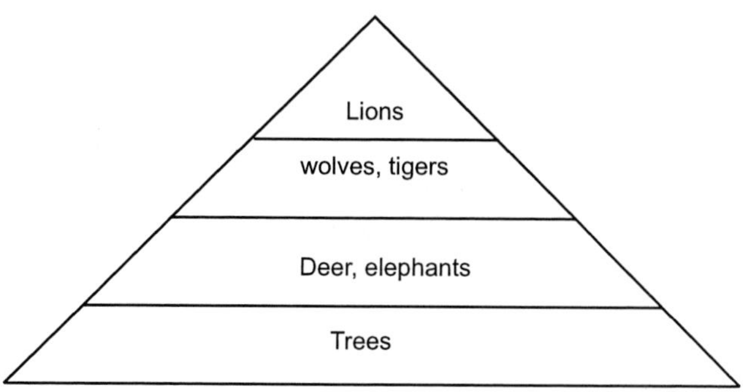

Upright pyramid of biomass of a forest

If the biomass of the producers is more, the biomass goes on decreasing at successive consumer levels. In such a case, the pyramid is erect. The best example is a grassland ecosystem. Pyramids of forests are also erect.

On the other hand, pyramid of biomass of a pond ecosystem is inverted. This is because phytoplanktons like diatoms and green algae which form the base of the pyramid are very light in weight. The small fish of consumer level 1 are heavier than producers and the big fish of consumer level 2 are the heaviest. Thus the pyramid is inverted.

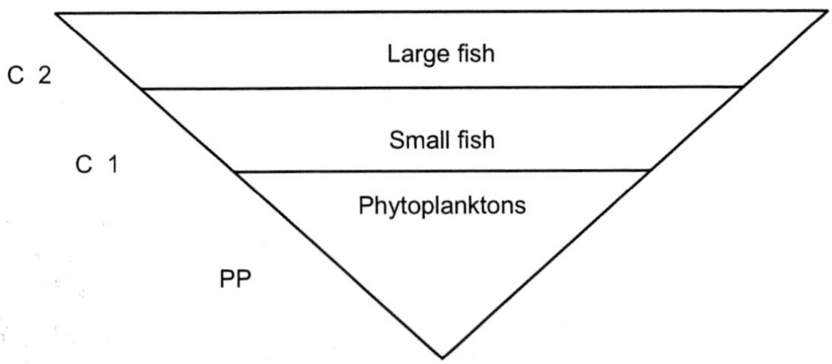

Inverted pyramid of biomass in a pond

Pyramids of Energy

Pyramids of energy show the total energy contained at the different trophic levels in ecosystems. The total energy in the producers is the highest.

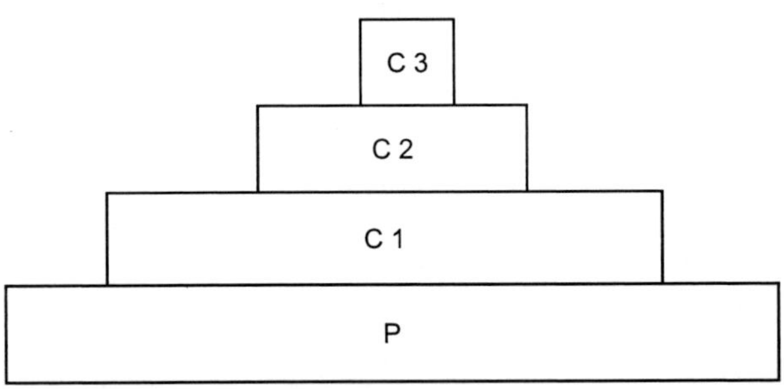

Pyramid of Energy

Herbivores contain lesser energy than the producers. The total energy in the carnivores is lesser than that in the herbivores. Thus, as we go up the trophic levels, the energy content goes on decreasing. This is because energy flows from the producers to the consumers at various levels. Therefore, the pyramids of energy are always erect in all ecosystems. Decomposers get energy from all trophic levels.

BIOGEOCHEMICAL CYCLES (CYCLING OF MATERIALS IN THE ECOSYSTEM)

In ancient Hindu scriptures, the concept of 'Panchatatva' is very clear. These are 'Prithvi', i.e. the earth (mineral nutrients); Jalam, i.e. water; Agni, i.e. fire (energy); 'Vaayu', i.e. gases and 'Aakaasam', i.e. the sky or space. This reflects the circulation of materials.

Living organisms obtain many material nutrients from the environment. These are gases like oxygen, carbon dioxide, etc., and minerals like calcium, potassium, iron, nitrogen, phosphorus, sulphur, etc. and water. These materials pass through the different trophic levels and are ultimately returned to the environment in many ways and are reutilized by the plants. Every nutrient atom enters the biotic component of the ecosystem from the abiotic component and finally returns to the environment. In this way, atoms are used over and over again, i.e. they are repeatedly recycled. This *circulation of materials between the biotic and the abiotic components of the ecosystem is called the biogeochemical cycle.* There are many types of biogeochemical cycles in nature. The more important of these are water cycle, carbon cycle, nitrogen cycle, sulphur cycle and phosphorus cycle.

Water Cycle (Hydrologic Cycle)

About 73% of the earth's surface is covered by water. Part of it evaporates and forms cloud. Rain returns this water to the earth. Plants absorb water from the soil and lose most of it to the atmosphere through transpiration. A part of it is passed on to the next trophic level. Animals get water from plants and also the environment. They also return water to the environment through evaporation. Thus the water cycle goes on continuously.

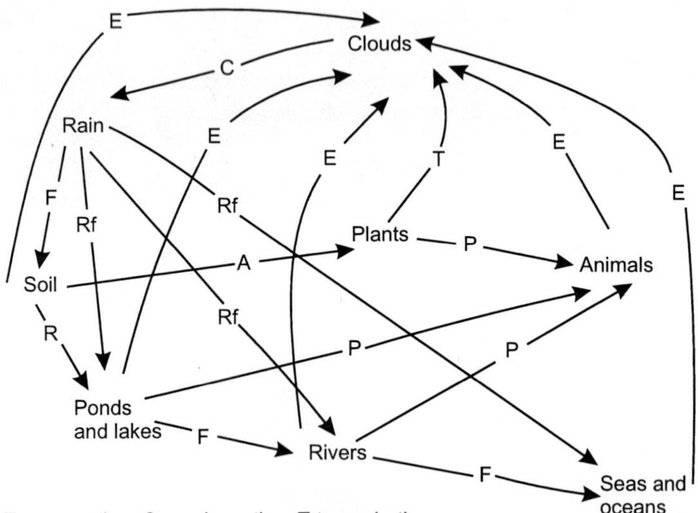

E-evaporation, C-condensation, T-transpiration,
F-filtration, R-runoff, A-absorption, R-rainfall
P-provide, F-flow

WATER CYCLE

Carbon Cycle

This is a gaseous cycle. 0.03% by volume of CO_2 in the atmosphere is the source of carbon for the biotic community. Carbon is also obtained from

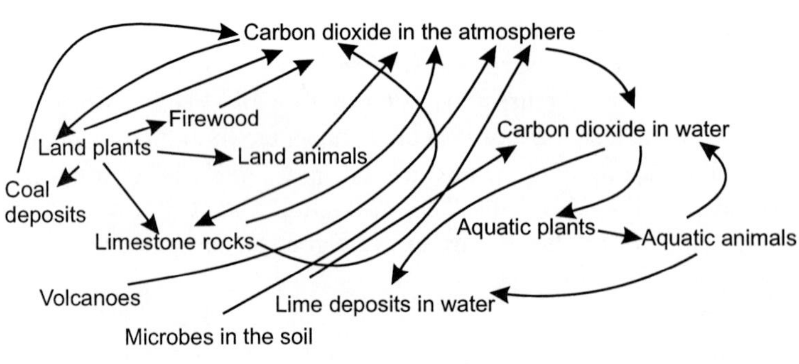

CARBON CYCLE

water where it is present in the soluble form. Carbon is the most important element of all organic compounds. Coal deposits, limestone rocks, volcanoes, lime deposits in the seas and oceans, etc. are also sources of CO_2 in the atmosphere. Other sources are from respiration, death and decay of plants, animals and microbes; weathering of limestone rocks, burning of coal and firewood, etc. In addition of this other sources of CO_2 are factories, automobiles, etc. Thus there are many sources of CO_2 in the atmosphere. But the only removal of CO_2 from the atmosphere is done by plants. They absorb it from the atmosphere and convert it into organic food during photosynthesis.

There is a separate cycle in the water bodies too, which is however connected to that on land. In short, the Carbon cycle is very complicated.

Nitrogen Cycle

The ultimate source of nitrogen for the living beings is the atmosphere. It contains 79% by volume of free nitrogen. Nitrogen is an important element for all forms of life, but it is not absorbed directly from the atmosphere. Plants get it from the soil as nitrates. Free nitrogen in the atmosphere is converted to nitrates in the soil by bacteria and blue-green algae and during lightning. Animals get nitrogen from plants. During decomposition after death of plants and animals, ammonia and free nitrogen are returned to the atmosphere. Part of ammonia in the atmosphere is converted by nitrifying bacteria in the soil into nitrates. At the same time, some nitrates in the soil are lost back into the atmosphere by the activity of denitrifying bacteria. Nitrogen cycle is thus a gaseous cycle.

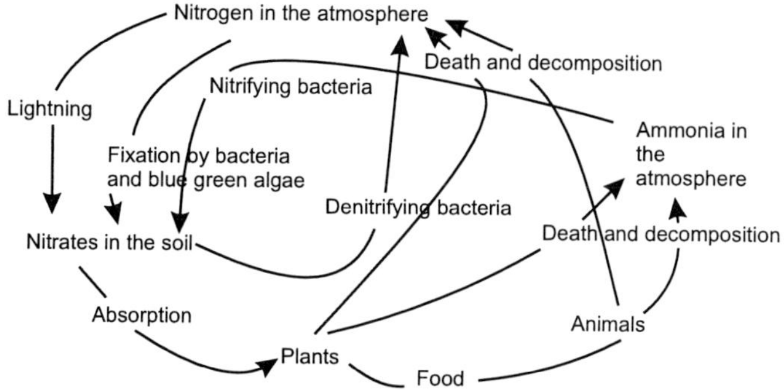

NITROGEN CYCLE

Sulphur Cycle

Sulphur is present mostly as sulphates in the soil. A small amount of hydrogen sulphide in the gaseous form is also present in the soil. Addition of fertilizers leads to increase in sulphates in the soil. Plants uptake sulphate from the soil and pass on part of it to animals. On death and decay, plants and animals return sulphates into the soil. Burning of fuel, automobiles and some industries release sulphur dioxide into the atmosphere. Part of this forms acid rain which falls to the earth and the water bodies. Some sulphates from the soil are washed into the water bodies. Aquatic plants and animals get their supply of sulphur from the water itself in the form of sulphates.

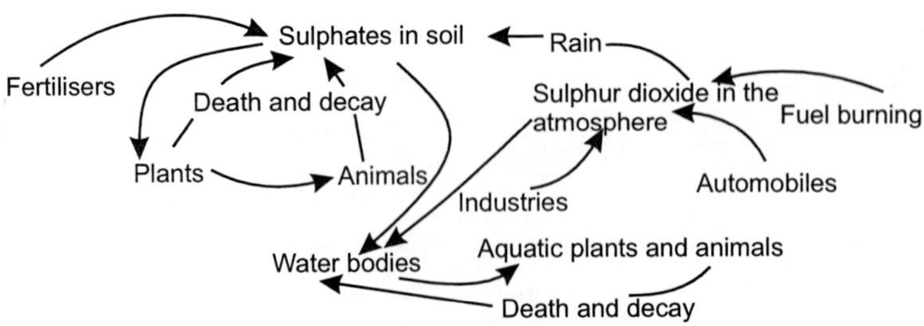

SULPHUR CYCLE

Phosphorus Cycle

Plants get Phosphorus in the form of phosphates from the soil. Sources of phosphorus in the soil are phosphate rocks and volcanoes. Plants uptake phosphates from the soil. Animals get it from plants. When plants and animals die and decompose, the phosphates are returned to the soil. Some phosphate is lost from the cycle by sedimentation in seas and oceans.

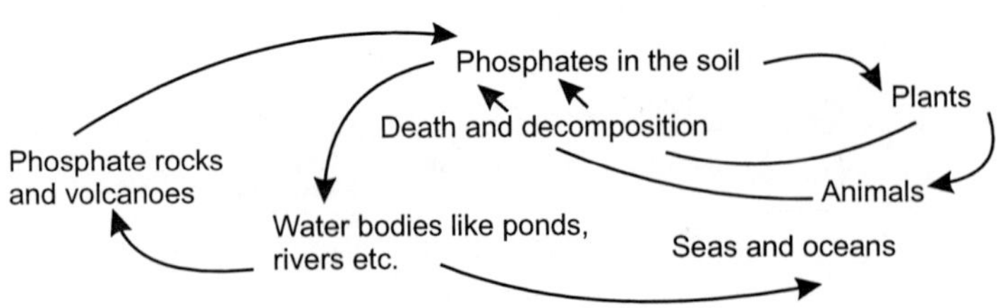

PHOSPHORUS CYCLE

In this way, all minerals are continuously circulating through the biotic and abiotic components of the ecosystem.

ENERGY FLOW IN THE ECOSYSTEM

If there are no plants, the light energy of the sun will be converted to heat energy which will heat up the earth. However, plants convert light energy into the chemical energy of the food during photosynthesis and it is this food that is used by the animals. In short, if there are no plants, there will be no life on earth. The life of all animals depend on the light energy of the sun (solar energy) trapped by the green plants. Thus, sun is the ultimate source of energy.

The sun is extremely hot. In the sun, nuclear rearrangements are taking place continuously, converting hydrogen to helium. This process releases huge amount of energy which radiates out in all directions in the form of electromagnetic waves. This is the solar radiation. It is only an extremely minute portion of this energy, i.e. about one out of fifty million parts, which reaches the earth's atmosphere.

When sunlight reaches the atmosphere, 30% is reflected back into space; almost 50% is absorbed by the ground and water as heat and 20% is absorbed by the atmosphere itself. This shows that only very little sunlight is used by the plants for photosynthesis. This is about 0.2%. Yet this is the source of energy for all life on earth. During photosynthesis, light energy is converted into chemical energy.

Solar energy travelling in the form of radiation is called radiant energy. Of this, rays having wavelength between 400mμ to 760mμ is the visible range or light energy. Light can easily be converted to heat. In the chloroplasts of green plants, light passes through the electron transport chain of the photosystems, and is fixed ultimately in carbohydrates. These are the immediate food sources.

The first law of thermodynamics proposed by Joule states that 'energy cannot be created or destroyed, but may be converted from one form to another'. This law is applicable in all cases.

About 90% of the total photosynthesis in the world is done by aquatic plants, especially diatoms and algae. Only the remaining is done by trees and other land plants. Among land plants, the largest contribution is by forests, followed by cultivated plants and grasses. Deserts contribute the least.

The unit of energy measurement is erg. One erg is the work done in lifting one gram of weight to a height of one centimeter against the force of gravity, which is 981 ergs. 10^7ergs = 1 joule. All forms of energy can be completely converted to heat energy. Heat is measured in calories (cal). 1 cal = 4.2 joules. 1000 cals = 1 kilocalorie (Kcal or Cal). In environmental studies, energy is therefore measured in calories or kilocalories; sometimes in kilojoules.

A large fraction of the solar energy received by the earth's atmosphere does not reach the ground, as it is scattered by suspended particles in the atmosphere. The constant rate at which solar energy reaches the earth's outer atmosphere is called solar constant. 25% of the incoming solar radiation is reflected by clouds and 9% by suspended dust. 9% is held by ozone, water vapour and other gases. 10% is absorbed by the clouds. Thus about 47% reaches the surface of the earth.

The quantity of solar radiation that is received in any place thus depends on the clarity of the atmosphere in that place. Latitude is another important factor. Thus regions near the equator receive maximum solar radiation, while it reduces gradually towards the poles.

In ecological energetics, studies are carried out mainly in three fields.

1. The quantity of energy reaching an ecosystem / unit area / unit time (for example, 1 sq metre / one hour or one day, etc.)
2. The quantity of energy absorbed by green plants and converted to chemical energy
3. The quantity and path of energy flow from green plants to different trophic levels in a known area for a known time period

The available energy decreases from producers to consumers of different levels. Green plants are the *primary producers*. The total energy fixed in food by green plants is called *Gross Production*. Some of this is utilized for respiration and other activities and the remaining is called *Net Production*. Some of this is consumed by herbivores, some is decomposed and some is not utilized and accumulate as sediments.

Of the total energy that herbivores get from the autotrophs, some is used in respiration, some part is used by carnivores, some is decomposed and some is not utilized.

Some herbivores in the ecosystem are eaten by carnivores. So the herbivores are called *secondary producers*. The carnivores get energy from herbivores. Of this, some is used in respiration, some is decomposed and some part is not utilized.

Thus, it is clear that energy flows only in one direction in the ecosystem. This is called the *unidirectional energy flow*. As food passes from one trophic level to the next in an ecosystem, the energy lost as heat at various levels of transfer does not return to the living organisms. The energy that goes to herbivores can not return to the autotrophs. The energy of the carnivores can not go back to herbivores or autotrophs. Thus the

energy flow in an ecosystem is always unidirectional. As it moves through different trophic levels, it is no longer available to the previous level. So if the primary source is cut off, the whole system will collapse. The one way flow or the single channel flow of energy is a universal phenomenon and it is true to the basic law of thermodynamics. To maintain the ecosystem, therefore, a constant supply of external energy is needed, which comes from the sun.

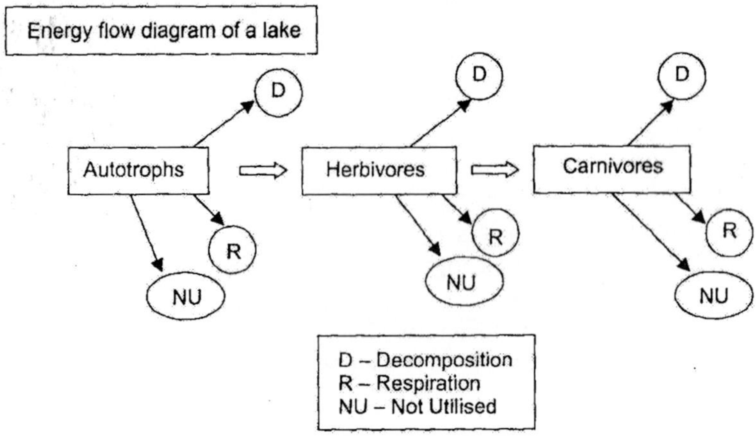

Energy does not flows freely, but through organic materials from one trophic level to the next. For this, inorganic material has to be converted to the organic form. This is done in primary production. The ecosystem is balanced as long as the energy utilization by animals is matched to the energy trapped by plants. Otherwise, food levels become short and a famine results. Consumers die if no food is available from outside the ecosystem.

In the ecosystem there is a progressive decrease in energy at each trophic level during energy flow. This is because of dissipation of energy as heat in different metabolic activities and also unused energy. This means that if a food chain is shorter, the available food energy will be more. The longer the food chain, lesser will be the food energy available for the organisms of the last trophic level.

Materials like carbon, nitrogen, oxygen, water, etc. are repeatedly used through the cycling of materials. All these pass on from living to non-living and back in a cyclic form. But energy flows in a unidirectional manner, in a single channel. This is one of the most important characteristics of energy flow in any ecosystem.

In any ecosystem, it is of great ecological significance to know the following:

1. The efficiency of primary and secondary production at different levels
2. The total input of energy
3. Efficiency of assimilation (photosynthesis)
4. Loss through excretion
5. Loss through respiration
6. Gross and net production at each level

A thorough understanding of the ecosystem is essential for better management of resources for human welfare. Every citizen should be educated in non-technical terms, the cycling of materials and the flow of energy in an ecosystem, of which he himself is a part.

In the history of mankind, we see many instances where man himself has upset the ecosystem which has resulted in great harm to man himself. Killing of some animals at a higher rate than their capacity to regenerate leads to an increase in the population of plants eaten by these animals. Such and other disturbances in the ecosystem will affect the cycling of materials and also the energy flow.

Similarly, indiscriminate felling of trees upsets the entire balance of water and other mineral cycles as well as plant and animal populations. Rajasthan was under forest vegetation a few centuries ago. But man removed plants faster than they could regenerate. As the primary production reduced, the fauna also gradually declined. Now, most of Rajasthan is desert. Efforts are being made at present on a large scale to stop further spreading of desert, by water conservation and planting trees.

Maintenance of life on earth is dependent on one energy source – the solar energy. This is fixed only by green plants. Therefore, if forests are destroyed, the net energy fixation capacity of the earth will fall. On the other hand, human population is increasing drastically. To support this ever increasing human population, the population of green plants or the efficiency of photosynthesis or both should be increased. Another possibility is to consume water plants directly, especially marine flora as fish resources are already being overexploited.

FOOD CHAINS

The sequence or order of trophic levels through which food energy passes from the primary producers to the top carnivores is called the food chain.

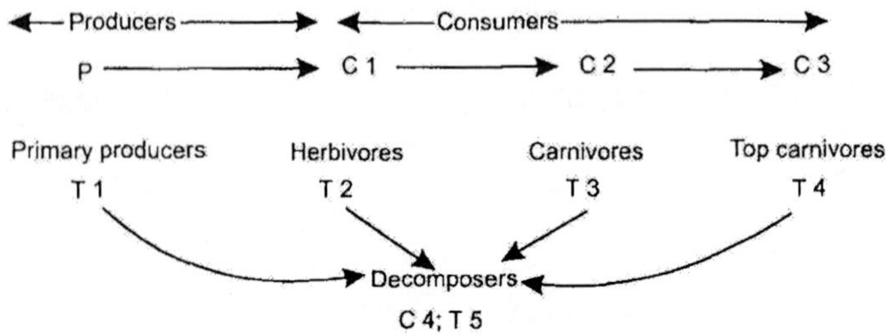

FOOD CHAIN

Decomposers get food from all trophic levels. Omnivores eat everything and are finally decomposed.

Scriptures of all old civilizations have references to food chains. The most ancient of these is the Mahabharata which mentions the importance of food chains and the interdependence of organisms for food. The Greeks also knew about food chain. In the modern times, it was Antonie van Leeuwenhoek who elaborated the principles of food chains at the end of the 17th century.

This classification of the living organisms of any ecosystem is based on their function and not species. Species that are very much different from each other may occupy the same trophic level. For example, in a pond, water lily and lotus which are flowering plants; non-flowering plants like water ferns and algae; and bacteria, all occupy T1.

In grasslands ecosystem, the food chain starts with the grasses, passes through grasshoppers, frogs, snakes, hawks, etc. In a pond, it starts with phytoplanktons and passes through water ferns, small fish, big fish, birds, large animals, etc.

Food chains are of two types.

1. *Grazing food chain: A grazing food chain starts from living plants and goes through herbivores to carnivores.* This type of food chain depends on photosynthesis. Most of the ecosystems in nature follow this type of food chain.

   Examples:

   Phytoplankton → Zooplankton → Fish

   Grass → Rabbit → Fox

2 *Detritus food chain: If a food chain starts with dead organic matter, moves into microbes and then into larger organisms, it is called a detritus food chain.* Such ecosystems do not depend directly on solar energy. For example, leaves in marshy vegetation

fall into water. These dead leaves are eaten by insect larvae, different types of worms, snails, etc. These are the detritus eaters. They are eaten by small fish, which form the food for big fish. Detritus food chains start from forest floors also.

Dead leaves in water ⟶ Worms and Snails ⟶ Small fish ⟶ Big fish

Dead leaves on the forest floor ⟶ Insects ⟶ Small birds ⟶ Snakes

FOOD WEBS

Food chains do not operate independently of each other. They are interconnected at some points so that an interlocking pattern is formed - this is called a food web.

Consumers always have alternative arrangements in nature. Snakes may eat mice or small birds; hawks may eat snakes, mice or rabbits. Besides these, vultures, foxes and human beings, etc. also make part of a grassland ecosystem. Thus food webs get interlocked at many points making the whole system very complicated.

In a pond, primary producers are the microscopic phytoplanktons and macrophytes (large plants). Whereas consumers of different levels are zooplanktons, different types of insects, snails, crabs, small and big fish, etc.

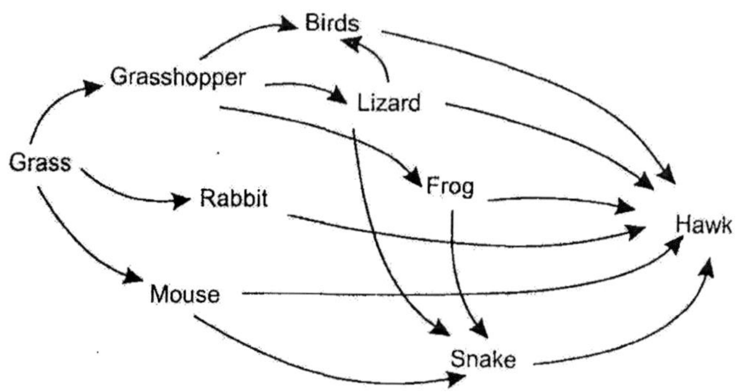

FOOD WEB IN A GRASSLAND ECOSYSTEM

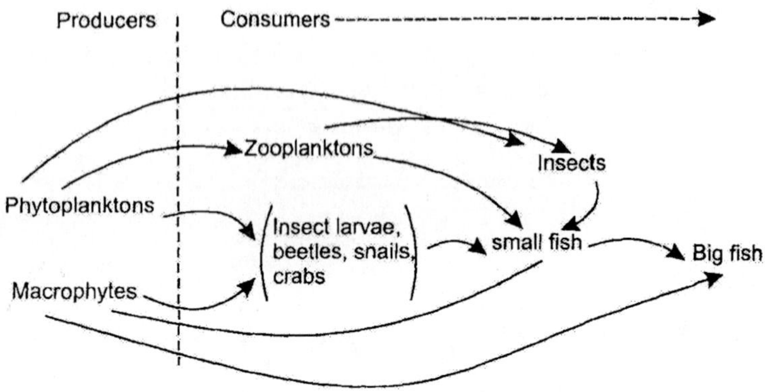

FOOD WEB IN A POND

Food webs are very important in maintaining the stability of ecosystems in nature. For example, in the grassland ecosystem, if the population of rabbits decreases due to some reason like a disease, population of alternative herbivore, i.e. mice increases. This may decrease the population of carnivores that prefer rabbits. Thus alternatives help to keep the ecosystems stable. This is necessary for the survival of all living things in nature. If herbivores are absent, the producers will multiply so much causing overcrowding and resulting competition. Same is the case of herbivores. Some of them should be consumed by the carnivores.

As we move further away from the primary source of food, i.e. the green plants, the energy available to the animals of successive food levels goes on decreasing. As a result, the population of animals that can be sustained also becomes smaller as we move up the food chain. As a herbivore, more energy is available to man than as a carnivore. If man starts taking more animal food, the available energy can support only fewer people. Then the human population will fall. Thus in nature, each species is under a check.

The more different the organisms in a system, the more complex is the food web. It depends on the length of the chain and the alternatives at different points. When alternatives are more, interlocking is also more. In seas and oceans, food webs are very complex.

Other than the producers and consumers, the ecosystem includes two more physiological groups – the parasites and the symbionts. *Parasites are organisms which depend on other living organisms totally or partially for food.* Examples of parasites are the disease causing bacteria and fungi which live in or on plant and animal bodies. *Symbionts are pairs of organisms which live in association with each other and benefit from each other.* Examples of symbionts are bacteria which live inside the nodules of roots of pulses, etc. When these parasites and symbionts are also included, food webs become more complex.

From this discussion on structure, components and various types of ecosystems, it is clear that each part of an aquatic or any other ecosystem has a definite role to play. Each part has its own function. The function of the producers is to produce food. The function of the consumers is to use part of the chemical energy trapped by the producers in the form of food for their various activities. They also function as a check against the overcrowding of the earth by the producers. The consumers act on the dead producers and consumers and release the elements bound in their dead bodies. This helps in the recirculation of the elements throughout the system.

Mother Nature has planned out everything to perfection. It is for us to maintain this equilibrium by not disturbing the different ecosystems in our biosphere.

QUESTIONS

A] Very short type questions:

a) Multiple choice questions – tick the right answer:

1. Pyramid of numbers in a pond ecosystem is always
   (a) Inverted (b) Upright
   (c) Irregular (d) None of these
2. Upper warm layer of water in a fresh water body is called
   (a) Hypolimnion (b) Metalimnion
   (c) Epilimnion (d) Profundal zone
3. Grasslands with trees here and there are called
   (a) Tundras (b) Estuaries
   (c) Plateaus (d) Savannas
4. Cycling of materials in the biosphere is called
   (a) Biochemical cycle (b) Biophysical cycle
   (c) Biogeochemical cycle (d) Geochemical cycle
5. The term 'Ecosystem' was coined by
   (a) Tansley (b) Odum
   (c) Warming (d) Blackmann

6. Pyramid of energy is always
   (a) Inverted (b) Upright
   (c) Rectangular (d) None of the above
7. Herbivores are
   (a) Primary Producers (b) Consumers
   (c) Decomposers (d) All of the above
8. Butterflies and bees are
   (a) Carnivorous insects (b) Detritus eaters
   (c) Herbivorous insects (d) Nocturnal insects
9. The sequence of organisms in a food chain is:
   (a) Consumer → Producer → Decomposer
   (b) Decomposer → Consumer → Producer
   (c) Producer → Decomposer → Consumer
   (d) Producer → Consumer → Decomposer
10. Which of the following is a top carnivore?
    (a) Tiger (b) Crocodile
    (c) Hawk (d) All of the above
11. One method of Nitrogen fixation in nature is by:
    (a) Thunder (b) Rain
    (c) Lightning (d) Storm
12. Energy flow in any ecosystem is always:
    (a) Bidirectional (b) Unidirectional
    (c) Multidirectional (d) None of the above

b) Define the following:
   1. Biome
   2. Ecosystem
   3. Autotrophs
   4. Heterotrophs
   5. Ecological pyramids
   6. Biogeochemical cycle
   7. Hypolimnion
   8. Epilimnion
   9. Metalimnion
   10. Standing state (Standing quality)
   11. Biomass
   12. Standing crop

c) Name the following:
   1. Three terrestrial ecosystems
   2. Three aquatic ecosystems

3. Three natural ecosystems
4. Three man made (artificial) ecosystems
5. Two lotic ecosystems
6. Two lentic ecosystems
7. Two freshwater ecosystems
8. Two marine ecosystems
9. The two components of an ecosystem
10. Three herbivorous animals
11. Three carnivorous animals
12. Three top carnivorous animals
13. Three herbivorous birds
14. Three carnivorous birds
15. Two top carnivorous birds
16. Two herbivorous insects
17. Two carnivorous insects
18. Two ecosystems with erect ecological pyramids of numbers

d) Write very brief notes on:
1. Ecological niche
2. Littoral zone
3. Limnetic zone
4. Profundal zone
5. Biomass
6. Detritus eaters
7. Predators

e) Differentiate between the following pairs:
1. Primary and secondary producers
2. Gross production and net production
3. Carnivores and top carnivores
4. Herbivores and carnivores
5. Producers and consumers
6. Autotrophs and Heterotrophs
7. Macroconsumers and microconsumers

B] Short answer type questions:

a) Write short notes on:
1. Inorganic components of an ecosystem
2. Organic components of an ecosystem
3. Physical components of an ecosystem
4. Decomposers
5. Pyramids of numbers

6. Pyramids of biomass
7. Pyramid of energy
8. Unidirectional energy flow
9. Pyramid of biomass in a pond
10. Pyramid of numbers in a tree ecosystem
11. Water cycle
12. Carbon cycle
13. Nitrogen cycle
14. Phosphorous cycle
15. Sulphur cycle
16. Importance of green plants on maintaining life on earth
17. Sun as the main source of energy on earth
18. Forest ecosystem
19. Grassland ecosystem
20. Desert ecosystem
21. Cropland ecosystem
22. Marine ecosystem

b) Explain the role of decomposers in the circulation of minerals

c) A fallen leaf is an ecosystem. Explain.

c) A roadside puddle is an ecosystem. Explain.

C] **Long answer type questions:**

a) Describe in detail the abiotic and biotic components of an ecosystem.

b) Give an account of the various types of ecological pyramids.

c) Write an essay on the cycling of materials in an ecosystem.

d) What do you understand by Unidirectional Energy Flow? Explain in detail.

e) What is the difference between food chains and food webs? Explain in detail with the help of suitable examples.

f) Give a detailed account of a pond ecosystem.

Unit 8

Biodiversity and its Conservation

CONCEPT OF BIODIVERSITY

Life originated on earth millions of years ago. It is believed that some organic molecules like simple nucleic acids were the first to evolve. Gradually, the simplest forms of life developed from them. The organisms to appear first were unicellular (an organism made up of one cell is called unicellular). After thousands of years, multicellular organisms (organisms made up of many cells) made their appearance. They evolved along so many lines that the whole pattern of evolution became a complicated web. The result of this is the biological diversity or biodiversity that we see today. *Biodiversity can be defined as the totality of genes, the species, the ecosystems in a given region of the world. In a broad sense, it stands for all life on earth.*

It was in 1992 that environmental issues came to be accepted as of global importance. From June 3-14, 1992, the Earth Summit or the UNCED (The United Nations Conference on Environment and Development) was held at Rio de Janeiro, capital of Brazil. It was a historic meet. Leaders from 170 countries participated in this summit. The definition formulated for biodiversity at this meet goes like this: 'Biodiversity is the variability of living organisms from all sources including terrestrial, marine and other aquatic ecosystems and the ecological complexes of which they are a part; this includes diversity within species, between species and of ecosystems'.

The number of species of known flowering plants alone is 200,000 at present. Add to this the number of species of non-flowering plants and those of animals — the total number of species which have been scientifically described will be about 2 million. The total number is much higher.

Levels of Biodiversity

Biodiversity is visible at three levels in nature.

1. Genetic diversity: In any organism, whether unicellular or multicellular, the cell contains some type of hereditary material. This is in the form of nucleic acids. Most of the organisms have Deoxyribo Nucleic Acid (DNA) as their hereditary material. Each DNA molecule is made of many genes. Thus *genes are the bearers of hereditary characters.* The characters that pass on from one generation to the next of any organism are controlled by genes.

The number of genes varies greatly in different organisms. Even a bacteriophage has about 100 genes. (Viruses are not organisms, as they are not made up of cells. However, even though they are made of only nucleic acids and proteins, once they enter living cells, they can multiply. Thus viruses show characters of both living and non-living things. Viruses cause many diseases in plants as well as animals. Bacteria on the other hand are unicellular organisms. Bacteria are of different types. Many of them are useful to us while many do cause plant and animal diseases. There are viruses which attack bacteria - these are called bacteriophages.) *E. coli*, a common bacterium has 4,000 genes while each cell of the common fruit fly (*Drosophila melanogaster*) has 13,000 genes. A cell of rice (*Oryza sativa*) plant contains 30 to 50 thousand genes while a cell of a human being (*Homo sapiens*) contains 35 to 45 thousand genes.

The genes differ from one another chemically. Therefore, the enzymes they code are also different. Chemical reactions in the cells depend on the enzymes. And the characters of any organism are the ultimate result of all the chemical reactions that takes place within their cells. This is how genes control the characters of an organism. In addition to this, many new characters can be developed by the interaction of genes, by the position effect of genes, etc. Thus, the number of characters controlled by genes runs into tens of thousands, based on the number of genes. During reproduction, the genes pass on from one generation to the next. This is how genetic characters pass on to succeeding generations.

However, genes are not immune to changes. Under certain conditions, they undergo changes and these are the mutations. The mutated gene will behave differently. When the mutated genes pass on to the next generation, changed characters will be seen.

In nature, many mutants (organisms with mutated genes) are seen. These along with the original wild type organisms are constantly affected by various environmental conditions. The result is natural selection. Those forms best suited to the environment are chosen by nature. These succeed in establishing themselves in nature. These may interbreed with the original forms and again give rise to new characters. *The sum total of all the genes in all the wild species make what is known as the gene pool or the gene reserve.*

The gene pool in nature is almost unlimited. The diversity in the genes in our crop plants and domesticated animals has developed over thousands of years. This gene reserve or gene pool is used by plant and animal breeders in their breeding experiments to give rise to better varieties of useful plants and animals. The gene pool is used by genetic engineers and biotechnologists also. They do research on producing genetically modified (GM) foods and other crops better suited for disease resistance, resistance to drought, salinity in the soil, etc., developing better medicines and other products for our industries.

2. Species diversity: When man studied more and more plants and animals on the earth, classification became necessary for convenience in studying. This led to establishment of different systems of classification during different periods of scientific study. Accordingly, many units of classification were decided upon, like order, family, genus, species, etc. Species is the smallest unit of classification. *Individuals resembling each other in all important external characters, both vegetative and reproductive, constitute a species.* From this it is clear that there are quite a large number of species of living beings on earth. The number of species of plants and animals present in a region constitutes its species diversity. Some regions have greater species diversity than others. For example, a rain forest has a more number of species than a desert. The species diversity is the least in a cropland, where only one crop is cultivated. Natural forests have great species diversity while plantation forests have only one or a few species planted and maintained by man. Thus though afforestation (planting of trees where there were no trees earlier) is a good practice, it can not compensate fully for deforestation (felling of trees in a forest and the resultant destruction of the forest). This is because of the vast difference in species diversity between the two. Natural forests give us a large variety of products in terms of timber as well as other forest produce. They sustain whole communities dependent on them for their livelihood. *Areas that are rich in species diversity are called hotspots*. The world has 15 nations which have hotspots; India is one of them.

3. Community and Ecosystem diversity: Ecosystems are of different types. Natural ecosystems are seen in the wild while man made ecosystems are seen in inhabited areas. As ecosystems are in general large sized, when we speak about ecosystem diversity, we refer to countries or states or districts. Ecosystems undisturbed by human activities have more chances of having greater species diversity. Biodiversity is the basis of the stability of ecosystems. Reduction in the number of individuals or species reduces its stability. Therefore protection of a single species at the cost of other species is not advisable. The whole ecosystem of which it is a part must be conserved. Overuse and also misuse of ecosystems will have long standing harmful effects on the environment.

Community and Ecosystem Diversity are of three types:

a) *Alpha* (α) *Diversity*: This is the diversity shown by the organisms living in the same community. These organisms live in more or less the same habitat. If a community has more Alpha diversity, it shows that the community is stable and that the inter relationships between the different organisms is healthy.

b) *Beta* (β) *Diversity*: This type of diversity is seen in gradients, i.e. in areas where two types of ecosystems meet. Here, the climate and other environmental factors show a gradual change from one ecosystem to the other.

c) *Gamma* (γ) *Diversity*: This is also known as regional diversity. Gamma diversity is the total diversity shown within a large zone, which may be a specific geographical area or a landscape. It indicates the total species richness in all the habitats of the zone put together.

Ecosystem diversity indicates the variety of ecosystems present in a large area. Each ecosystem has its own food chains and food webs, its own trophic levels, nutrient cycles, energy flow mechanisms, ecological niches, etc.

BIOGEOGRAPHICAL CLASSIFICATION OF INDIA

India is the seventh largest country in the world. In Asia, it is the second largest. Its total land area is 3,287,263 sq km. The mighty Himalayan ranges are located to the north, north-west and north-east of our country. The southern part of India is a peninsula with Arabian Sea on the west, Bay of Bengal on the east and the Indian Ocean on the south. Pakistan is situated on our north-west. China, Tibet and Nepal are the countries on our north-east. Myanmar (previously known as Burma) is located on our east.

India has climates varying from the equatorial to the alpine in character. The chief controlling factors of the climate are rainfall and temperature. Depending on rainfall, India can be divided into four regions — wet zone with rainfall above 80″, intermediate zone with 40-80″ rainfall, dry zone with 20-40″ rainfall and arid zone with less than 20″ of rainfall. Temperatures vary greatly. Highest recorded temperature is 50.6° C in Alwar while lowest recorded temperature is - 45° C in Kashmir. There are four major seasons in India — winter from December to February, summer from March to June, rainy season from July to mid-September and autumn from mid-September to November. As per our traditional calendar, we have 6 seasons — Spring (Vasant), Summer (Greeshm), Monsoon (Varsha), Early Autumn (Sharad), Late Autumn (Hemant) and Winter (Shishir), each of two months.

We have a rich variety of climates and vegetation. There are 10 biogeographical regions in our country. These are:

1. Trans-Himalayan zone: This is the very cold region of Ladakh. It is a snow covered mountainous region. The Trans-Himalayas come under the alpine zone. 15,000′ is the snow line. Above this there is usually no vegetation. There is only snow. Animals of this zone include wild sheep, goat, yak, snow leopard, Tibetan asses and wolfs, etc.

2. The Himalayan ranges and the valleys of Kashmir, Himachal Pradesh, Utharkhand, Assam and other North-Eastern States: This zone starting from the north-west of Kashmir extends eastwards and covers the north-west, central and eastern Himalayas. Vegetation-wise, this biogeographic zone is divided into the western Himalayas (from Kashmir to Kumaon) and the eastern Himalayas (Kumaon eastwards, up to Assam). Rainfall is more in the eastern part while the climate is dry towards the west. Going up the mountain, climate changes as if from traveling from the equator to the poles. Due to these latitudinal variations, there are definite belts of vegetation here. Transition from one zone to the next is rapid. 3 climatic zones each with its own characteristic vegetational belts are recognized in the Himalayan ranges. These are the submontane (lower or basal) zone, the montane (temperate) zone and the alpine zone.

A) *The submontane zone* extends from the foot of the Himalayas up to 5,000′ height. Climate is tropical and subtropical. In the lower belts, continuous belt of sal (*Shorea robusta*) forms a tropical deciduous forest. Many other trees like mango (*Mangifera indica*), amaltash (*Cassia fistula*), etc. are seen in the western Himalayas and kadam (*Anthocephalus cadamba*), *Terminalia* species, etc. in the eastern Himalayas. Riverian (forests near rivers) and mangrove forests and savannas (tall grasses with trees here and there) are also seen in this zone.

B) *The Montane zone* extends from 5,000′ to 12,000′ . The lower belt here is made of coniferous forests while the upper belt has deciduous forests. Common trees are different species of Pine (*Pinus*), deodar (*Cedrus deodar*), etc. in the western Himalayas and white oak (*Quercus incana*), champa (*Michelia champa*), jamun (*Eugenia jambolana*), etc. in the eastern Himalayas. The alpine zone lies above 12,000′ , which is the timber (tree) line, i.e. no trees grow here. The lower alpine zone has some shrubs like *Juniper, Rhododendron,* etc. in both west and east Himalayas.

C) *The Alpine zone* has herbs with brightly coloured flowers in the western Himalayas. Some examples are *Ranunculus* with yellow flowers, *Polygonum* and *Pedicularis* with red flowers, *Epilobium* and *Geranium* with pink to red flowers, *Primula*, *Aster*, *Delphinium*, *Gentiana,* etc. with blue flowers, crucifers with white flowers, *Potentiella* with scarlet flowers, *Clematis* with black flowers…the colour festival can be visualized in this beautiful part of India. Diversity of animals is more in the Himalayas than in the Trans-Himalayas. It includes sambar, musk deer, wild bear, snow leopard, brown bear, pheasants, etc. Sikkim dog which used to live here earlier has now become extinct.

3. Desert zone: Rajasthan, Punjab, Haryana, Gujarat, Delhi, Rann of Kutch, etc. are included in this zone. The Thar of Rajasthan is a famous desert. Rainfall is scanty, and winter temperatures are low, often nearing the freezing point. Summers are hot and dry with scorching dry winds and sometimes dust storms. Soil is sandy. Except in irrigated fields, land dries up very quickly after rain. Vegetation is therefore poor. Tree species are very few. They are seen only near the rivers, forming dry forests. Examples of trees are poplar (*Populus euphratica*), shisham (*Dalbergia sissoo*), *Cassia,* etc. Some shrubs and many short lived annuals are seen. Most common shrubs are the cacti, Indian ber (*Zizyphus jujuba*), etc. Common herbs are prickly poppy (*Argemone mexicana*), mako (*Solanum nigrum*), etc. South-west of Punjab merges into the hot dry desert of western Rajasthan, west of the Aravalli hills. In the south-eastern part of Aravalli, dry thorny forests of *Acacia*, *Zizyphus*, *Euphorbia,* etc. are seen. The fauna of the desert zone includes wild ass, fox, desert cat, sand grouse, the great Indian bustard, etc.

4. Semi-desert zone: This zone includes Madhya Pradesh, Chattisgarh, parts of Orissa and parts of Gujarat. Rainfall ranges from 25″ in the west and centre to 70″ in the east. Depending on the rainfall, forests of different types grow here. Sal (*Shorea robusta*) forms forests in eastern parts of M.P. and Orissa. Mixed deciduous forests of teak (*Tectona grandis*), shisham (*Dalbergia sisso*), amaltas (*Cassia fistulosa*), palash

(*Butea frondosa*), amla (*Phyllanthes emblica*), etc. are seen in the central parts of M.P. and Chathisgarh. Thorn forests of *Acacia, Zizyphus,* etc. occupy the drier parts. The endemic (characteristic to a region and not found elsewhere) animal of the semi-desert zone is the Gir lion.

5. The Gangetic plain: The Gangetic plain extends from Eastern Punjab to the Sunderbans in West Bengal. It passes through Haryana, Delhi, U.P., parts of Bihar and Orissa. Land is very fertile. Rainfall is less in U.P. but more in Bengal. Vegetation is mostly tropical moist and dry decidous forests. Northern regions have sal forests. Other trees are mango, peepal (*Ficus religiosa*), neem (*Azadirachta indica*), etc. Forest floor is covered with shrubs and herbs. At the eastern end are the mangrove tidal forests of the Sunderbans. These occupy the tidal areas close to the Gangetic delta. Marsh plants form swampy vegetation here. Fauna includes gazelle, elephant, rhinoceros, crocodile, black buck, barasingha, turtle, etc.

6. The Western Ghats: This zone comprises the Malabar Coast and the Western Ghats. It extends from South Gujarat to Cape Comorin. Rainfall is very heavy. Vegetation is of four types — tropical moist evergreen forests, mixed deciduous forests, subtropical temperate evergreen forests and the mangrove forests on the sea shores. Tropical evergreen forests are multi-storied with tall trees like teak, mango, shisham, etc. Bamboo (*Bambusa arundinacea*) is common. Growth is luxuriant. Forest floor is densely covered with shrubs. Climbers and epiphytes (plants growing on trees) are also many. Mixed deciduous forests occur high on the mountains, up to 5,000′ altitude. Teak and shisham trees growing here are not very tall. Several bamboos grow here. Temperate evergreen forests are seen on the Nilgiris at 5,000′ and above. They are known locally as 'sholas'. Prominent trees are champa, jamun, etc. Mangrove forests form large areas on the coast of Mumbai and its suburbs. Examples are *Rhizophora, Phoenix,* etc. The Western Ghats have a very large number of endemic flora and fauna. Threatened species of animals are crane turtle and Travancore tortoise.

7. Deccan peninsula: The Deccan peninsula includes the Deccan plateau in the south, the central and eastern plateaus, Chotanagpur plateau and central highlands. Average altitude is 2,500′ . Rainfall is less; it is 20 to 40″ as the area lies in the rain shadow region of the Western Ghats. Common plants are acacia, siris (*Albizzia siris*), sandalwood (*Santalum album*), teak, etc. The fauna include sloth bear, leopard, wild boar, barasingha, chital, wild buffalo, sambar, wild elephant, etc.

8. Coasts: The coastline of our country is on both sides of the Southern

Peninsula. Dry evergreen forests are seen in the Eastern Coromandel Coast. Rainfall here is 40 to 50′ . Besides the plants of the Deccan plateau, trees like *Strychnos*, jamun, etc. are also seen here. In the drier parts, *Zizyphus*, date palm (*Phoenix*), *Casuarina*, etc. grow. The spinous grass *Spinifex* is common in the sandy shores. Halophytes grow in the marshes.

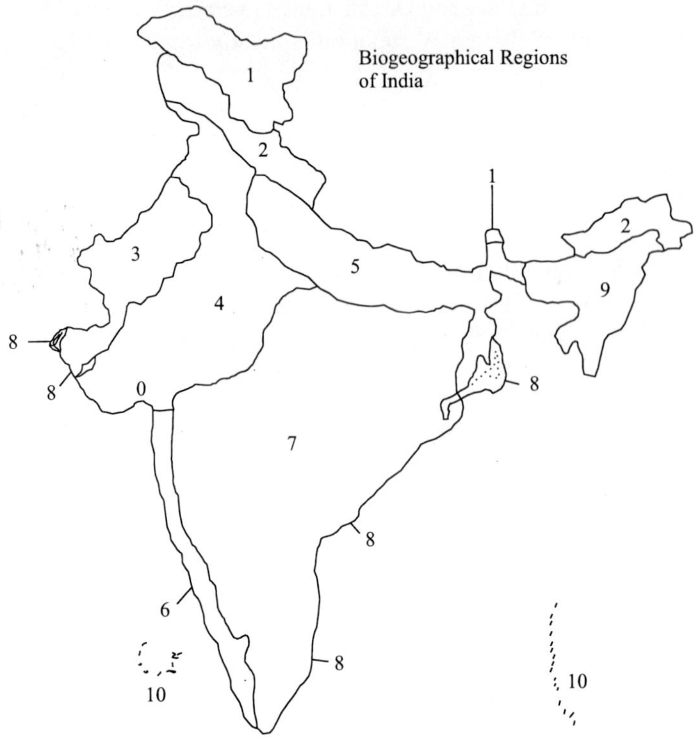

1. Trans - Himalayan Zone
2. Himalayan Ranges
3. Desert Zone
4. Semi desert Zone
5. The Gangetic Plain
6. The Western hats
7. Deccan Peninsula
8. Coasts
9. North East Zone
10. Islands

Ratnagiri in Maharashtra and Pichavaram near Chennai have typical estuarine (lower tidal) vegetation. Animals characteristic to the region are inshore dolphins, estuarine turtles and crocodiles, etc.

9. North-east zone: This zone has a very heavy rainfall. Annual rainfall is more than 80″. Cheerapunji has more than 400″ annual rainfall.

This heavy rain is responsible for dense tropical evergreen forests. In the valleys of Brahmaputra, outside the Himalayas, common plants are tropical chestnuts (*Sterculia*), champa (*Michelia*), mulberry (*Morus*), Fig. (*Ficus*), etc. *Dendrocalamus* is the common bamboo. limbers and evergreen shrubs are very common. On the riverbanks, siris, silk cotton (*Bombax*), insectivorous pitcher plant (*Nepenthes*), etc. grow. Sal forests occur in the Garo Hills. The fauna includes elephants, rhinoceros, swamp deer, pygmy hog, buffalo, etc.

10. Islands: The islands include Andaman and Nicobar in the Bay of Bengal and also the Lakshadweep in the Arabian Sea. The flora of Andaman and Nicobar resembles that of Myanmar and Malayasia. The Andaman has a wide range of spreading coastal vegetation like mangroves, beech forests and in the interior, evergreen forests of tall trees. Some dry pockets are also seen in the interior parts. Mangrove forests comprise *Rhizophora*, and *Sonneratia*. Common trees are betel nut (*Areca catechu*), muga (*Erythrina indica*), etc. Many bamboos also occur. Several animals and birds are endemic to this region.

VALUES OF BIODIVERSITY

There are numerous benefits of preserving biodiversity on our planet. Biodiversity is useful to us in many ways at the local, national and global levels. The maximum biodiversity is seen in the forests. In any country, at least 33% of the land should be under forests to run smooth economy of the country. Forests are a great national wealth. Trees purify the atmosphere by releasing large quantities of oxygen and by absorbing a same amount of carbon dioxide. They help in maintaining the water cycle. Trees afford excellent protection to the soil against erosion. They prevent early siltation (falling of the soil at the edges into water bodies) of lakes and dams and the formation of gullies (water channels formed by the flowing of rain water on slopes) which result in loss of fertile top soil. Forests prevent floods as well as droughts. They maintain the flow of rivers and streams throughout the year. Fixing and recycling of nutrients in nature is done maximum by forests. They provide us with a variety of products necessary for our food, clothing, housing, medicines and many other essential things and also give many people employment. Tribal people depend on forests for most of their needs.

Without thinking of the importance of forests to mankind as a whole, great damage has been and is being done to our forest wealth by unplanned lumbering by man without replacement. Valuable timber is cut up to many times the permission given by governments. This has damaged the whole environment in many parts of the world. Floods and droughts and soil erosion have been increased many fold. Valuable wildlife is gradually moving towards extinction. As forest covers are depleted and the number of polluting industries and automobiles increases, there is an increase in global warming which is extremely dangerous to the world environment. It also causes many changes in our atmosphere, and raise temperatures.

Moreover, wild plants and wild animals in forests constitute a never ending gene pool for our plant and animal breeders. From this vast gene reserve, they select useful genes for introducing into cultivated plants and domesticated animals.

Each species has a specific function in the ecosystem — it has an ecological niche. The ecological niche of an organism concerns basically three aspects: 1.The space where it lives; 2.What it does, i.e. how and where it gets energy from, how it behaves with other members of the system, how it responds to the environment, how it modifies the environment, etc. and 3. How is it constrained by other species. Thus the importance of any species in nature may not be known to us, but it is an essential part of nature. One of the main benefits of preserving biodiversity is thus maintaining the balance or stability of the ecosystems in nature.

The values of biodiversity can be divided into six categories.

1. *Consumptive use value*: Local communities all over the world make direct use of different plants and animals available in their respective areas for various purposes. The biodiversity in the ecosystem provides the local people with most of the things they need on a daily basis. Rural people get the grains, pulses, oils, vegetables and fruits, milk, meat, eggs, fish, etc. from their own villages or nearby villages. The rural folks depend on other plant and animal produce like medicinal herbs, fodder for their animals, building materials, firewood, honey, etc. on the local ecosystems. Fishermen catch fish from ponds, lakes, rivers and seas and oceans. If diversity was not there in nature, people would have had to be content with the same produce over and over again.
2. *Productive use value*: It is because of the genetic diversity present in nature that scientists, farmers and biotechnologists are able to selectively improve plants and animals. All the species of plants cultivated for various uses at present had their origin from wild species. *The process of cultivating wild species for human use is called domestication of plants*. This has played a vital role in

human civilization. Origin of agriculture dates back to 7,000 to 13,000 years ago. It started in the fertile lands near Indus, Nile, Euphrates and Tigris. Cultivation started with plants propagating vegetatively by buds, tubers, etc. but soon, man came to know how seeds give rise to new plants. Cereals were brought into cultivation first. These were unipurpose plants. Later, multipurpose plants were domesticated. Soon, plant breeding was devised as an important technique for plant improvement. Different types of selection were the first methods tried for plant improvement. Later, the method of hybridization (crossing two types of plants) was found out. Tissue culture is a modern method. This is based on a phenomenon called *totipotency* which *is the total capacity of any cell to give rise to a new plant.* More recent is genetic engineering. *In genetic engineering, one or more genes from a desirable variety is a introduced into or deleted from another variety which needs to be improved.* Genetic engineering is a branch of biotechnology.

Those plants into which one or more foreign genes have been introduced permanently are called transgenic plants or genetically modified (GM) crops. Such plants have been produced in tobacco (*Nicotiana tabacum*), tomato (*Lycopersicum esculentum*), etc. and many others. For best results in these fields, selection of the best variety among many is most important so that it can be still further improved by human effort. This is provided by the biodiversity in nature. For this, plant materials or genes of as many varieties in nature as possible are collected (in the form of seeds) and stored in live form. Tubers, rhizomes, bulbs, etc. can also be preserved. *Institutions where plant materials (genes) are stored are called gene banks*. Where seeds are not produced, tissue culture techniques are used for storing germplasm (reproductive cells or the germ cells and the cells from which they are formed).

The discovery of new types of plants and animals is never complete. It is an ongoing process. Also, among the known species, only very few have been studied till now for their use as medicines, industrial potential, etc. Loss of such organisms will also be a great loss to humanity.

3. *Social values*: In the earlier days, population was less and the needs of humans were limited. Those days, nature was not tampered with. But as human population grew and human beings

became more and more demanding, natural resources started getting depleted faster and faster. Coupled with their excessive usage, when replacement was not done properly, nature's resources started dwindling further. Biodiversity, one of the most valuable gifts of nature to man was no exception. However, even now, we can see that traditional societies are preserving nature better than the modern society. Therefore, we need to learn a lesson from them.

The Vedic scriptures make explicit references as to how forests and other natural resources should be maintained along with best utilization. Trees like peepal (*Ficus religiosa*), neem, mango, vat (*Ficus bengalensis*) and many others, herbs like Tulsi (*Ocimum sanctum*), grasses like Nagarmotha (*Cyperus rotundus*), Doob (*Cynodon dactylon*), animals like cow, ox, tortoise etc. and birds like peacock, owl etc. have been given place of respect in the scriptures. Medicinal plants and trees (vanaushadhis) were personified as goddesses of forests. All this inspired and motivated the people of those times to conserve biodiversity.

Traditional agriculture systems in India always followed multiculture (cultivation of different crops) as against monoculture practiced in some places today. Moreover, some governments in India are giving incentives to those farmers to start cash crops for sale in national or international markets. This has lead to local shortage of food, unemployment, draught, flood, etc. For the prosperity of local people, farm produce made locally must be available to them at reasonable prices. Otherwise, there will be unrest in society. Cultivating various crop plants and rearing domestic animals provide employment to lakhs of people in India. Thus conserving biodiversity has great social value.

4. *Ethical and moral values*: Protecting all forms of life was one of the principles on which life was based in the Vedic period in India. The Vedas preach, "Aatmavat sarvabhootheshu" — *Regard every being like your own self.* The Smritis, Puranas and other ancient texts integrates the civil and social duties with religious faith. It was believed that all forms of life have a right to live on earth. Man is only part of nature, just like the other organisms in nature. All ancient Indian scriptures propagated this idea.

   The attitude of respect towards earth as mother is widespread among the Indian society. Moral and ethical norms had been designed to ensure whole hearted participation of the masses in sustaining the biodiversity of nature. Many cultural, moral and ethical values are associated with nature even today in our country. One example is the sacred groves. These are small or large undisturbed forest areas in villages with tiny places of worship in the centre. In Kerala, sacred groves along with their associated ponds constituted a unique network of ecosystems. These are intertwined with the life and culture of the people of Kerala. They are dedicated mainly to goddesses/snakes. Most of them are private properties, owned by families. Some are owned by communities or temples. Due to neglect, the sacred groves with their immense biodiversity have been dwindling of late. Realizing the importance of such groves, local community members have formed committees to manage and protect sacred groves in some places. Pattuvaom Gram Panchayat was in news recently for preparing the first Community Biodiversity Register in which the medicinal plants and their uses were listed. Such efforts from the people will go a long way in creating awareness regarding the ethical and moral values of biodiversity among the general public.

5. *Aesthetic value*: Biodiversity adds to the aesthetic value of any place. Variety is the spice of life. How monotonous life would have been if there was no variety in nature. It is due to biodiversity that we have so many different varieties of plants and animals on earth. We live amidst varying ecosystems of forests, rivers, lakes, ponds, grasslands, agricultural pastures, deserts, snow bound peaks…even urban areas with some remains of vegetation and water bodies. All these add to the aesthetic beauty of nature. India is a vast country with varying types of climates and the resulting immense variety of plants and animals. These are a great attraction for national as well as international tourists. The more the biodiversity in a country, the more will be the number of tourists that visit that country. In this age of eco-tourism, preserving biodiversity will therefore make a government earn more revenue. This will boost up economy. Tourism has become a big industry in India. Conserving our biodiversity and thereby the places of aesthetic value in our country is therefore very beneficial.
6. *Option value*: We never know which species of animals or plants will become useful to us in future. Only very little has been found

out as yet. Before the discovery of penicillin, who would have thought that the lowly fungus *Penicillium* has such great therapeutic value? Nothing is impossible. Not only plants and animals, but also many microorganisms are of use to us in different ways. There is no guarantee that those which are not used at present will remain useless in future. Any day, any use of any organism may be found out. So it will be wise to preserve all forms of life present today in the best possible way. To get improved plants and animals, we go back to the wild types very often. So the wild dominant genes and genomes (a single set of genes is a genome) must be preserved. This is the option value of biodiversity.

BIODIVERSITY AT GLOBAL, NATIONAL AND LOCAL LEVELS

The number of known and described species of living organisms in the world today is about 1.8 million. According to some scientists, more than this number remains undiscovered. This shows how vast is the extent of biodiversity on the earth.

Life evolved on earth more than 600 million years ago. Many species have evolved and many others have become extinct over these long years. This is proved by fossils. By studying fossil records, we can know how the species evolved, and when these forms dominated the earth. The fossil records also show that there have been periodic cycles of mass extinction in the history of the earth. Fossil history shows that over the last 150 million years, 1, 50, 000 species of birds have evolved and become extinct. Yet, as per the information we have on birds, 9,000 species of birds are living worldwide today. This is a good number of species, we might think. But the problem is that human induced extinctions are much more rapid than those by natural processes. For example, even in the 19th century, 2,000 million individuals of passenger pigeons were living. But due to hunting and habitat destruction, the species became extinct by 1914.

The developed nations have a lesser biodiversity than the developing ones. The economically developed nations have exploited their diversity for centuries. The developing nations, including India, should therefore take a lesson from this. India as well as Brazil, Malaysia and Indonesia have a vast biodiversity at present. The species composition of these countries varies greatly. We must therefore preserve our own diversity to the maximum extent possible. As we have advanced much in biotechnology and genetic engineering, we can make better use of this natural resource too, compared to others.

The importance of preserving biodiversity has come to be realized globally. Many international agreements have been signed these past few years to protect this resource. Some of them are Washington Convention on International Trade in Endangered Species of wild flora and fauna (CITES), 1973; Bonn Convention on the Conservation of Migratory Species of wild animals (CMS), 1979; Convention on Biological Diversity (CBD), 1992, etc. These treaties gain importance in a situation when many animals and plants are nearing extinction world wide. For example, the global tiger population is only about 7,000; giant pandas are only 1000 in the wild today while only 1,700 individuals of horned parakeet are alive today, world wide. Ginseng root grows only in small areas of some countries and may disappear in a few decades.

India has signed many international conventions on biodiversity including CITES. The Government of India has also declared many areas as protected. Project Elephant, Project Tiger, etc. are actively being implemented in India.

Jharkhand is one of the forest states of India. Forest cover of the state is a healthy 31% at present. Tree plantation is being done on a large scale. Saranda forest, with sal as its major tree, is the home of the endangered flying lizard and elephants. Betla National Park boasts of gaur, chital, panther, wild bear and mouse deer. It was the site of the first tiger census in the world. In the Dalma wildlife sanctuary, one can see elephants, barking deer, sloth bear, leopard, tiger, porcupine, pangolins, etc. Many rare medicinal plants grow in the forests of Jharkhand.

MEGA BIODIVERSITY ZONES

Not all countries in the world are rich in biodiversity. Some countries do not have climates suitable for the development and sustenance of diverse forms of life. In some others, extinction rates have been rapid. At present, twelve countries have been listed as mega diversity countries in the world. They are: 1. Mexico 2. Columbia 3. Brazil 4. Peru 5. Ecuador 6. Zaire 7. Madagascar 8. Indonesia 9. Malaysia 10. India 11. China and 12. Australia. These countries together contain 60 to 70% of the world's biodiversity.

INDIA AS A MEGA DIVERSITY NATION

India is one of the mega diversity centres of the world. 320 species of wild crop relatives and several species of domesticated animals have originated here. The genetic diversity within these species is astounding. Examples from domesticated species given here are from National Bureau of Plant Genetic Resources and other sources.

- Rice ………..50,000 varieties
- Mango………....1,000 varieties
- Sorghum……..5,000 varieties
- Pepper………....5,00 varieties
- Cattle……………....27 breeds
- Goats……………22 breeds
- Sheep……………40 breeds
- Poultry…………...18 breeds

This amazing diversity is not only natural, but also a result of careful

Biodiversity Element	Range Native to India	Destroyed / Under threat
Ecosystems		
Forests	200 types, scrub to rainforests	Approximately 50% wiped out over last century; efforts on to increase forest land by Afforestation and Re-forestation.
Wetlands	8 types, seasonal floodplains to lakes	One third drained out, 70% polluted
Agro-ecosystems	20 agro-ecological zones	Mass homogenization across plains
Coasts	Several types of beaches, mangroves, coral reef systems	40% of mangroves wiped out; major portion of coral reefs bleached or silted
Wild Species	47,000 plants and 89,000 animal species	
Flowering plants	17,500 species	At least 20% extinct; 10% threatened
Mammals	390 species	2 or 3 extinct; 20% threatened
Birds	1,232 species	2 extinct; 5 to 10% threatened
Insects	57,000 species	No estimates

(Information from Zoological Survey of India, Botanical Survey of India and others)

selection and cross breeding done by India's farmers and scientists over the years. Cultural customs have helped conserve this diversity against the forces of destruction. Livelihoods of millions in India depend on the forests, ponds, rivers, coasts, fields and mangroves. Therefore people should not be removed from an area to protect wildlife.

Many villages in the country are engaged in self-initiated community forest protection. The villagers will destroy the biodiversity of an area only if their means of livelihood are snatched away. Main reasons for destruction lie elsewhere.

India is a land of diversity in ecosystems and wild species too. Some species have been destroyed, or are under threat.

Geographical history of the world shows that about 70 million years ago, the single giant continent of the earth was, split into two - the southern and the northern continents. India was part of the southern landmass, called Gondwanaland. Australia, Africa and Antarctica were also part of the southern landmass. Later, India moved to the north to join the northern Eurasian continent. The sea between these two landmasses was called Tethys Sea. When it dried up, plants and animals that had evolved in Eurasia and also the Far East migrated into India before the Himalayas were formed. A final lot came from Africa. These were plants adapted to dry regions. Thus India got the advantage of being in the centre, among 3 different centres of plant and animal origin. These geological events are the single largest cause of the biodiversity we see in our country today. High percentage of endemism is also seen here. About 18% of our species are endemic, seen nowhere else. For example, 62% of our amphibians (animals that can live on land and water, like frogs) are endemic, while 50% of our lizards are endemic. Many marine worms, insects, centipedes, etc. are also endemic to different parts of India.

India has a good world ranking in species diversity.

Species Diversity	India's world ranking
Mammals	8th
Birds	8th
Reptiles	5th
Amphibians	15th
Angiosperms (Flowering, fruiting plants)	15th to 20th

Moreover, our gene banks have collected more than 34,000 cereals and 22,000 pulses and many others grown in India. Research works in genetic engineering and biotechnology are going on in many laboratories in our country using materials from the gene banks. It is expected that better varieties of many cultivated species will be released sooner than later.

HOTSPOTS OF BIODIVERSITY

Of the more than thousand major ecoregions in the world, 200 are very rich. These are called the 'Global 200'. Most of these are under some threat, but some are under immediate threat and need immediate attention. Such *regions in the world that have rich biodiversity and high endemism and are under threat of extinction and habitat destruction are called hotspots for preservation.* Norman Myers (1988) identified 12 hotspots of rain forests. These hotspots cover only 0.2% of the total land surface, but contain 14% of the world's plant species. Later, in 1991, Myers expanded his study and declared eight non-forest areas as hotspots — four in the tropics and four outside the tropics. 50,000 endemic plants comprising 20% of global plant population live in these 20 hotspots. As per different environmentalists, the total number of hotspots is 18, 24, etc. Such variations are due to international priorities and overlapping.

Of these twenty four hotspots of the world, twelve are rainforest hotspots as identified by Myer in his first list. They are seen in different countries or regions like North and South America, Philippines, Australia, the Himalayas, etc.

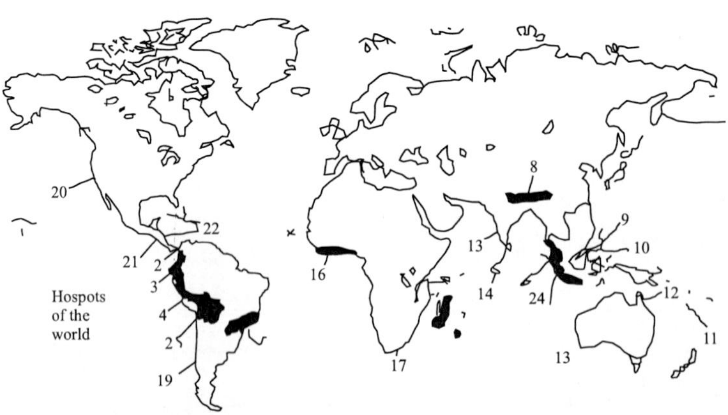

1 to 12 - Rainforest hotspots; 13 to 20 - Eight hotspots in other climates; 21 to 24 - Four more zones recognized as hotspots

The 12 rainforest hotspots are:

1. Hawaian Islands, west of North America
2. Colombian Choco
3. Western Ecuador
4. Western Amazonia
5. Atlantic forests of Brazil
6. Eastern Madagascar
7. Peninsular Malayasia
8. Eastern Himalayas
9. Philippines
10. Northern Broneo
11. Queensland, Australia
12. New Caledonia (Oceania)

The 8 hotspots in other climatic ecosystems are:

13. South-West Australia
14. Sinharaja Forest, Srilanka
15. Western Ghats, India
16. Eastern Arc Forests, Tanzania
17. Cape Floral Province, South Africa
18. Upper Guinean Forests (Ivory Coast)
19. Central Chile
20. California Floral Province

In addition to these twenty hotspots, four more hotspots have been recognized by many. These are:

21. Central American Highland Forests
22. Central American Lowland Forests
23. Tropical Andies, and,
24. Indonesia

As mentioned earlier, opinions differ regarding the number of hotspots. So the regions also may differ slightly. However, the fact is that over the world, biodiversity is under threat in many regions, spread in many countries. In India, we have two hotspots - the Western Ghats and the Eastern Himalayas.

The Western Ghats lie parallel to the west coast of peninsular India. Its length is about 1,600 kms. It covers parts of Maharashtra, Karnataka, Tamil Nadu and Kerala. Evergreen and semi-evergreen forests are found here. Altitude is up to 1,500 meters. The main centre of biodiversity is the Silent Valley in Kerala.

Eastern Himalayas of the North-Eastern India was an active centre of evolution. Many endemic species (species growing specifically in some regions and not anywhere else) grow here. Mountains as well as valleys are rich in flora and fauna. Altitude is up to 3,500 meters. Many primitive angiosperms are found in this region. (See Bio geographic classification of India).

THREATS TO DIVERSITY

Biodiversity of any region is threatened in different ways. The most serious threat is from the human beings. Destruction done to nature and natural systems by us is of the worst kind.

1. Habitat destruction: Habitat means the place where an organism lives. In nature, we see that organisms live in a variety of habitats. If these are destroyed due to any reason, the organisms will have to move to places more suitable for them or perish. Thus habitat destruction is a very serious threat to biodiversity. Humans destroy habitats in different ways.

One is by *deforestation.* Valuable timber is cut up to many times the permission given by governments. Cutting down trees by stealth is common in some forests. Both these take place where protection is not up to the mark. In poor countries around the world, the rural poor are pushed onto marginal, often hilly lands from which they have to cut trees and clear land for cultivation just to survive. In such cases, poverty plays a major role. Forest fires — natural or induced by man, for cultivation — often destroy vast areas of forests. Urbanization is the cause in some cases. Villagers encroach into nearby forests as population increases. Roads and sometimes railways are constructed through forests. For dams and hydroelectric projects on rivers flowing through forests, large areas are cleared. Thus there are many reasons for deforestation. Deforestation results in floods, droughts, reduced rain, soil erosion, pollution, etc. in addition to flight of or death of many animals and birds, reduction in the yield of river fish, etc. The damage caused by Orissa's killer cyclone would have been much less if the mangrove forests that once protected the coastline had not been cut for 'coastal development'.

Draining up of marshes is another reason for habitat destruction. Marshes are like sponges. They absorb rain water in the monsoon and hold it for the whole year. In summer, when the myriads of organisms living in the marshes are in need, water is released. Moreover, estuarine and wetland ecosystems are the most productive ecosystems in the world. Highest agricultural production is in rice fields which are wetlands. If these are drained to make way for human settlements or prawn farming, one can very well imagine the level of damage this will inflict on the natural habitats.

Bleaching and destruction of coral reefs: Polyps are small aquatic, mostly marine animals, usually fixed and colonial. These belong to a group of animals called Polyzoa. Coelenterata is another group of similar animals, simpler in structure. These animals secrete skeletons of calcium carbonate around themselves. The skeletons of numerous animals are joined together to form what we see as corals. The animals live inside the corals. *Corals in continuous populations are called coral reefs*. They are very beautiful and may be white or of different colours. They grow in unpolluted waters of the sea. Innumerable small fish of various species go to coral reefs for breeding as big fish cannot enter the reefs and so the reefs are safe and secure. Thus, coral reefs are ecosystems by themselves. Around the world, coral reefs are disappearing from seas and oceans due to pollution, siltation from the seacoasts and a rise in temperature of the surface water. In many places, sewage is just released into the sea, agricultural runoff containing pesticides flows into the sea, sedimentation is increased due to erosion caused by deforestation.

Soil degradation: *The fall in the quality of soil is called soil degradation*. This is caused by many factors. Addition of fertilizers to an undesirably high extent leads to the death of soil microbes, many of which are useful. Biological nutrients do not get replenished as before. This has led to the degradation of millions of hectares of agricultural land in Punjab, Haryana, Eastern Uttar Pradesh, etc. The practices of monoculture in U.P., shifting cultivation in Nagaland, etc. also degrade once fertile land. Other reasons for land degradation are water logging caused by floods resulting from deforestation, mining, dam building, increased salinity, acidity etc.

Pollution: Air, water and soil pollution are causing untold damage to habitats all over the world. Dust, smoke, sulphur dioxide, fluorides, etc. in the air are injurious to plants. Many airborne microbes cause diseases to human beings, animals and plants. Water is polluted in different ways by us. Water pollution decreases or even destroys entire aquatic ecosystems. Soil is also polluted in many ways. Vast fields are threatened as the natural enemies of their pests have been killed off by pesticides. All these cause great harm to biodiversity. (For details, please see chapter on Pollution).

2. Habitat fragmentation: Before the advent of urbanization and industrialization and the like, before the formation of thousands of settlements all over the land, natural ecosystems were vast expanses. Now these habitats have been fragmented by roads, rail lines, canals, power lines, etc. This has disturbed the ecosystems by affecting dispersal of fruits and seeds of plants, by changing territoriality shown by animals, etc.

3. Introduction of alien species: Species native to any region are adapted to the environmental conditions in that region perfectly. Alien species brought to any geographical region may harm the indigenous ones. They may change the habitat also to such an extent that the native ones gradually disappear. This is true of plants and animals. Some fish, on being introduced into ponds and lakes eat the native fish so much, beyond their capacity of multiplication that the latter disappear completely. Plants like *Eupatorium ayapana* have become so wide spread in many regions in India that it has already replaced many other plants. *Parthenium hysterophorus*, an alien plant introduced into India is seen growing almost everywhere; its pollen are powerful allergens; people who are unaware of this do not destroy the plants and get affected by lung diseases. Disease causing microbes introduced into new localities may cause havoc not only by their own multiplication, but also by modifying the indigenous microbes genetically.

4. Desertification: Effect of wind erosion is very clear in the deserts of Rajasthan. The land west of the Aravali Mountains is a sandy waste. Over the years, the desert has spread due to reckless tree felling by man and overgrazing by camels, sheep and cattle. Large towns near the desert are threatened by the advancing sand dunes.

5. Wrong policies: In many developing countries, the policies formulated by the governments are faulty. Environmental issues are not given the importance due to them. This is mainly due to lack of knowledge on the part of the policy makers and also due to their indifferent attitude while planning the economic system to be adapted by the country.

6. Extinction of species: This is the most serious threat to biodiversity. Biological extinction had taken place in nature many times, in many places. *Natural extinction* is a gradual slow process because of environmental changes. If a large number of species become extinct together due to some natural calamity, it is called *mass extinction*. But *anthropogenic extinction*, i.e. the extinction due to man's activities is the worst as it is extremely fast and is going on and on. Thousands of species are becoming extinct. As species in the tropics show more variety, greatest losses are in the tropics. Selective and excessive exploitation of species is going on continuously in many countries. In addition to this, proper management practices are not done for their regeneration. When a species becomes extinct, that particular combination of gene pool is permanently lost. Along with each disappearing species, its dependent species of higher animals, insects, etc. also disappear. Therefore conservation of endangered species is a must. The world may lose half of its biodiversity if extinction is not checked now.

As far as our country is concerned, at least 10% of our plants and animals are threatened with extinction. Much of the crop and livestock diversity characteristic of Indian farming is on the decline. Some areas have lost all their seed diversity and all the 18 breeds of poultry are threatened by breeds introduced from outside India. Thus extinction is taking place not only in the wild, but also among the crop plants and the domesticated animals.

The Red Data Book is published by IUCN (International Union for Conservation of Nature and natural resources). It lists plants and animals which are in different levels of extinction. Threatened species have been classified into three categories — rare, vulnerable and endangered. If the rare species are not protected, they will become vulnerable. If the vulnerable species are not protected, they will become endangered. If the endangered species are not protected, they will become extinct.

In the Red data book prepared by the Govt. of India, 435 species of endangered plants have been listed. Examples - *Rauwolfia serpentiana* (medicinal), *Atropa acuminata* (medicinal), *Dioscorea* species (source of steroids), *Santalum album* (sandalwood or chandan), etc.

CONSERVATION OF BIODIVERSITY

Conservation of biodiversity can be done by conservation of species and ecosystems. It is not advisable to protect a single species, though this is done in special cases when it is protected for aesthetic or medicinal or other purposes. The whole ecosystem of which it is a part must be conserved. Natural areas need protection as wild stocks cannot be maintained in zoos and parks. The attention they receive in zoos and park results in a rapid loss of variability. In addition to the protection and conservation of threatened species, habitats that are particularly interesting can be given the special status of National Park or Nature Reserve to prevent further deterioration. Areas of outstanding scientific importance can be conserved in this way for future generations.

In Situ Conservation

Complete protection of one species may result in its overpopulation. This shows that management of a reserve is not easy, especially as man has greatly disturbed the natural balance by eliminating predators. A Nature Reserve is an area where a particular habitat is completely protected together with its flora and fauna. Human activity is reduced and scientific workers' access is also restricted. Industrial and commercial activities are prohibited.

Such *conservation of biological diversity in their natural surroundings is called in situ conservation.* These are of different kinds.

1. *Protected areas:* Protected areas may be of land or water. These are protected legally by the Government or by community efforts. Government action may be at the national or local levels. Sometimes, land is bought and conservation carried out by private organizations. Frequently, the governments of developing countries may tie up with international organizations, multinational banks or governments of developed countries for funds and training of conservation experts.

S. No.	Type of Protected Area	Characteristics of the Protected Area
1.	Scientific Reserves and Strict Nature Reserves	Strictly protected areas. Maintained for education, environmental monitoring and scientific study. Undisturbed
2	National Parks	Large areas. Picturesque. Maintained for educational and scientific use. Not exploited for commercial use.
3	National Monuments and Landmarks	Smaller area. Preserves unique areas of national heritage
4	Wildlife Sanctuaries and Nature Reserves	Limited human population and harvesting allowed
5	Protected Landscapes	People residing here can use the environment, but no destruction allowed. Tourism and recreation permitted
6	Resource Reserves	Controlled resource use allowed, in accordance with the law of the country
7	National Biotic Areas and Anthropological Reserves	Resources can be used. Hunting of common animals permitted
8	Multiple-use Management Areas	Sustained use of all natural resources like timber, water, wildlife, etc. allowed. Activities like fishing, tourism, grazing permitted

System of Classification of Protected Areas developed by IUCN (1985) The first five categories in this classification by International Union for Conservation of nature and natural resources (IUCN) are fully protected areas. The main purpose of their maintenance is conservation of biodiversity. However, in the last three categories also, biodiversity is observed.

India has about 580 protected areas. These cover only 4.7% of the land area while the recommended figure is 10%. Yellow Stone National Park of US is very famous. Important National Parks in India are Kaziranga National Park in Assam, Sundarbans in West Bengal, Corbett National Park in Nainital, etc. Wildlife sanctuaries protect animals only. Animals are allowed to live in natural surroundings. India has 492 sanctuaries. Important ones are Keoladeo Ghana Bird Sanctuary in Bharatpur, Rajasthan; Periyar Sanctuary, Kerala; Chilka Lake Bird Sanctuary, Orissa; Manas Wildlife Sanctuary, Assam, etc.

2. *Biosphere Reserves:* Biosphere reserves are special protected areas where people form a component. In the centre of the reserve is the undisturbed *Core Zone*. Outer to this is the *Buffer zone* where educational activities and research are allowed. In the outermost *Transition Zone*, management of the reserve interacts with the local people.

3. *Sacred forests, lakes, groves, etc.:* These are maintained by local communities. They are areas where many rare and endangered plants and animals are protected. Aquatic flora and fauna are protected in sacred water bodies. Such centres are seen in many states in India like Kerala, Karnataka, Maharashtra, Meghalaya, etc.

4. *Legal protection of species and habitats:* This can be done at local, national and international levels. In many countries, private conservation organizations buy land and conserve species. At the international level, CITES (Convention on International Trade in Endangered Species) was established in 1973. Countries which signed the Convention agreed to restrict trading in animals, plants and their parts. There are international agreements for protection of habitats, etc. One example is the Ramsar Convention on Wetlands of International Importance, 1971, signed by 123 countries.

Chilka Lake in Orissa is a Ramsar site. This lagoon is situated along the east coast of India. It is a mixture of marine and freshwater ecosystems, an area of great biodiversity where many endangered species listed in the Red Data Book live. To restore the lagoon to its original status, the Government of Orissa created Chilka Development Authority (CDA). The

CDA decided to involve various institutions and the local communities in the restoration work. The result has been amazing. Chilka Lake (lagoon) stands completely restored now. For wetland conservation and sustainable use in any part of the world, CDA got the Ramsar Wetlands Award, which is given only once in 3 years, in 2002. This is also an example of *in situ* conservation.

Ex Situ Conservation

As *in situ* conservation is not possible in many cases, *ex situ conservation* is done. *Conservation of genetic resources outside the natural habitats of organisms by maintaining sample populations is called ex situ conservation.* Genetic drift and inbreeding, environmental variations, habitat loss, competition, diseases, overexploitation, etc. lead to decline and extinction of species in the wild. So species have to be maintained in artificial conditions under human supervision. This *ex situ* conservation is therefore also called *off site conservation.* Zoos, aquaria, game farms, etc. in the case of animals and botanical gardens, seed banks, etc. in the case of plants are established for this purpose. Tissue culture and genetic engineering are very helpful in *ex situ* conservation.

Ginkgo biloba (maiden hair tree) is a gymnosperm (plants with naked seeds, i.e. seeds not covered by fruits are called gymnosperms). It was once known as 'living fossil' as it was seen only in some countries and was about to become extinct. However, serious efforts were made to save the species all over the world. Now Ginkgo is a common plant grown in many gardens. This is a very good example of conserving genetic resources by *ex situ* conservation.

Gene Banks: Plants with desirable characters have to be stored in a live condition for use by plant breeders and biotechnologists. Institutions where such materials are preserved are called gene banks. In gene banks, living materials are stored in the form of seeds or genetic materials or both. It can be used by scientists in any part of the world.

Seed is the precursor of plant. Inside seeds, embryos remain in an inactive state for variable lengths of time periods and later germinate into plants. The process of germination needs many external and internal factors. The main external factors are water, air and suitable temperature. The internal factors necessary for seed germination are the required enzymes and hormones, and release of energy through respiration. If these factors are denied to seeds artificially, they do not germinate; at the same time, they remain alive too. This is done by storing seeds in a dry condition (about only 5% moisture) and at low temperatures (-10 to -20° C).

Orthodox seeds and recalcitrant seeds: Not all seeds can tolerate drying and freezing conditions. *Seeds of some species are not damaged by drying and freezing; such seeds are called 'orthodox seeds'*. Cereals and legumes come under this category. *Seeds that get killed by drying and freezing are called recalcitrant seeds.* Examples are coconut, litchi, jackfruit and many other fruits and timber plants. They can be stored only for short periods, after giving them fungicides, required moisture and oxygen. Such plants can be conserved using tissue culture.

Tissue culture and storage of germplasm: *Growing isolated cells or tissues or organs or embryos in an artificial medium under aseptic conditions outside the parent organism is called tissue culture.* In cases where seeds are recalcitrant or are not formed at all, tissue culture is very helpful. Many plant organs are used for tissue culture, but shoot tips are the best. Shoot tips are easy to regenerate and are free of viruses. The main steps in tissue culture are as follows:

1. Mother plant is selected.
2. Growing tissue (eg., Shoot tip) is dissected out.
3. Surface is sterilized
4. Tissue is introduced into culture medium(nutrients dissolved in distilled water)
5. Tissue grows and forms a lump, called callus; it is cut into pieces
6. Pieces are grown in many tubes containing culture medium (subculturing)
7. Process repeated many times
8. Shoots are then rooted by applying hormones
9. Transferred to pots
10. Transferred to fields

Thus, tens of thousands of plants are produced in a very short time from one shoot tip or embryo. In this way, plants in the danger of extinction can be multiplied and propagated. This helps in the conservation of germplasm and biodiversity.

Cryopreservation: Storing germplasm at very low temperature of –196° C *(temperature of Nitrogen) is called cryopreservation.* No biological activities take place at this temperature. Therefore genetic changes are not possible. Animal cells, sperms and ova, tissues of embryos, whole embryos, etc. can all be preserved in this manner.

Zoos as ex situ conservation centres: Zoos are very good examples of *ex situ* conservation. In India, it was in the 1960's that scientists realized the gravity of the problem of rapid biodiversity loss in the country. This led to zoos being converted to conservation centres in addition to their original role of education and entertainment. Thus, zoos came to have 3 additional functions

1. Captive breeding
2. Conservation, and
3. Research

Indian zoos have successfully bred some species like tiger, wild dog, leopard, crocodile and many such other animals. Managers of zoos have come to realize the role of zoos in *ex situ* conservation.

However, it is no use of protecting some areas while the rest of the country is allowed to deteriorate. This can happen especially outside the protected areas. In such areas, conservationists are engaged in restoring the degraded environment. *Scientific measures taken by experts in conservation leading to restoration of degraded or damaged areas surrounding protected areas is called Ecological Restoration. The branch of study dealing with conservation of species outside protected areas is called Restoration ecology.* In fact, the protection of nature must be extended to cover the whole country. It is possible to protect nature only with public awareness. If the diversity of nature is conserved, man can become fully integrated with nature.

Questions

A] Very short answer type questions:

a) Multiple choice - tick the right answer

1. Maintaining biodiversity in natural habitats is called:
   (a) *Ex situ* (b) *In situ*
   (c) *In vitro* (d) *En masse*
2. Cryopreservation is preserving germplasm at:
   (a) High temperatures (b) Intense light
   (c) Very low temperature (d) High pressure
3. Endemic plants are those which:
   (a) Grow in cold countries
   (b) Grow in tropical countries
   (c) Live exclusively in some areas
   (d) Are distributed in all types of localities
4. Sundarbans is the name of a place in:
   (a) Assam (b) West Bengal
   (c) Karnataka (d) Orissa

5. Mangrove forests are seen in:
   (a) Mumbai (b) Nicobar Islands
   (c) West Bengal (d) All of the above
6. Forest cover of Jharkhand at present is:
   (a) 18% (b) 31%
   (c) 45% (d) 12%
7. There are no hotspots in:
   (a) Canada (b) India
   (c) America (d) Australia
8. Western Ghats are located in:
   (a) North East India (b) Maharashtra
   (c) Peninsular India (d) Gujarat
9. Red Data Book was prepared by:
   (a) IUCN (b) UNEP
   (c) IBP (d) UNESCO
10. Earth Summit was held in:
   (a) 2000 (b) 1996
   (c) 1995 (d) 1992

b) Define the following:
   1. Biodiversity
   2. Species
   3. Hotspots
   4. Domestication of plants
   5. Totipotency
   6. Transgenic (Genetically Modified or GM) plants
   7. Ecological restoration
   8. Cryopreservation
   9. Soil degradation
   10. *Ex situ* conservation
   11. Tissue culture
   12. Ecological restoration
   13. Orthodox seeds
   14. Recalcitrant seeds

c) Write very brief notes on:
   1. Earth Summit
   2. Trans-Himalayan Zone
   3. Desert zone of India
   4. Semi-deserts in India
   5. Gangetic Zone

6. Deccan Plateau
7. Coasts of India
8. North East zone of India
9. Islands of India
10. Consumptive use value of biodiversity
11. Productive value of biodiversity
12. Social value of biodiversity
13. Ethical and moral value of biodiversity
14. Aesthetic value of biodiversity
15. Option value of biodiversity
16. Megabiodiversity zones of the world
17. Draining of marshes
18. Habitat fragmentation
19. Red Data Book
20. Biosphere Reserves
21. Sacred groves
22. Cryopreservation
23. Tissue culture

d) Differentiate between the following pairs:
1. Natural extinction and Anthropogenic extinction
2. Orthodox seeds and Recalcitrant seeds
3. Habitat destruction and habitat fragmentation

B] Short answer type questions:

Write short notes on:
1. Species diversity
2. Community or ecosystem diversity
3. Deforestation
4. Himalayan ranges
5. Megadiversity nations
6. Destruction of coral reefs
7. Extinction of species
8. Gene banks

C] Long answer type questions:

a) Write an essay on the biogeographical regions of India
b) Give a detailed account of the values of biodiversity
c) Write long notes on:
1. Genetic diversity
2. *In situ* conservation

Appendix 1

Annual Rainwater Harvesting Potential (ARHP)

To calculate ARHP, the following data are essential:

1. Average rainfall and frequency of rainfall; 2.Catchment area with details of paved and non-paved areas; 3.Runoff coefficient of the catchments.

The formula is: ARHP = R × Rc × A, where,

R - Annual rainfall in metres; Rc - Runoff coefficient; A - area in metres

The runoff coefficient is assumed as follows for different materials:

Tiled roof – 0.8-0.9
Corrugated metal sheet roof – 0.7-0.9
Concrete ground surface – 0.6-0.8
Brick pavement – 0.5-0.6
Compacted and smoothened soil – 0.3-0.5
Rocky areas – 0.2-0.5

To calculate the maximum amount of rainfall that can be harvested from a rooftop, suppose the following data is available:

R = 611mm (0.61m)
Rc = 0.85
A = 100sq.m.
ARHP = 0.6 × 0.85 × 100 = 51cu.m. (51,000 litres)

SETTLEMENT TANK

To know the optimum capacity of the settlement tank, three aspects must be considered: 1.Size of the catchments 2. Intensity of rainfall 3. Rate of recharge.

Capacity of the tank should be enough to retain runoff occurring from conditions of peak rainfall for at least 15 minutes.

Suppose the following data is available:

Area of rooftop catchment = 100 sq m

Peak rainfall in 15 minutes = 25 mm (0.025m)

Runoff coefficient = 0.85

Capacity of settlement tank = $A \times R \times C = 100 \times 0.25 \times 0.85$

= $2.125\ m^3$ (2,125 litres)

Any container with adequate capacity of storage can be used as a settlement tank. Generally, masonry or concrete underground tanks are preferred as they do not occupy any surface area. For over ground tanks, prefabricated PVC or ferrocement tanks can be used. Prefabricated tanks are easier to install.

Appendix 2

Answers to Multiple Choice Questions

UNIT – 1

1 (d) 2 (c) 3 (a) 4 (d) 5 (b) 6 (c)

UNIT – 2

1 (b) 2 (c) 3 (c) 4 (c) 5 (a) 6 (c) 7 (d) 8 (a) 9 (d) 10(d) 11 (b) 12 (c) 13 (a) 14 (d) 15 (c) 16 (c) 17 (d) 18 (c) 19 (b) 20 (d)

UNIT 3

1 (d) 2 (c) 3 (a) 4 (d) 5 (c) 6 (d) 7 (d) 8 (a) 9 (c) 10 (c) 11 (d) 12 (a)

UNIT -4

1 (b) 2 (a) 3 (d) 4 (d) 5 (c) 6 (a) 7 (d) 8 (d)

UNIT – 5

1 (b) 2 (d) 3 (b) 4 (d) 5 (d) 6 (a) 7 (c) 8 (a) 9 (a) 10 (b) 11(d) 12 (d) 13 (d) 14 (c) 15 (b) 16 (d)

UNIT – 6

1 (c) 2 (d) 3 (d) 4 (c) 5 (b) 6 (d) 7 (a) 8 (c) 9 (c) 10 (b) 11 (d) 12 (a) 13 (b) 14 (d) 15 (d)

UNIT – 7

1 (b) 2 (c) 3 (d) 4 (c) 5 (a) 6 (b) 7 (b) 8 (c) 9 (d) 10 (d) 11 (c) 12 (b)

UNIT – 8

1 (b) 2 (c) 3 (c) 4 (b) 5 (d) 6 (b) 7 (a) 8 (c) 9 (a) 10 (d)

Appendix 3

Websites and Search Engines

Suppose you want to get some information on a particular topic and would like to use the internet for the same. After getting connected to the internet, type in the name of any search engine in the long box named address, and press enter. Google.com, yahoo.com, msn.com, ask.com, lycos.com, alltheweb.com, altavista.com, etc. are all search engines.

As soon as the page of the search engine appears, type the most important words from the topic in the Search box. Search Engines are websites that help you find information located on other websites based on the words you supply. Thus, as soon as you type the words, the search engine gives you a list of web pages to visit.

Now you can click on any website you want. Within a very short time, mostly within seconds, you will reach a world of unlimited knowledge. After you finish with the first site, you can click on 'back' to return to the previous page and click on another website. The links are unlimited.

There are different types of websites. A website ending with .edu is often the most reliable, as it is the website of any educational institution. Similarly, websites ending with .org are those of organizations. One ending with .gov is the website of the US government. Indian government websites have .nic.in at the end. Thus the last letters in the name mostly indicate the organization / institution / government, etc.

If you know the name of the website, you do not need to go through the search engine. You can type the site name directly in the address box. For example, if you want to go to the site of the World Health Organisation (WHO), type Who.org in the address box and press enter. The site of United Nations Environment Programme is unep.org.

Similarly, if you want to visit the Ministry of Environment and Forests (MoEF), Government of India, type in envfor.nic.in and press enter.

Some websites endings are country specific. For example, the website of the Government of Australia ends in .gov.au. A website ending with .uk is that of the United Kingdom while .pk is the website of Pakistan.

Encyclopedias and dictionaries are also available on the net. For example, wikipedia is a web-based encyclopedia just like dictionary.com is a web-based dictionary. Wikipedia is an excellent encyclopedia. It appears almost in the beginning on opening many sites. The information you get in wikipedia is not obtained by searching through various other websites – it is information that experts on various topics have already entered in wikipedia itself.

Some of you must be aware of all this and more. This Appendix is meant for those students who would like to know the basics of browsing (surfing) the internet. In these days of technical advancement all over the world, knowledge of computers has become a necessity. Those who do not know how to search the net are losing out a lot. This is the main reason for dealing with this topic briefly here. Hope it will be of help to you. Happy surfing!!

Appendix 4

Some Tips to Students

Following are some tips to students. By following these recommended suggestions, the students can improve their memory, concentration power and confidence. This will help them perform better in classrooms and examinations. It will be in fact helpful throughout life. Tips are given on how to answer examinations also. Facts have been presented in an informal fashion.

- *Importance of attending classes*: Hope you have started attending the third year classes right from day 1. Attendance is important not only for the sake of procuring the required percentage of attendance. Listening to your teacher with full attention in the classroom will help you a lot in the examination hall.
- *Interest in the subject*: When you are seeing a movie, I am sure you will watch the scenes and listen to the dialogues and songs with rapt attention. It is because you want to know what is happening. Similarly, if you want to really know what is being taught in class, nothing can stop you from paying attention. Develop interest in the subject consciously, in case you did not have it earlier! Enjoy the subject. There is no subject that is mundane.
- *Need of single minded attention*: Remember the story of Arjun. Once, he and his brothers were asked by their archery teacher to hit the eye of a wooden bird placed sidewise on the branch of a tree. The boys stood in a line ready to shoot their arrows. As each boy came forward and took aim, the teacher asked him what he was seeing. Some answered they were seeing the tree; some said the branch on which the bird was kept and yet some others said they saw only the bird. Only Arjun gave the right answer. He said, "I see only the bird's eye". The teacher was happy and Arjun hit the aim with great accuracy.... When the teacher is explaining

Environmental Science, do not think of anything else, about home or anything that happened in the bus or of Chemistry, Economics or Commerce. Forget everything else and concentrate on Environmental Science. In Environmental Science, think of only that chapter, only that part of the chapter, that is being taught. Suppose the teacher is explaining Ecosystem. Think of Biotic factors when that part is being dealt; within that, think of producers when then teacher is talking about that, if it is primary producers that is being discussed, think of primary producers. In this way, keep your mind glued to each sentence, each word being spoken in class. This is possible. It comes with practice.

- *Yoga to remove tension, improve concentration power and memory*: Follow some yoga practices every day. Yoga has become popular in many foreign countries. It is high time it became popular in its land of origin, our very own India.
  Each asana (posture) has multiple benefits. Those of you who have learnt yoga know that inverted and balancing poses are especially good to improve concentration. As inverted poses are more difficult, and you can learn them later, let me tell you about the balancing poses. There are many different balancing poses. Let me explain the simplest of them, Pranamasana. It is also called Eka Pada Pranamasana, meaning the one legged prayer pose.
  Pranamasana: Stand erect, with feet together and arms on your sides. Lift one foot and place its sole on the inside of your other thigh. Fold your hands in front of your chest. Look straight ahead, fixing your eyes on one point, like a black dot made on a wall. Do this asana for as long as possible. For balancing your body, you have to remain concentrated. Therefore, it improves your power of concentration. It also removes stress, anxiety and worries. If you become tense at any time due to any reason, you can do Pranamasana for some time. Additional benefits are strengthening the leg, foot and ankles.
  Trataka: Practice of Trataka can help to improve concentration to any extent. There are many variations to this technique. I shall explain the simplest form here: Place a lighted candle on a table or stool in such a way that the flame will be at the level of your eyes when you sit down for the practice, about one and a half to two feet away. Sit in a convenient pose and make yourself comfortable. Any broad based pose will be ideal for this. Keep

your back straight. Relax your whole body and close your eyes. Soon, you will become aware of your body only and you will be still like a statue. Now open your eyes and look at the brightest spot of the flame, just above the tip of the burning wick, with total concentration. Do not blink or move your eyeballs. Do not change the position of your body even slightly.

When the eyes become tired, close your eyes. Now you can see the image of the flame in front of your eyes. Practice Trataka on this image. When the image begins to fade, open your eyes and gaze at the real flame again.

Trataka can be done for about 15 minutes. In the beginning, your eyes may pain. So increase the time to 15 minutes only gradually, within a few days or weeks. Soon, you will be practising Trataka with ease.

Additional effects of Trataka are that it stabilises nerves, relaxes troubled minds and corrects weakness as well as some defects of eyes like short sightedness.

Pranayama: Pranayama is excellent to increase the power of concentration. When we breathe normally, the lowermost part of our lungs is never emptied. Therefore some stale air always remains there. By forcefully expelling this stale air, like during the Kapalbhati Pranayama, we can increase the efficiency of our lungs. Pure air filling the lungs cleans all cells in the body during blood circulation. Brain cells also become more active. This leads to improved memory and concentration.

It is best to do Pranayama early in the morning. However, it can be done at other times also, but not immediately after meals. A gap of about 3 to 4 hours can be left. There are various types of Pranayama. Of these, Anulom Vilom, Nadishodhan etc. are very good to improve memory. For more details on pranayama, various balancing asanas (postures), other asanas for general health and well being, and to know about exercises for eyes, refer any good yoga book.

- *Eye care*: Reading or writing for a long time continuously, especially before examinations, tires the eyes a lot. Therefore, give your eyes some rest in between. Close the eyes and just relax, physically and mentally for a minute or so after about every hour of reading. You can also rub your palms to make them warm and place them on your closed eyes during this period. Putting one or two drops of pure rose water (gulab jal) in each eye once or twice a day also refreshes the eyes.

- *Regularity and planning*: Study daily. Do not keep everything for the last minute. Make a plan daily and try to follow it. Every night, before going to sleep, check how much of the syllabus you have been able to finish. This will help you to know your strengths and weaknesses. You will be able to make better plans for the next day. In the weekends, revise what was taught in the whole week. Similarly, in long vacations, revise everything taught in the term.
- *Display method*: Write equations, formulae, important points, etc. on big sheets and display them in prominent places. You can read them while walking past them.
- *Revising while relaxing*: Do not think that you are studying only when you have a book in your hands. You can try to recollect what you read recently while taking bath or brushing your teeth or walking to the market. This is also studying.
- *Drawing and writing*: If there are pictures, formulae, etc. in the chapter, draw or write them down while studying. This can be done after studies also, as a self-test.
- *The importance of getting concepts clear*: Do not just mug up facts without understanding the meaning. Studying only to get good marks is not advisable. Study to learn new things. That makes your concepts clear. This will invariably translate into good marks. When your concepts are clear, you will answer the examination with confidence.
- *Thinking positive*: A positive mindset is crucial in every walk of life. As far as examinations are concerned, if you know the lessons, you will be in a positive frame of mind. That helps a lot in the examination hall.
- *Be relaxed*: Do not tensed up at all while you enter the examination hall. You will feel that you have forgotten all that you have learnt. This happens to everybody. There is no need to worry. Once you start writing, you will remember everything.
- *Carefulness in the examination hall*: While writing your roll number and other particulars on the answer book, be very careful. If you forget to write your roll number or enter a wrong number, you may fail.
- *Reading and understanding the questions*: Read the question paper

thoroughly. Understand fully what has been asked. See the marks allotted to each question. Decide which questions you plan to answer. Find out how much time is available for each answer. Try to follow this plan as much as possible. Time management is very important. Many students write the first and sometimes second answers for a very long time and then get very little time to answer the rest. This reduces the overall marks drastically.

- *Order of writing*: Write the one you know best in the beginning. This will boost up your morale. However, follow the time schedule while answering.
- *Revision*: If there is time for revision, do that. Otherwise, just check the entries on the first page before handing over the paper. Go home happily. Come prepared for the next examination. Remember, 'God helps those who help themselves". Wish you all the best! God bless you!

Index